# Econophysics of Systemic Risk and Network Dynamics

# New Economic Windows

*Series Editors*
MARISA FAGGINI, MAURO GALLEGATI, ALAN P. KIRMAN



For further volumes:
www.springer.com/series/6901

Frédéric Abergel • Bikas K. Chakrabarti •
Anirban Chakraborti • Asim Ghosh
Editors

# Econophysics of Systemic Risk and Network Dynamics

Springer

*Editors*
Frédéric Abergel
Chaire de Finance Quantitative
Laboratoire de Mathématiques Appliquées aux Systèmes
École Centrale Paris
Châtenay-Malabry, France

Anirban Chakraborti
Chaire de Finance Quantitative
Laboratoire de Mathématiques Appliquées aux Systèmes
École Centrale Paris
Châtenay-Malabry, France

Bikas K. Chakrabarti
Theoretical Condensed Matter Physics Division
Saha Institute of Nuclear Physics
Kolkata, India;
Economic Research Unit
Indian Statistical Institute
Kolkata, India

Asim Ghosh
Theoretical Condensed Matter Physics Division
Saha Institute of Nuclear Physics
Kolkata, India

ISSN 2039-411X ISSN 2039-4128 (electronic)
New Economic Windows
ISBN 978-88-470-2552-3 ISBN 978-88-470-2553-0 (eBook)
DOI 10.1007/978-88-470-2553-0
Springer Milan Heidelberg New York Dordrecht London

Library of Congress Control Number: 2012939162



Printed on acid-free paper

Springer is part of Springer Science+Business Media (www.springer.com)

# Preface

Systemic risk has long been identified as a potential for financial institutions to trigger a dangerous contagion mechanism from the financial economy to the real economy itself. One of the commonly adopted definitions of systemic risk is: "risk of disruption to the flow of financial services that is

(i) caused by an impairment of all or parts of the financial system; and
(ii) has the potential to have serious negative consequences for the real economy".

Evident from this definition, or from any of its variants that one can find in the growing literature on the subject, are two characteristic aspects. The first one being that such a risk takes place at a much larger scale than that of an individual institution. The second one being that it eventually spreads to the real economy outside the financial system through various "leakage" mechanisms, of which the last crisis has given some examples: liquidity shrinkage, fire sale of assets, drop in market value of derivatives...

This type of risk, long confined to the monetary market, has spread widely in the recent past, culminating in the subprime crisis of 2008. The understanding and control of systemic risk has therefore become an extremely important societal and economic question. Such problems are now extensively being studied by people from disciplines like economics, finance and physics. The contributions by physicists are relatively new.

The Econophys-Kolkata VI conference, the 6th event in this series of international conferences, held during October 21–25 last year, was dedicated to address and discuss extensively these issues and the recent developments. Like the last event in the series, this one was also organized jointly by the École Centrale Paris and the Saha Institute of Nuclear Physics, and was held at the Saha Institute of Nuclear Physics, Kolkata.

This proceedings volume contains the written versions of most of the talks and seminars delivered by distinguished experts from all over the world, participating in the meeting, and accepted after refereeing. For some completeness in the cases of one or two important topics (like in the case Many-agent Games), some reviews, by experts who could not attend, were invited and incorporated in this volume.

These Proceedings volume is organized as follows: Part I dedicated to the study of systemic risk, network dynamics and other empirical studies. Part II devoted to model-based studies. We have also included Part III for "miscellaneous reports", to present some on-going or preliminary studies. Finally, we have summarized in a brief "discussion and comments" Appendix, some of the remarks made by the participants during the various interesting and animated exchanges that took place during the panel discussion in the conference.

We are grateful to all the participants of the conference for their participation and contributions. We are also grateful to Mauro Gallegati and the Editorial Board of the New Economic Windows series of the Springer-Verlag (Italia) for their support in getting this Proceedings volume published as well, in their esteemed series.[1]

The editors also address their thanks to the Centre for Applied Mathematics and Computational Science at Saha Institute, and École Centrale Paris for their support in organizing this conference. They would also like to thank Gayatri Tilak for providing invaluable help during the preparation of the manuscript.

Châtenay-Malabry, France — Frédéric Abergel
Kolkata, India — Bikas K. Chakrabarti
Châtenay-Malabry, France — Anirban Chakraborti
Kolkata, India — Asim Ghosh
April, 2012

[1] Past volumes:

(i) Econophysics of Order-driven Markets, Eds. F. Abergel, B.K. Chakrabarti, A. Chakraborti, M. Mitra, New Economic Windows, Springer-Verlag, Milan, 2011.
(ii) Econophysics & Economics of Games, Social Choices and Quantitative Techniques, Eds. B. Basu, B.K. Chakrabarti, S.R. Chakravarty, K. Gangopadhyay, New Economic Windows, Springer-Verlag, Milan, 2010.
(iii) Econophysics of Markets and Business Networks, Eds. A. Chatterjee, B.K. Chakrabarti, New Economic Windows, Springer-Verlag, Milan, 2007.
(iv) Econophysics of Stock and other Markets, Eds. A. Chatterjee, B.K. Chakrabarti, New Economic Windows, Springer-Verlag, Milan, 2006.
(v) Econophysics of Wealth Distributions, Eds. A. Chatterjee, S. Yarlagadda, B.K. Chakrabarti, New Economic Windows, Springer-Verlag, Milan, 2005.

# Contents

# Part I
# Systemic Risk, Network Dynamics and Other Empirical Studies

# Chapter 1
# Diffusion of Defaults Among Financial Institutions

**Gabrielle Demange**

**Abstract** The paper proposes a simple unified model for the diffusion of defaults across financial institutions and presents some measures for evaluating the risk imposed by a bank on the system. So far the standard contagion processes might not incorporate some important features of financial contagion.

## 1.1 Introduction

Financial institutions use the interbank market to develop relationships that protect them against liquidity risk. These interbank claims are an important concern for regulators as some argue that they have played a large role in the dissemination of the financial crisis starting in 2007. However this is still a controversial issue in part because of the dual role played by these claims as risk sharing and risk spreading instruments.

Results are rather inconclusive at both theoretical and empirical levels. On one hand, the theoretical studies about the role and rationale of these relationships, though pointing to important phenomena such as the insurance against liquidity shocks and the limitation of bank runs, are conducted in rather simple frameworks (see e.g. Allen and Gale [2], Freixas et al. [8]). Given the complexity of the existing architecture, the robustness of the analysis can be questioned. On the other hand, empirical studies based on simulation methods have so far been unable to reproduce the extent of the crisis, even though they do not introduce the frequent bail-out interventions observed in practice (for a detailed recent survey, see Upper [15]). Simulations however shed some light on the role of the network in systemic risk and on how the assessment of the 'systemic' importance of a bank varies with the chosen measure.

Systemic risk can be understood as a diffusion process of defaults. This paper, building partly on previous studies on diffusion processes, presents a simple model of diffusion of defaults to discuss some questions about measuring systemic risk.

G. Demange (✉)
Paris School of Economics, EHESS, Paris, France
e-mail: demange@pse.ens.fr

F. Abergel et al. (eds.), *Econophysics of Systemic Risk and Network Dynamics*,
New Economic Windows,
DOI 10.1007/978-88-470-2553-0_1, © Springer-Verlag Italia 2013

Indeed, diffusion processes have been introduced in various areas, ranging from sociology to study the diffusion of innovation or ideas through social networks, in epidemiology, viral marketing and so on.

The paper is organized as follows. Section 1.2 presents the model, gives examples, and defines the diffusion processes. Section 1.3 discusses some measures of systemic risk and analyze in more details one measure.

## 1.2 The Framework

I describe a simple model in which financial institutions draw some risky revenues from their activities, are endowed with capital and portfolios, and are linked through claims on each other.

Consider $n$ financial institutions, called banks for simplicity and denote $N = \{1, \ldots, n\}$. A bank $i$ is endowed with some capital $e_i$. Let $\tilde{z}_i$ represent[1] the (risky) revenue that $i$ expects from its activities excluding the interbank relationships. In the sequel $z_i$ is called the *net worth*. Examples are described below. From a balance sheet perspective, $\tilde{z}_i$ is equal to the asset values (stocks + loans to consumers) minus the consumers' deposits. The interbank liabilities are described by $(\omega_{ij})$ where $\omega_{ij}$ represents the magnitude of $i$'s nominal debt obligation towards $j$.

When dealing with a large number of banks, the pattern of their relationships is quite stable and specific, with some banks having regular and large relationship while others having none. In such a situation, the interpretation of financial interlinkages as a network, where banks are nodes and bilateral exposures are the links, is very compelling. It may be useful to think of the graph $G$ formed with the set of links $(i, j)$ where $i$ has an obligation toward $j$, $\omega_{ij} > 0$.

Let us describe the timing. The contagion process takes place ex post once the net worth values $\mathbf{z} = (z_i)$ are realized. In the process described below, the creditors to a defaulting bank receive nothing but the repayment from their not-in-default debtors. The loss incurred by the defaulting bank can engender defaults among its creditors. Defaults can then spread sequentially through the system, and affect, perhaps, a significant number of banks. The resulting set of defaulting banks will be denoted by $D(\mathbf{z})$ and called defaulting set. Thus, given $\mathbf{z}$, the process is well-defined and deterministic. Ex ante however, the defaulting set is random given by $D(\tilde{\mathbf{z}})$, hence with a distribution driven by the distribution of $\mathbf{z}$.

The risk of contagion is affected by various factors such as the magnitude of the inter-bank linkages, the risk distribution of the banks' net worth values, the sensitivity of the asset prices to distress sales. Let us clarify this by considering three examples with increasing complexity.

(i) The *pure network* model.
The elements of $\tilde{\mathbf{z}}$ are independent. Defaults can spread only through the interbank liabilities.

[1] In the following, a $\tilde{}$ on variable $a$ means that the variable is random.

(ii) *Aggregate shocks.*
The presence of aggregate shocks induces correlation in the $\tilde{z}_i$ hence the possibility of simultaneous defaults. This is described by

$$\tilde{z}_i = \tilde{y}_i + \beta_i \tilde{\eta},$$

where the $(\tilde{y}_i)_{i \in N}$ are independent across banks and independent of $\tilde{\eta}$. $\tilde{\eta}$ is interpreted as a common macroeconomic factor and $\beta_i$ the sensitivity of bank $i$'s assets to that factor. If the value of the macroeconomic shock is known at the beginning of the process, and is not affected subsequently by the contagion process, the analysis of the contagion process itself is unaffected since it takes place for each realization of the $\mathbf{z}$.

(iii) *Amplification effects.*
It has been argued that the impact of 'distress' sales from defaulting banks amplifies crises in case of illiquidity. 268 positions and the sensitivity of prices to sales determine the strength of this effect. A simple description captures this effect. Letting $x_i$ be $i$'s asset holding in the market portfolio, $i$'s net worth is

$$\tilde{z}_i = \tilde{y}_i + \beta_i \tilde{\eta} + x_i p_0$$

in which $p_0$ is the asset's price when no default occurs. As previously, the $(\tilde{y}_i)_{i \in N}$ are independent across banks and independent of $\tilde{\eta}$. For example, taking $\beta_i = x_i$, the macroeconomic shock $\tilde{\eta}$ is interpreted as the unexpected variation in the asset's price. The liquidation of assets by the defaulting banks triggers a decrease in asset's price, hence in the asset value of the balance sheets of *all* banks. Thus, defaults generate a correlated change in the banks' net worth levels through their assets' positions along the contagion process.[2]

### 1.2.1 Contagion Process

We assume that there is no possibility of partial default.[3] Let us describe the process by which the default of some banks propagates to other banks, first in the pure network model.

**The Pure Network Model** Once $\mathbf{z} = (z_i)$ are realized, each bank faces a solvency constraint. Given the bank's endowment, the net worth and the amount of its incoming and outgoing liabilities, the constraint for bank $i$ assuming no default from its debtors is:

---

[2]In a model with partial default as in [6], Cifuentes et al. [4] introduce a different mechanism in which the non-defaulting banks have to sell in order to satisfy some solvability ratio constraints.

[3]Eisenberg and Noe [6] introduce a model in which default can be partial, represented by a default level on liabilities. For a measure of the threat index of a bank in that model, see Demange [5].

$$z_i + e_i + \sum_{j \in N} \omega_{ji} - \sum_{j \in N} \omega_{ij} \geq 0. \tag{1.1}$$

The left hand side is referred to as $i$'s *net equity*. The initial assumption on the absence of default is indeed satisfied if the net equity of each bank is non-negative, namely if the solvency condition (1.1) holds for each $i$. To simplify notation, let us write it as $z_i \geq v_i$ where $v_i$ is defined as the initial net value of interbank liabilities minus the capital

$$v_i = \sum_{j \in N} \omega_{ij} - \sum_{j \in N} \omega_{ji} - e_i. \tag{1.2}$$

If some defaults occur, non-defaulting banks suffer a loss and the solvency condition is modified. Let $D$ denotes the set of defaulting banks at some point in time. We assume no recovery, so that a bank that has not failed incurs a loss that amounts to the total of its loans to the banks in $D$, $\sum_{j \in D} \omega_{ji}$. Thus, given that banks in $D$ have failed, the solvency condition for bank $i$ not in $D$ is:

$$z_i \geq v_i + \sum_{j \in D} \omega_{ji}. \tag{1.3}$$

The process of contagion follows from the initial defaults. Here we simply assume that all the banks that are insolvent at some step are declared defaulting (this assumption is relaxed afterwards). Given the realized values for the $\mathbf{z}$, the initial set of defaulting banks is

$$D_0 = \{i, \text{ for which } z_i < v_i\}.$$

At the beginning of step $t$, $t = 1, 2, \ldots$, for which the process has not stopped, there is a set $D_{t-1}$ of banks that have failed. The solvency conditions (1.3) for the other banks, given the total loss they have incurred due to the failures in $D_{t-1}$, determine the set of defaulting banks at $t$:

$$D_t = \left\{ i, \text{ for which } z_i < v_i + \sum_{j \in D_{t-1}} \omega_{ji} \right\}.$$

The sequence $\{D_t\}$ is increasing, $D_{t-1} \subset D_t$, until no additional failure occurs, that is $D_{t-1} = D_t$ (there are at most $n - 1$ steps).

The solvency condition (1.3) on a bank only depends on its defaulting debtors, that is, the $j$ with a positive $\omega_{ji}$. Hence, starting with a single default, the defaulting banks along the process are connected.[4]

Let us now allow for correlations in the net asset values and amplification effects.

[4]The process fits in the class of linear 'threshold' models as introduced by Granovetter [9]. More precisely, the linear model assumes a uniform distribution on the threshold, here the $\tilde{\mathbf{z}}$, and an influence of $j$ on $i$, here $\omega_{ji}$. A node not yet 'active' at step $t$ becomes active in step $t$ if the influence of its neighbors in step $t - 1$ is larger than its threshold, here if the sum of their liabilities, $\sum_{j \in D} \omega_{ji}$ to $i$ is larger than the threshold $z_i$.

**Macroeconomic Risk** In the presence of a macroeconomic risk, the diffusion process is defined for each realized value of $\mathbf{z}$ where $z_i = y_i + \beta_i \eta$ as in the pure network model. There is a change in the ex ante evaluation of the defaulting set because the correlation in the distribution of the $\tilde{z}_i$ affects the distribution of the initial defaulting set.

**Amplification Effects** The presence of an illiquid asset needs some explanation. Given the realizations of the $y_i$ and $\eta$, assuming no default, let $p_0$ denote the asset's price in that case (recall that the variation in the prices is incorporated in the factor $\eta$, so $p_0$ can be interpreted as the expected value under normal conditions). Denote by $z_i$ the known value $y_i + \beta_i \eta + x_i p_0$. The solvency condition for $i$ writes

$$(y_i + \beta_i \eta + x_i p_0) + e_i + \sum_{j \in N} \omega_{ji} - \sum_{j \in N} \omega_{ij} \geq 0 \tag{1.4}$$

or

$$z_i \geq v_i \quad \text{where } v_i = \sum_{j \in N} \omega_{ij} - \sum_{j \in N} \omega_{ji} - e_i. \tag{1.5}$$

If the solvency condition (1.5) holds for each $i$, the initial assumption of no default is justified. If instead it is not met for a bank, this bank will be unable to fulfill its obligations, even by selling its asset at price $p_0$ and a fortiori at any lower price (we use here that $x_i$ is non-negative). The bank defaults and liquidates its asset.

Thus, the default of banks in $D$ results in an aggregate amount of asset's sales equal to $x_D = \sum_{i \in D} x_i$. The sale triggers a decrease in the asset's price equal to $p_0 - P(x_D)$. Accounting for this decrease, the solvency condition (1.3) for non-defaulting banks is replaced by

$$z_i \geq v_i + \underbrace{\sum_{j \in D, (j,i) \in G} \omega_{ji}}_{\text{direct loss}} + \underbrace{x_i \big(p_0 - P(x_D)\big)}_{\text{indirect loss}}. \tag{1.6}$$

Bank $i$'s incremental loss incurred by the default of $D$ is composed of the direct loss due to the non-reimbursement of its defaulting neighbors and the indirect loss due to the variation in prices. To simplify notation, the solvency condition (1.6) on bank $i$ under defaulting set $D$ writes

$$z_i \geq V_i(D) \quad \text{where } V_i(D) = v_i + \sum_{j \in D, (j,i) \in G} \omega_{ji} + x_i \big(p_0 - P(x_D)\big). \tag{1.7}$$

Thus, with amplification effects, the default of a bank is influenced not only by its defaulting neighbors as in a pure network but also by a term reflecting the whole set of defaulting banks, neighbors or not: the value $V_i(D)$ is strictly increasing in $D$ for each $i$ when each bank holds a positive quantity of the asset and the asset price is sensitive to sales. As a result, each bank is affected by the default of any other banks, even they are not its debtors. The network is made 'complete' by the presence of the price effect.

A general formulation that encompasses previous examples is described as follows.

**Activation Model** *For each $i$ there is an activation function $V_i$ defined over subsets $D$ of $N$, nondecreasing in $D$, so that the solvency condition on bank $i$ under defaulting set $D$ writes*

$$z_i \geq V_i(D). \tag{1.8}$$

*Denote $v_i = V_i(\emptyset)$.*

Given $\mathbf{z}$, a diffusion process is defined as follows. At $t = 0$, define

$$D_0 = D_0(\mathbf{z}) = \{i, \text{ for which } z_i < v_i\}.$$

At the beginning of step $t$, $t = 1, 2, \ldots$, for which the process has not stopped, there is a set $D_{t-1}$ of banks that have failed. Similarly at time $t$, the set of defaulting banks $D_t$ is updated from $D_{t-1}$ by checking the solvency conditions (1.8) for banks not in $D_{t-1}$:

$$D_t = \{i, \text{ for which } z_i < V_i(D_{t-1})\}.$$

If $D_{t-1} = D_t$, each bank not in $D_t$ is solvent and the process stops. The reached set is called the defaulting set.

Alternative processes can be contemplated in which all the banks that are insolvent at some step are not necessarily eliminated but at least one of them must be. Specifically a process is said *without stop* if at a step $t$ for which a 'new' bank is not solvable, that is, a bank $i$ not in $D_{t-1}$ for which $z_i < V_i(D_{t-1})$, then at least one such a bank is declared defaulting, i.e. $D_t$ strictly includes $D_{t-1}$. The process surely stops in at most $n - 1$ steps, though it may involves more steps than in the one described previously.

### *1.2.2 Characterization of the Defaulting Set*

The defaulting set is characterized by a property, which is independent of the process (provided the process is without stop). Hence, whatever the speed and the order of eliminations, the same defaulting set is reached.

The characterization is based on the notion of closed set.

**Definition 1.1** A set $C$ is said to be *closed* at $\mathbf{z}$ if no $i$ outside $C$ defaults in case of a failure of each bank in $C$. Specifically $C$ is closed at $\mathbf{z}$ if

$$z_i \geq V_i(C) \quad \text{for each } i \text{ not in } C. \tag{1.9}$$

The condition requires banks outside to be solvent when they incur losses corresponding to the failure of $C$. Observe that no condition bears on elements in $C$ and some banks in $C$ can be solvent at $\mathbf{z}$. By construction, a process without stop can only settle at a set that is closed and contains $D_0(\mathbf{z})$. Furthermore,

**Claim 1** *Let the solvency conditions be described by* (1.8). *Given* $\mathbf{z}$, *the defaulting set is the smallest set that is closed and contains* $D_0(\mathbf{z})$, *whatever the process without stop. Denote it by* $D(\mathbf{z})$.

*Proof* Let us first show that there is a smallest set among those that are closed and contain $D_0(\mathbf{z})$. Since the value $V_i(C)$ does not increase with $C$, the non-empty intersection of two closed sets is closed as well.[5] This readily implies that there is a smallest set among those that are closed and contain $D_0(\mathbf{z})$.

Given $\mathbf{z}$ denote by $D_\infty$ the defaulting set reached by a process without stop. $D_\infty$ is closed at $\mathbf{z}$ and contains $D_0(\mathbf{z})$. It thus suffices to show that $D_\infty$ is a subset of any $C$ that is closed at $\mathbf{z}$ and contains $D_0(\mathbf{z})$. Since $D_\infty$ coincides with $D_t$ for $t$ large enough, it suffices to show that $D_t$ is included in $C$ for any $t$. We prove this by induction on $t$.

The induction assumption is true at $t = 0$ by the assumption on $C$: $D_0(\mathbf{z}) \subset C$.

Let us assume $D_t \subset C$ at $t$. By monotony of the $V_j$ this implies that $V_j(D_t)$ is less than $V_j(C)$. By definition, a bank in $D_{t+1}$ defaults at the level given $V_j(D_t)$ hence *a fortiori* each one of them defaults at level $V_j(C)$:

$$\text{for each } j \in D_{t+1}, \quad z_j < V_j(D_t) \le V_j(C).$$

Since $C$ is closed, $z_j < V_j(C)$ implies that $j$ belongs to $C$: $D_{t+1}$ is a subset of $C$, which proves the induction assumption. □

## 1.3 Measuring Losses and Externalities

Now that a diffusion process has been defined, it remains to evaluate the impact of a bank on the risk of the system, in particular, how it influences the reached defaulting set. Alternative measures of the impact of a bank on the risk of the system have been proposed (see Elsinger et al. [7], Tarashev et al. [14], Adrian and Brunnermeier [1], the survey of Upper [15] among a few). They differ in the following three dimensions at least.

(i) The time—and the available information—at which the expected loss imposed by a bank is evaluated: this loss can be evaluated ex ante, thereby accounting for the probability of initial insolvency of the bank or ex post, conditional on its insolvency. Also the loss can be computed conditional on some macroeco-

[5] For closed sets $C$ and $C'$ and $i$ not in $C \cap C'$, surely $z_i \ge V_i(C)$ or $z_i \ge V_i(C')$. Since both $V_i(C)$ and $V_i(C')$ are larger or equal to $V_i(C \cap C')$, $z_i \ge V_i(C \cap C')$: $C \cap C'$ is closed.

nomic events. For example, in the model with macroeconomic shocks, one can use the distribution conditional on the macroeconomic variable $\eta$ being smaller than some value. This is in spirit with proposals to compute a VaR measure (or an expected shortfall) conditional on 'systemic events' (to be made precise).

(ii) The risk that is measured, as an initiator of the default, or as a propagator of defaults, or both as in the 'contribution' approach. Some use the term top-down versus bottom-up.

(iii) The cost evaluation associated to the reached defaulting set. One may take the point of view of stockholders and evaluate the equity of the defaulting banks or rather be concerned with the loss to the non-financial creditors.

I describe very roughly some measures in the context of this paper and then study in more detailed one ex-post measure.

### *1.3.1 Some Measures*

The measures are built on the cost associated to the set of defaulting banks. Let $C(D)$ be the cost associated to the defaulting set $D$.[6] For example, taking the point of view of stockholders, one has $C(D) = \sum_{j \in D} e_j$ to evaluate the loss in capital. If one is concerned with the loss to the non-financial creditors, the loss possibly includes some measurement of the impact on economic activity. In such a case, it is reasonable to assume that the cost $C$ has some form of increasing returns to default (super-modularity as defined in Sect. 1.3.2).

In the sequel I focus on the loss in capital to fix the idea. In some cases it simplifies the presentation owing to the linearity of $C$ in that case.

**Loss and Externality Conditional on a Bank's Default (Bottom Up)** The expected loss conditional on a bank $i$'s default is assessed by assuming that $i$ is the only bank to initiate default. Thus, it is computed by considering the (random) defaulting set following the initial insolvency of $i$ only. Specifically, the loss conditional on the failure of $i$ is

$$L(\{i\}) = E\big[C\big(D(\mathbf{z})\big) | D_0(\mathbf{z}) = \{i\}\big] \tag{1.10}$$

in which the expectation is taken over the distribution of the defaulting set $D(\mathbf{z})$ following $i$'s default. The external loss is obtained by considering the loss imposed on other banks, that is

$$L^*\big(\{i\}\big) = L\big(\{i\}\big) - e_i.$$

Let us illustrate the computation in a simple hierarchical graph, as depicted in Fig. 1.1 in which an arrow represents a liability, that is, a bank is indebted to a

[6]This assumes that the cost only depends on the set $D$ and not on the precise values of $\mathbf{z}$.

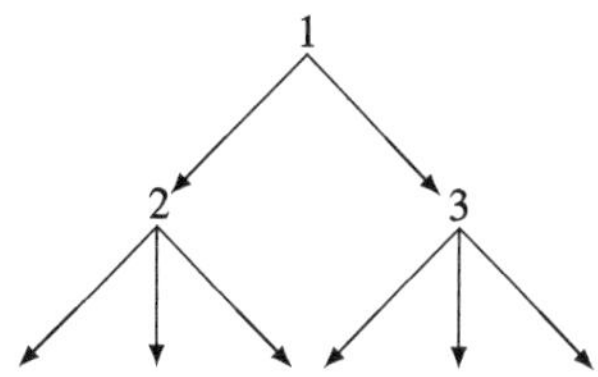


**Fig. 1.1** Hierarchical structure

successor. $L$ is given by a simple recursive expression:[7]

$$L(\{i\}) = e_i + \sum_{(i,j)\in G} q_j L(\{j\})$$

where $q_j$ is the probability that $j$ fails if its unique debtor $i$ fails. It is thus given by $q_j = \frac{F_j(v_j+\omega_{ij})-F_j(v_j)}{1-F_j(v_j)}$ where $\omega_{ij}$ is the loan made by $j$ to $i$. In the particular case where each bank has the same capital value $e$, the same liability to each of its creditor $\omega$, and the same distribution $F$ one obtains: $L(\{i\}) = e(1 + q\cdot$ number of $i$'s creditors $+q^2\cdot$ number of creditors of $i$'s creditors $+\cdots)$.

The smaller the probability of inducing default, the closer the measure is to the number of creditors. When $q$ is larger, indirect default starts to be determinant. In the case depicted in Fig. 1.1, $L(\{2\})$ is larger than $L(\{1\})$ for small enough $q$ and the reverse for $q$ large enough.

For more complex networks with cycles, there is not such a simple recursive expression. One may also consider simultaneous defaults. The (equity) loss to the initial default of a subset $A$ is defined by

$$L(A) = E\left[\sum_{j\in D(\mathbf{z})} e_j | D_0(\mathbf{z}) = A\right] \tag{1.11}$$

and the externality (external loss) induced by $A$ as

$$L^*(A) = L(A) - \sum_{i\in A} e_i. \tag{1.12}$$

These computations are performed according to some distribution of the payoffs of the non-initially defaulting banks. As said previously one may want to account for some information say on macroeconomic events.

**Top Down Ex Ante Evaluation** I present here a simple form of the contribution approach.[8] The basic idea here is first to evaluate the total risk in the system and second to define a contribution of a particular bank or of a subset of banks to that risk.

---

[7] Here the expression is similar to some measures of 'power' or 'prestige' developed in sociology such as the Katz prestige index [11].

[8] [14] also consider the loss imposed by a bank to each subsystem and derives the contribution to risk of that bank by taking an average (the Shapley value).

**Fig. 1.2** A simple example

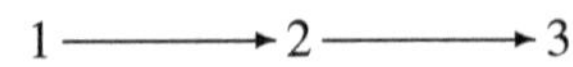


The total risk is evaluated as the expected loss accounting for all possibilities of failure. Let $\Pi(A)$ denote the probability that $A$ is the initial insolvent set. In the pure network model for example $\Pi(A) = \prod_{k\in A} F_k(v_k) \prod_{k\notin A}(1 - F_k(v_k))$. The total risk is

$$T(N) = \sum_{A\subset N} \Pi(A)L(A). \tag{1.13}$$

With macroeconomic shocks, as said above, the cost can be evaluated conditional on a low enough value for the macroeconomic variable.

The cost that $i$ imposes on the system, also called the 'contribution' of $i$ to systemic risk, is defined as the ex ante benefit of making $i$ totally safe. It is given by the difference

$$T(N) - T\big(N - \{i\}\big)$$

where $T(N - \{i\})$ is the risk in the system with $i$ totally safe. Let us examine the cost in more detail. Specifically let $L_{-i}(A)$ denote the loss due to $A$ defaulting in the system where $i$ is totally safe; we have $T(N - \{i\}) = \sum_{A\subset N-\{i\}} \Pi(A)L_{-i}(A)$. We may thus write

$$T(N) - T\big(N - \{i\}\big) = \underbrace{\sum_{A\subset N, i\in A} \Pi(A)L(A)}_{\text{direct cost}} + \underbrace{\sum_{A\subset N-\{i\}} \Pi(A)\big[L(A) - L_{-i}(A)\big]}_{\text{indirect cost}}.$$

The cost imposed by $i$ is composed of two terms: a direct one associated to all the events in which $i$ defaults, and an indirect one that reflects the role of $i$ as a vector in propagating defaults. Indeed the indirect cost only accounts for events in which $i$ defaults but is not as an initiator; making it totally safe prevents $i$ to spread default in those events. To illustrate, consider the simple example in Fig. 1.2. Take the same capital levels $e$, an identical probability for each of becoming insolvent alone, and an identical probability $q$ of triggering default on a creditor. To simplify the presentation, let us assume that the events with several banks simultaneously initiating defaults are negligible. The direct cost is of course decreasing going down in the hierarchy ($L(\{1\}) = (1 + q + q^2)e$, $L(\{2\}) = (1 + q)e$, $L(\{3\}) = e$) but the indirect cost is larger for 2 than for 1 because 1 never spreads default. Taking the sum of the direct and indirect impact, simple computation shows that the system with 2 safe is safer than with 1 safe.

### *1.3.2 Some Properties of the Conditional Loss*

I concentrate here on some properties of the measure $L$, still assuming the cost to be given by equity loss. The diffusion processes defined in the previous section fit in known classes in some special cases, such as the linear 'threshold' models (see footnote 4). The aim of this section is to draw insights from previous works on diffusion processes, and to see whether their main assumptions and results are adapted to our context. The focus is on properties referred to as sub-modularity or super-modularity. They describe concavity or convexity properties for a function defined over subsets.

**Definition 1.2** Let $\Phi$ be a function defined over the subsets of $N$. $\Phi$ is *sub-modular* if for each $i$ in $N$,

$$\Phi\big(A+\{i\}\big)-\Phi(A)\geq\Phi\big(B+\{i\}\big)-\Phi(B)\quad \text{for any sets } A\subset B\subset N. \tag{1.14}$$

$\Phi$ is *super-modular* if the reverse inequality in (1.14) holds.

Sub- or super-modularity on a function bears on its *incremental variation* and not on its level, and indicates decreasing or increasing incremental variations. The sub-modularity of $\Phi$ is interesting from a computational point of view in problems of maximization[9] of $\Phi$.

Applied to the loss function $\Phi = L$, subsets are the initial sets of defaulting banks. Sub-modularity (resp. super-modularity) indicates decreasing (resp. increasing) returns to default: the incremental effect of a bank initially defaulting on others' defaults is decreasing with the set of banks that have already failed. Consider the problem of finding a subset $A$ that maximizes $L(A)$ under some constraints, say the cardinality of $A$ less than a number $m$. Since $L$ is the loss conditional on the initial default of $A$, the solution to the problem may be thought as the set of most 'dangerous' $m$ banks in terms of the level of the losses (using $L^*$ instead of $L$, the problem is to find the subset of $m$ banks that inflict the largest external loss on others). This type of problem has been investigated in the context of diffusion to find the nodes in a network that are the most important in terms of spreading some property. In viral marketing for example, these nodes are useful to spread the adoption of a new product.

In our context, however, the following situation looks more relevant. Let a regulation agency have some information on the value of $\mathbf{z}$ and in particular knows that the banks in $D_0$ will fail without intervention. Contemplating the possibility of rescuing a subset $A$, it evaluates the benefit according to $\Phi(A) = L(D_0) - L(D_0 - A)$. $\Phi$ inherits the modularity properties of $L$. Hence starting rescuing $j$, the value of rescuing $i$ as well is decreasing or increasing depending on whether $L$ being super-modular or sub-modular. Thus, here also, the modularity property of $L$ matters.[10]

---

[9]Consider for example the problem of finding a subset $A$ that maximizes $\Phi(A)$ under some constraints, say the cardinality of $A$ less than a number $m$. Under sub-modularity, a fast algorithm provides an approximation for the problem. The algorithm is called 'greedy': it first looks for $i$ with the largest value for $\Phi$ among the singletons, say $i_1$ that maximizes $\Phi(\{i\})$, then for $j$ with the largest incremental change over $i_1$, say $i_2$ that maximizes $\Phi(\{i, i_1\}) - \Phi(\{i_1\})$, and so on $m$ times. The exact problem is known to be NP-hard in the size of $n$.

[10]Intuitively, there should be a link between the sub-modularity or super-modularity of $L$ and the properties of the diffusion process itself, in particular the distribution of $D(\tilde{z})$ given $A$. Sub-

The next proposition directly follows from Kempe, Kleinberg, and Tardos [12]. They consider the expected size of the contaminated set (or any positive linear function of it) and prove its sub-modularity in a linear threshold model. The result directly applies to the pure network model when each net worth $z_i$ is uniformly distributed on some interval and the loss incurred by the default of a single debtor can trigger default.

**Proposition 1.1** *In the pure network model, assuming the $\tilde{z}_i$ uniformly distributed on $[a_i, b_i]$, $L$ and $L^*$ are sub-modular if each bank $i$ defaults with a positive probability when one of its debtor fails*:

$$z_i < v_i + \omega_{ji} \quad \textit{for each } j, \textit{ with } \omega_{ji} > 0. \tag{1.15}$$

Observe that sub-modularity is a fortiori true if the cost associated with a default set is itself sub-modular, and not additive as in the case of equity. However, sub-modularity in the cost function is surely not an appropriate assumption in our context. For example, if the cost represents the loss to creditors outside the financial system, it is more plausible that the converse assumption holds: the larger the loss, the more costly it is to absorb an additional loss.

Also, it is important to note that Proposition 1.1 may not apply in the presence of information effects, for example due to correlation in the net worth values. To be more specific, when $D$ has defaulted and a bank has not yet defaulted, the only relevant information is that it has survived the loss induced by $D$, i.e., that $z_i$ is larger than $V_i(D)$. No information is drawn from the fact that $D$ has defaulted, as would be the case if the net worths were correlated.[11]

Condition (1.15) says that each node can contaminate each one of its neighbor.[12] This assumption is taken somewhat implicitly in the standard linear threshold model (which assumes the threshold to be uniformly distributed on [0, 1] and the influences factors of neighbors to be positive) and not emphasized. However it cannot be dispensed with as shown in the example below.

**Example of Non-sub-modularity** A simple example illustrates why sub-modularity may fail. There are three nodes, with 2 linked with 1 and 3. The influence weights of 1 and 3 on 2 are equal to 1 and the threshold value for 2 is bounded below by 1.5. Hence, if only 1 or 3 fails, 2 remains safe, whereas if both fail, 2 fails with some positive probability. We thus have $L^*(\{1\} + \{3\}) > 0$ and $L^*(\{1\}) = L^*(\{3\}) = 0$.

---

modularity of $L$ suggests a non-explosive dynamics of contagion. However I do not know of any result of this kind.

[11] What matters is the correlation that is unknown at the time of the evaluation.

[12] The condition is satisfied if there is a chance for the bank to default alone. This occurs for example if its initial values of the capital and liabilities are set so that a Value at Risk requirement is just binding. Specifically, given a level $\alpha$, say 99 %, the level of capital, $e_i$, and the interbank assets and liabilities are set so that the probability of default is equal to the level $1-\alpha$: $F_i(v_i) = 1-\alpha$. But VaR does not make much sense in a context with a uniform distribution.

This contradicts sub-modularity: adding 3 to 1 increases more the expected size of the defaulting set than adding 3 to the empty set.

The example is not pathological and extends to any situation in which the default of a single neighbor is not enough to threaten bank $i$ because the loss of its default inflict to $i$ is smaller than $i$'s minimum payoff. This explains the condition (1.15).

The assumption of a uniform distribution is of course not adequate for net worth. I examine here what are the conditions on a distribution that favor or deter sub-modularity. Basically, sub-modularity may fail if an additional loss to a bank may have a sudden large impact on the incremental probabilities of failure conditional on the fact that it has not yet defaulted as suggested in the example. This is avoided by assuming the concavity of the distribution functions, as stated is the following proposition (see our comment later on). Proposition 1.2 considers the more general formulation for a diffusion process through activation functions, as described by (1.8). It is derived by using an extension of the sub-modularity property of the expected contaminated size when the activation functions are not linear but sub-modular, still under a uniform distribution (Mossel and Roch [13]). The proof uses a transformation of the variable $\tilde{z}_i$.

Let us say that $j$ *directly influences* $i$ if $V_i(\{j\}) > v_i$. In other words, $i$ incurs a loss under the single default of $j$. In a pure network model $i$ is directly influenced by its debtors and in a model with price effects by all other banks (assuming all positions $x_j$ positive).

**Proposition 1.2** *Let the solvency conditions be described by* (1.8). *Assume that, for each* $i$,

- *the functions* $V_i$ *are sub-modular.*
- *the cumulative distribution function of* $\tilde{z}_i$ *is strictly increasing on its support, concave and*

$$F_i\left(V_i(\{j\})\right) > 0 \quad \text{for each } j, \text{ that directly influences } i. \tag{1.16}$$

*Then* $L$ *and* $L^*$ *are sub-modular.*

The sub-modularity of $V_i$ extends the linearity of the $V_i$ in the pure network model. In the model with amplification effects, intuitively, sub-modularity should be satisfied if the decrease in the price due to some sales, $p_0 - P(x_D)$, diminishes with the amount of sales. This is indeed the case, as stated in the next claim

**Claim 2** *Consider the model with amplification effects in which the activation functions are given by* (1.7): $V_i(D) = v_i + \sum_{j \in D, (j,i) \in G} \omega_{ji} + x_i(p_0 - P(x_D))$. *If the drop in the price* $p_0 - P(x)$ *is concave in the sales* $x$, *the functions* $V_i$ *are sub-modular.*

When there are panic effects however, the drop in prices might be convex.

Consider now the assumptions on the distributions. Condition (1.16) has the same interpretation as Condition (1.15). It says that $i$ fails with positive probability under the single default of a bank that has a direct influence on $i$. The concavity of a distribution however is a strong assumption. The larger the amount of losses that have already been incurred, the less likely it is that an additional amount of loss generates default. Most standard distributions are not concave.[13] For example, a normal distribution is not concave but convex for values below the mean. These values are precisely the ones that matter when considering the diffusion of the defaults.

The reverse assumptions, a convex distribution and a super-modular activation functions, do not guarantee super-modularity of the functions $L$. The reason is the following, that I call the ground effect. Starting with a larger initial default set, there are less banks susceptible to default. This effect is true not only at the beginning of the diffusion but also all along the process. This effect is a force towards sub-modularity, independently of the distribution. A formulation of this idea is the following proposition, stated in the pure network model for simplification.

**Definition 1.3** Consider the pure network model. $D$ is said to be a threat if for each $i$ not in $D$

$$Proba\big(z_i < V_i(D)\big) > 0 \quad \text{each } i, \tag{1.17}$$

where $V_i(D) = v_i + \sum_{j \in D} \omega_{ji}$.

**Proposition 1.3** *Let $D$ be a threat. Assume in addition the distribution $F_i$ to be concave for $z_i > V_i(D)$ for each $i$. Then $L$ is sub-modular on larger sets, that is* (1.14) *holds for any sets $A$ and $B$ containing $D$.*

Condition (1.17) says that each bank not in $D$ has some chances to fail if all banks in $D$ default. If whatever situation the bank never fails, no set $D$ is a threat. Otherwise, each bank is threatened when all its neighbors fail, so (1.17) is satisfied for 'large' enough sets. Observe also that a superset of a threat is also a threat. The second condition, which requires the concavity of the upper tail of the distributions, is satisfied by most distributions.

The proof is easy: It suffices to apply Proposition 1.2 to the diffusion process assuming that $D$ has already defaulted. The distribution for the net worth of the banks that have not yet defaulted is adjusted only conditional on their absence of default, that is conditional on $z_i > V_i(D)$.

**Some Concluding Remarks** In conclusion, the properties of the diffusion processes studied so far in the (non-financial) literature might not be relevant to financial crises, and there is a need for further development. They focus on some properties on the distribution functions that are not appropriate. The analysis here

---

[13]Most standard distributions have a log-concave density. Then the ratio $\frac{F(x+\delta)-F(x)}{1-F(x)}$, related to the hazard rate (see for example Bergstrom and Bagnoli [3]) is increasing. Though this is compatible with the concavity of $F$, this suggests that concavity is far from being guaranteed.

distinguishes two influences of the distributions of the net worth on the diffusion of risks. On one hand, the lower tails of the distributions of the net worth matter for determining the initial defaults. On the other hand, the reaction of the diffusion to additional defaults, as embodied by the incremental values of the loss $L$, is instead affected by the shape of the distributions for larger values of the net worth, and this holds especially if interbank liabilities are large. Under standard assumptions on the distributions, this second effect might play an important role in shaping the loss. Finally, diffusion processes do not account for information effects.

**Acknowledgements** I would like to thank Yuan Zhang for research assistance.

## References

1. Adrian T, Brunnermeier MK (2011) CoVaR. NBER Working Paper No. 17457
2. Allen F, Gale D (2000) Financial contagion. J Polit Econ 108(1):1–33
3. Bergstrom T, Bagnoli M (2005) Log-concave probability and its applications. Econ Theory 26:445–469
4. Cifuentes R, Ferrucci G, Shin HS (2004) Liquidity risk and contagion. J Eur Econ Assoc 3(2–3):556–566
5. Demange G (2012) Contagion in financial networks: a threat index. PSE WP 2012-02. http://halshs.archives-ouvertes.fr/halshs-00662513
6. Eisenberg L, Noe TH (2001) Systemic risk in financial systems. Manag Sci 47(2):236–249
7. Elsinger H, Lehar A, Summer M (2006) Risk assessment for banking systems. Manag Sci 52(9):1301–1314
8. Freixas X, Parigi BM, Rochet J-C (2000) Systemic risk, interbank relations, and liquidity provision by the central bank. J Money Credit Bank 32(3):611–638
9. Granovetter M (1978) Threshold models of collective behavior. Am J Sociol 83(6):1420–1443
10. Gai P, Kapadia S (2010) Contagion in financial networks. Proc R Soc A 466(2120):2401–2423
11. Katz L (1953) A new status index derived from sociometric analysis. Psychometrika 18(1):39–43
12. Kempe D, Kleinberg J, Tardos É (2003) Maximizing the spread of influence through a social network. In: Proceedings of the ninth ACM SIGKDD international conference on knowledge discovery and data mining (KDD '03). ACM, New York, USA, pp 137–146. doi:10.1145/956750.956769
13. Mossel E, Roch S (2007) On the submodularity of influence in social networks. In: Proceedings of the thirty-ninth annual ACM symposium on theory of computing (STOC '07). ACM, New York, USA, pp 128–134. doi:10.1145/1250790.1250811
14. Tarashev NA, Borio CEV, Tsatsaronis K (2010) Attributing systemic risk to individual institutions. BIS Working Paper No. 308. Available at SSRN: http://ssrn.com/abstract=1631761
15. Upper C (2011) Simulation methods to assess the danger of contagion in interbank markets. J Financ Stab 7:111–125

# Chapter 2
# Systemic Risk and Complex Systems: A Graph-Theory Analysis

**Delphine Lautier and Franck Raynaud**

**Abstract** This chapter summarizes several empirical studies in finance, undertaken through the prism of the graph theory. In these studies, we built graphs in order to investigate integration and systemic risk in derivative markets. Several classes of underlying assets (i.e. energy products, metals, financial assets, agricultural products) are considered, on a twelve-year period. In such a high dimensional analysis, the graph theory enables us to understand the dynamic behavior of our price system. The dimension of the fully connected graph being high, we rely on a specific type of graphs: Minimum Spanning Trees (MSTs). Such a tree is especially interesting for the study of systemic risk: it can be assimilated into the shortest and most probable path for the propagation of a price shock. We first examine the topology of the MSTs. Then, given the time dependency of our correlation-based graphs, we study their evolution over time and their stability.

## 2.1 Introduction

This chapter summarizes several empirical works in finance undertaken since 2009: [8, 10, 11]. These works share two common points: first, they all focus on systemic risk and the integration in organized derivative markets. Second, they provide for a large-scale analysis and rely on the graph theory.

Integration and systemic risk are linked: the former is a favorable condition for the second to appear in. Concerns about systemic risk have recently grown in derivative markets, notably among energy commodities. These markets are supposed to be more and more integrated, both in regard to each other and to other markets. For some months now, fluctuations in the prices of energy products have often been

D. Lautier is also member of the Fime laboratory and associate research fellow at MinesParisTech.

D. Lautier (✉)
DRM-Finance, UMR CNRS 7088, University Paris-Dauphine, Paris, France
e-mail: delphine.lautier@dauphine.fr

F. Raynaud
Laboratory of Cell Biophysics, EPF Lausanne, Lausanne, Switzerland
e-mail: franck.raynaud@epfl.ch

F. Abergel et al. (eds.), *Econophysics of Systemic Risk and Network Dynamics*, New Economic Windows,
DOI 10.1007/978-88-470-2553-0_2, © Springer-Verlag Italia 2013

invoked to explain corresponding fluctuations in soft commodities like soy, corn, or wheat. Furthermore, because commodities are nowadays considered a new class of assets, investors use them for diversification purposes. Therefore, the price fluctuations recorded in commodity markets might be, at least partially, explained by external events like the fall in stock prices or in interest rates [2, 4, 7].

In our studies, we propose a holistic approach for systemic risk, which examines it simultaneously in three dimensions: space, time, and the maturity of the transactions. Such an analysis accounts for the eventuality that a price shock that occurs on a specific asset's physical market can spread, not only through its own futures market, but also into other physical and/or paper markets, and *vice versa*.

A full comprehension of systemic risk can only be made through a large scale analysis that requires the manipulation of a huge amount of data. In our most extensive study [10], we work on the basis of 14 derivative markets (six energy commodities, four agricultural commodities, and four financial assets), over a 12-year period. This leads us to setup a database containing more than 750, 000 futures prices. To perform such large scale analyses, we rely on methods initially designed for statistical physics, aiming at understanding the behavior of complex systems. They incite us to consider all prices, quoted in different places and with different maturities, as a complex dynamic system. Moreover, this consideration leads us to a set of tools that proves very useful for the study of systemic risk: graph theory.

Through this prism, the nodes in our graph are the daily price's returns and the links stand for distances, the latter being computed as a function of the correlations between the returns [12, 14, 15]. This representation allows us to analyze the linkages between the markets and their evolution, thanks to the structure of the connections between the futures contracts. What is especially interesting here is that we can consider, simultaneously, all possible pairs of assets.

The size of the fully connected graph being high, we rely, most of the time, on a specific type of graphs: Minimum Spanning Trees (MSTs). A MST provides a way to extract the most important information contained in the initial graph. It is unique and corresponds to the shortest path covering all the nodes of the graph without loops. Such a tree is thus especially interesting for the study of systemic risk: it can be assimilated into the shortest and most probable path for the propagation of the price shock. To the best of our knowledge, it is the first time that this tool has been used this way.

The visualization of the MST and the computation of some specific measures, like allometric coefficients, make possible the analyzation of the organization of the trees. Two extreme configurations are used as references. A chain-like organization signifies that, when it appears at one extremity of the price system, only one way exists for the price shock to propagate: before reaching the other extremity of the graph, the shock will have to cross each node. On the other hand, in a star-like organization, the paths for the transmission of fluctuations are less easy to predict. Here, the node located at the center of the star is of crucial importance: whenever a shock arises at this point, it might disseminate to the whole system! We first examine the MST according to these two ideal types of organizations. Then, given the time dependency of correlation-based graphs, we study their evolution over time and their stability.

Our first main results lie in the economic meaningfulness of the graphs. In the spatial as well as in the 3-D analyses, the trees are organized into sub-trees corresponding to the three sectors of activity under examination: energy commodities, agricultural products, and financial assets. In the maturity dimension, as a result of arbitrage operations, the trees are ordered according to the maturity of the contracts. The second set of results, interesting for regulatory purposes, shows that energy products promote the connection between the different sectors. Moreover, crude oil stands at the center of the energy complex. A third category of results concerns the evolution of integration over time. In commodity markets, both spatial and maturity dimensions tend to be more integrated. Thus, the conditions for the appearance of systemic risk increase.

Section 2.2 of this chapter explains the data. Section 2.3 gives insights into the methodology retained. In Sect. 2.4, we present the empirical results: we summarize the main conclusions reached in previous studies (more particularly in [10]), and we add some other results, still unpublished until now. Section 2.5 concludes.[1]

## 2.2 Data

We select futures markets corresponding to three sectors: energy, agriculture, and financial assets. On the basis of the Futures Industry Association's reports, we retain those contracts whose characteristics are large transaction volumes over long time periods. We rearrange the futures prices in order to reconstitute daily term structures, that is, the relation linking, at a specific date, several futures contracts with different delivery dates. In order to obtain continuous time series, we remove some maturities from the database. We also take away all observation dates that are not shared by all markets. Once these selections carried out, our database still contains more than $750,000$ prices. Table 2.1 summarizes the characteristics of this database.[2]

## 2.3 The Graph Theory: From Full Connected Graphs to Minimum Spanning Trees (MST)

Among the different tools provided by the graph theory, we select those that allow us to analyze market integration and systemic risk by using a 3-D approach. We first focus on the synchronous correlations of price returns. Having transformed these correlations into distances, we are able to draw a fully connected graph of the price system, where the nodes (vertices) of the graph represent the time series of futures prices. In order to filter the information contained in the graph, we then rely on

[1] The review of the literature related to this chapter can be find in [10].

[2] Another study, including the freight rate, can also be find in [8].

**Table 2.1** Main characteristics of the collected data: nature of the underlying asset, trading place of the futures contract, localization of the exchange, time period, longest maturity (in months) and number of records per maturity. CME stands for Chicago Mercantile Exchange, ICE for Inter Continental Exchange, US stands for United States and Eu for Europe

| Underlying assets | Exchange–Zone | Period | Maturities | Records |
|---|---|---|---|---|
| Light crude | CME–US | 1998–2011 | up to 84 | 3343 |
| Brent crude | ICE–Eu | 2000–2011 | up to 18 | 2923 |
| Gasoil | ICE–Eu | 2000–2011 | up to 12 | 2950 |
| Nat. gas (US) | CME–US | 1998–2011 | up to 36 | 3336 |
| Nat. gas (Eu) | ICE–Eu | 1997–2011 | up to 9 | 3698 |
| Wheat | CME–US | 1998–2011 | up to 15 | 3412 |
| Soy bean | CME–US | 1998–2011 | up to 14 | 3370 |
| Soy oil | CME–US | 1998–2011 | up to 15 | 3447 |
| Corn | CME–US | 1998–2011 | up to 25 | 2960 |
| Eurodollar | CME–US | 1997–2011 | up to 120 | 3689 |
| Gold | CME–US | 1998–2011 | up to 60 | 3060 |
| FX rate USD/EUR | CME–US | 1999–2011 | up to 12 | 3239 |
| Mini SP500 | CME–US | 1997–2011 | up to 6 | 3611 |

Minimum Spanning Trees [12]. Such a tree can be defined as the one providing the best arrangement of the network's different nodes.[3]

### *2.3.1 Synchronous Correlation Coefficients of Prices Returns*

The synchronous correlation coefficients of price returns are defined as follows:

$$\rho_{ij}(t) = \frac{\langle r_i r_j\rangle - \langle r_i\rangle\langle r_j\rangle}{\sqrt{(\langle r_i^2\rangle - \langle r_i\rangle^2)(\langle r_j^2\rangle - \langle r_j\rangle^2)}}, \tag{2.1}$$

where $i$ and $j$ correspond to two different time series of futures returns. The daily logarithm price differential stands for price returns $r_i$, with $r_i = (\ln F_i(t) - \ln F_i(t - \Delta t))/\Delta t$, where $F_i(t)$ is the price of the futures contract at $t$. $\Delta t$ is the lag between two consecutive trading days.

For a given time period and a given set of data, we thus compute the matrix of $N \times N$ correlation coefficients $C$ for all the pairs $ij$. $C$ is symmetric with $\rho_{ij} = 1$ when $i = j$. Thus, $N(N-1)/2$ coefficients characterize $C$.

[3] For more details on the methodology used and on the graph theory, please refer to [11].

### 2.3.2 From Correlations to Distances

In order to use graph theory, we need to introduce a metric. The correlation coefficient $\rho_{ij}$ cannot be used as a distance $d_{ij}$ between $i$ and $j$, because it does not fulfill the three axioms that define a metric [6]: (1) $d_{ij} = 0$ if and only if $i = j$; (2) $d_{ij} = d_{ji}$, and (3) $d_{ij} \leq d_{ik} + d_{kj}$.

A metric $d_{ij}$ can be extracted from the correlation coefficients through the following non linear transformation:

$$d_{ij} = \sqrt{2(1 - \rho_{ij})}. \tag{2.2}$$

A distance matrix $D$ is thus extracted from the correlation matrix $C$ according to (2.2). Both, $C$ and $D$ are $N \times N$ dimensional. Whereas the coefficients $\rho_{ij}$ can be positive for correlated returns or negative for anti-correlated returns, the quantity $d_{ij}$ that represents the distance between price returns is always positive. This distance matrix corresponds to the fully connected graph: it represents all the possible connections in the price system.

### 2.3.3 From Fully Connected Graphs to Minimum Spanning Trees

A simple connected graph represents all the possible connections between $N$ points with $N(N-1)/2$ links (or edges). The graph can be weighted in order to represent the different intensities of the links and/or nodes. In our case, these weights represent the distances between the nodes. For a weighted graph, the MST is the one spanning all the nodes of the graph without loops. This MST also has less weight than any other tree.

Through a filtering procedure that reduces the information space from $N(N-1)/2$ to $N-1$, the MST highlights the most relevant connections in the system. In our study, the MST provides the shortest path to linking all nodes and discloses the underlying mechanisms of systemic risk. Thus, because this tree is unique, it can be considered the easiest path for the transmission of a price shock.

## 2.4 The Topology of the Trees

The first information that a Minimum Spanning Tree provides is the kind of arrangement that exists between the vertices: its topology. We thus focus on this topology and its consequences for systemic risk. We present the results obtained with static MST, (i.e. we consider the whole time period—or some sub-sets of this period—as a single window, and we perform a static analysis).

The first step in studying the topologies lies in the visualization of the trees in the three dimensions under consideration. In a second step, we separate the whole time

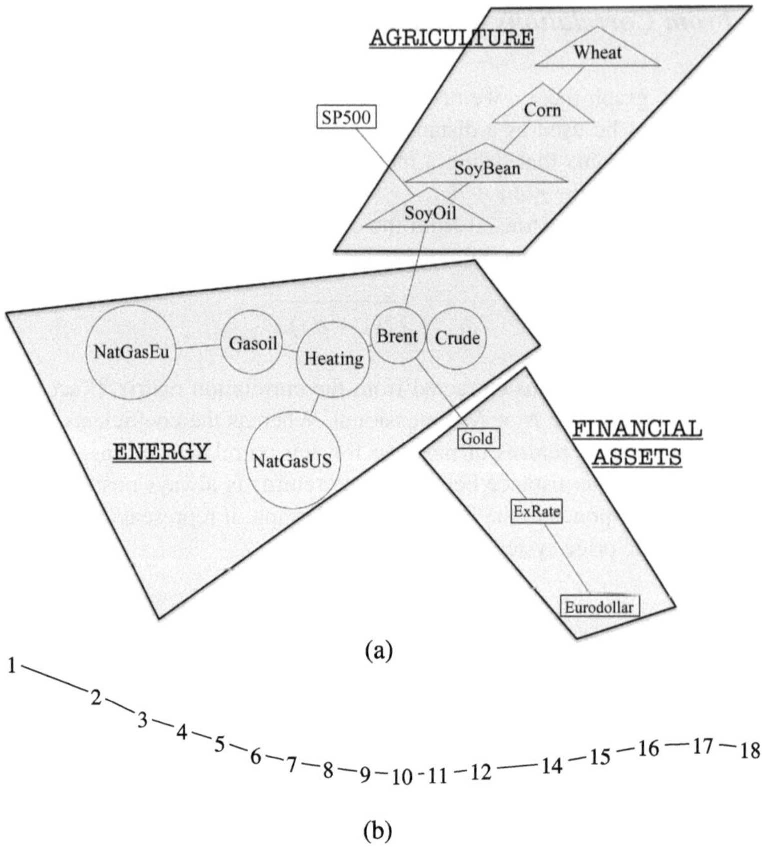


**Fig. 2.1** Static Minimum Spanning Trees built from the correlation coefficients of the prices returns. (**a**) MST in the spatial dimension (April 2001–April 2011). (**b**) MST of the Brent crude in the maturity dimension (April 2000–April 2011). The curvature only eases the visualization

period into three sub-periods and show how the topology evolves in the maturity dimension. Finally, we use allometric coefficients to determine whether the MST are totally organized, totally random, or are situated somewhere between these two extreme topologies.

### *2.4.1 The Emerging Taxonomy in the Three Dimensions*

Figure 2.1 presents the MST obtained for the spatial dimension. As far as the spatial dimension is concerned, all three sectors can be identified. Energy comprises American as well as European markets and is situated between agriculture (on the top) and financial assets (on the bottom). The most connected node in the graph is the Brent, which makes it the best candidate for the transmission of price fluctuations in the tree (actually, the same could have been said for the Crude (Light crude), as the distance between these products is very short). Further, the energy sector is the most integrated of the three sectors because the distances between the nodes are short. The link between the energy and agricultural products passes through soy oil, which can be used for fuel. The link between commodities and financial assets passes through gold, which can be seen as a commodity but also as a reserve of value. The only surprising link comes from the S&P500, which is more correlated to soy oil than to financial assets.

Such a star-like organization leads to specific conclusions regarding systemic risk. A price move in the energy markets, situated at the heart of the price system, will have more impact than a fluctuation affecting peripheral markets such as interest rates or wheat.

Things are totally different in the maturity dimension. The results are illustrated by the example of the Brent crude, depicted by Fig. 2.1b. For all contracts, the MSTs are linear and the maturities are regularly ordered from the first to the last delivery dates.

The results obtained in the maturity dimension give rise to three remarks. Firstly, the linear topology is meaningful from an economic point of view, as it reflects the presence of the Samuelson effect [17]. In derivative markets, the movements in the prices of the prompt contracts are larger than the other ones. This difference results in a decreasing pattern of volatilities along the price curve and leads to higher correlations between the maturities that are the closest to each other. Secondly, this type of organization impacts the possible transmission of price shocks. The most likely path for a shock is indeed unique and passes through each maturity, one after the other. Thirdly, the short part of the curves are less correlated with the other parts. This phenomenon can result from price shocks emerging in the physical market with the most nearby price being the most affected; it could also reflect noises introduced on the first maturity by investors in the derivative market.

Figure 2.2 represents the 3-D static MST. Its shape brings to mind the spatial dimension. However, it is enhanced by the presence of the different maturities available for each market. These maturities have a clear, linear organization. Again, the tree shows a clear separation between the sectors. Three energy contracts, the crude oil (Light crude), the Brent and the Heating oil, are at the center of the graph. They are the three closest nodes in the graph. Whereas the maturities of each market primarily have a linear organization, the American natural gas behaves differently and displays an atypical topology with numerous ramifications.

It is interesting to see which maturities connect two markets or sectors. Economic reasoning suggests that two kinds of connections should exist: with the shortest and/or with the longest part of the curves. In the first case, the price system would be essentially driven by underlying assets; in the second, it would be dominated by derivative markets. However, a closer analysis of the 3-D trees does not provide evidence of either kind. Furthermore, the analysis of the trees at different periods does not lead to the conclusion that there is something like a pattern in the way connections occur.

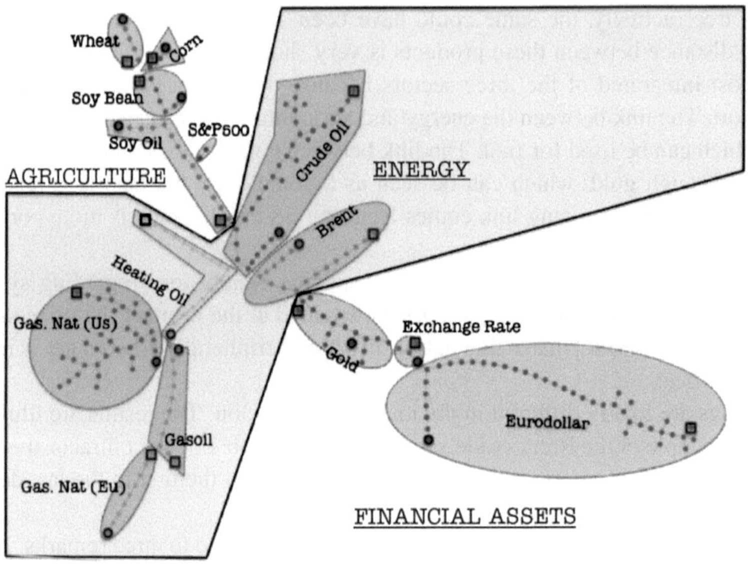


**Fig. 2.2** Static MST in 3-D (2000–2011). Each futures contract is enclosed in a shaded area with its name. The first and last maturities are respectively represented by a bold circle and a bold square. The distance between the nodes is set to unity

## *2.4.2 The Evolution of the MSTs Topology Through Time, in the Maturity Dimension*

As far as the topology in the maturity dimension is concerned, we observed, through different studies, that the linear topology exhibited by Fig. 2.1b is the result of a maturation process of the derivative market [13]. In such a market, indeed, the maturities of the futures contracts usually rise through time: the growth in the transaction volumes indeed results in the introduction of new delivery dates and extends the time horizon of arbitrage operations.

When we examine the topologies of the MST in the maturity dimension in a dynamical way, we find an illustration of this maturation process in the evolution of the topologies of the trees. Figure 2.3 gives an example of such a process through the case of the Eurodollar futures contract: in the beginning of the study period, the market is less integrated and the tree is not perfectly linear: there are branches on the latest maturities. This result is consistent with economic intuition, as the latter are always characterized by lower transaction volumes. As time goes on, however, the linearization progresses, reflecting the intensification of arbitrage operations towards longer time horizon. Moreover, the shortest maturities become perfectly ordered. As far as this maturation process is concerned, we found only one exception among the 14 markets under scrutiny: the American market for natural gas. Here, conversely to what can been observed on other markets, the MST becomes less and less linear as time goes on. More precisely, as shown by Fig. 2.4, a cyclical pattern emerges. As

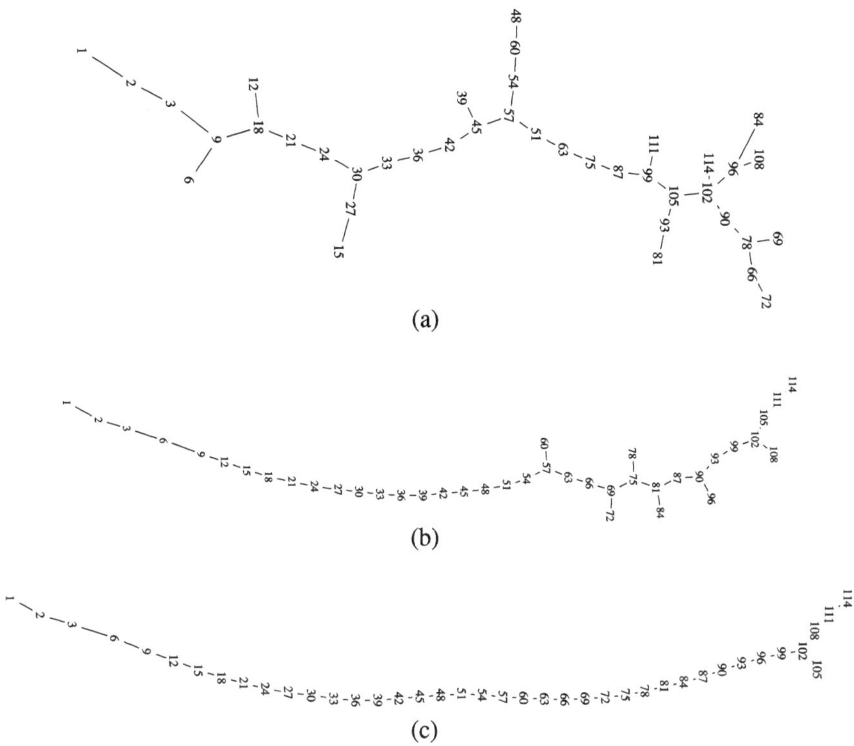


**Fig. 2.3** Time evolution of the Eurodollar futures contract. (**a**) 1998–2001. (**b**) 2001–2004. (**c**) 2005–2009

the natural gas market is characterized by a strong seasonality, we first thought that this pattern reflects this seasonality. The maturities of connections between the futures contracts, however, do not make sense with such an interpretation. Moreover, the European natural gas market does not behave similarly. A possible answer lies in a possible disorganization of the market after the difficulties encountered in 2006 with the hedge fund Amaranth. We leave such investigation for later studies.

## *2.4.3 Allometric Properties of the MST*

The computation of the allometric coefficients of a MST provides a means of quantifying where this tree stands between two asymptotic topologies: star-like trees that are symptomatic of a random organization, and chain-like trees that show a strong ordering in the underlying structure.

[1] developed the first model for the allometric scaling of a spanning tree. The first step of the procedure consists of initializing each node of the tree with the value of one. Then the root or central vertex of the tree must be identified. In what follows,

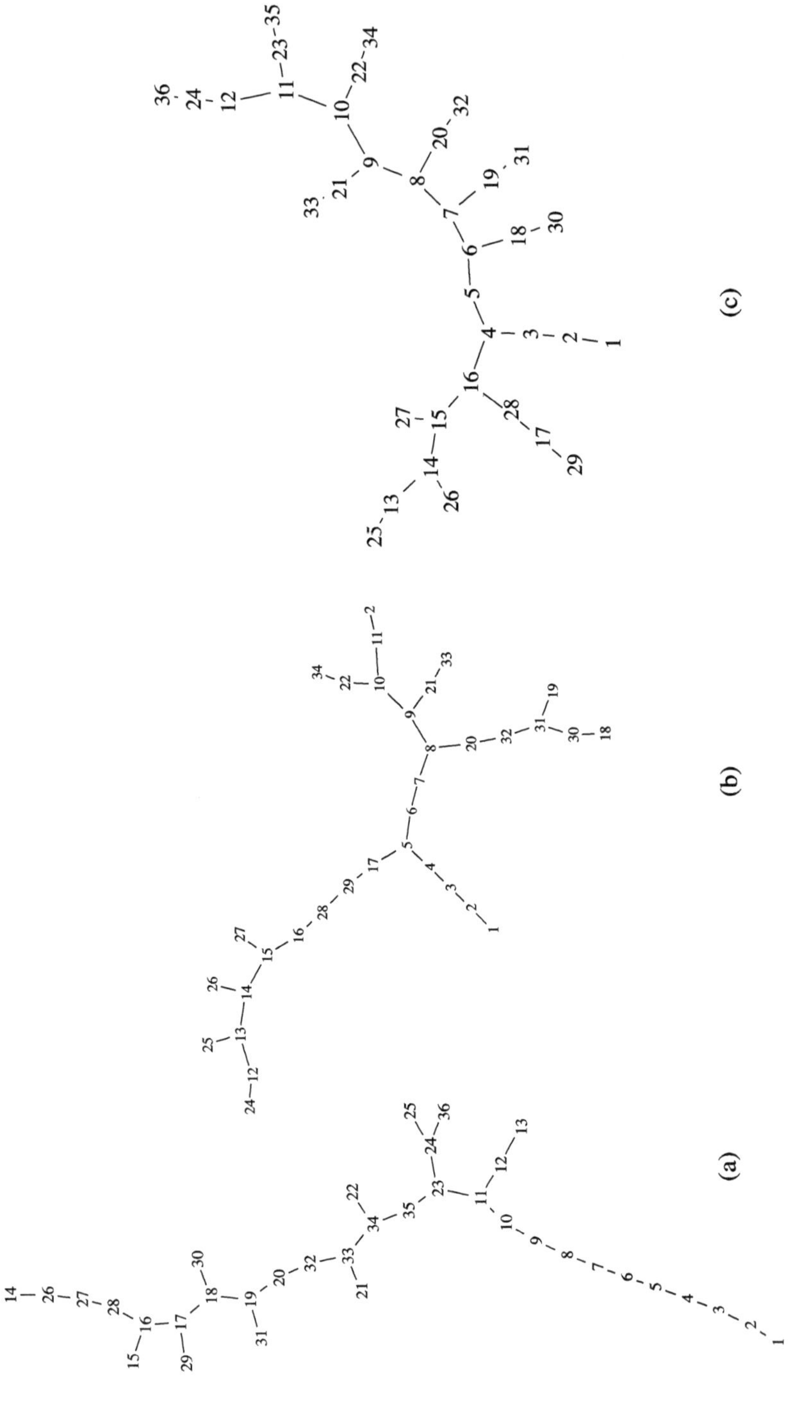


**Fig. 2.4** Time evolution of the American Natural Gas futures contract. **(a)** 1998–2001. **(b)** 2001–2004. **(c)** 2004–2009

the root is defined as the node that has the highest number of links attached to it. Starting from this root, the method consists of assigning two coefficients $A_i$ and $B_i$ to each node $i$ of the tree:

$$A_i = \sum_j A_j + 1 \quad \text{and} \quad B_i = \sum_j B_j + A_i, \tag{2.3}$$

where $j$ stands for all the nodes connected to $i$ in the MST. The definition of the allometric scaling relation is the relation between $A_i$ and $B_i$:

$$B \sim A^{\eta}, \tag{2.4}$$

where $\eta$ is the allometric exponent estimated after removing the leaf nodes $A = C = 1$ [5]. It represents the degree or complexity of the tree and stands between two extreme values: $1^+$ for star-like trees and $2^-$ for chain-like trees [16].

The main results obtained with this measure are the following:[4] within the maturity dimension, the coefficients tend towards their asymptotic value: $\eta = 2^-$, for all markets under investigation. However, they are a bit smaller than 2, due to finite size effects (there is a finite number of maturities). Such a result is probably due to arbitrage operations. When performed on the basis of contracts having the same underlying asset, such operations are easy and rapidly undertaken, thus resulting in a perfect ordering of the maturity dates. Such a scaling is appealing from an empirical as well as a theoretical point of view and suggests a possible universal behavior of the topologies of derivatives networks. Additionally arbitrage operations have also a deep impact on the prices dynamic leading to ubiquitous statistical properties of price returns along the term structure [9].

With concern for the spatial dimension, the exponents indicate that even if Fig. 2.1a exhibits a star-like organization, the shape of the MST is rather complex and stands exactly between the two asymptotic topologies. There is an ordering of the tree, which is well illustrated by the agricultural sector, which forms a regular branch. Finally, even if the topologies of the spatial and 3-D trees seem similar, they are quantitatively different. The allometric exponent for the 3-D tree is higher: the best fit from our data gives an exponent close to 1.757 as compared to the value of 1.493 in the spatial case. Thus, the topology in 3-D merges the organization in sectors induced by the spatial dimension and the chain-like organization arising from the maturity dimension.

## 2.5 Dynamical Analysis

Because it is based on correlation coefficients, our study of market integration is intrinsically time dependent. On the basis of the fully connected graph, we first examine the dynamic properties thanks to the node's strength, which provides in-

[4]For more detailed results, please see [10].

formation on how close a given node is to all others. We then turn to the MSTs. In order to study the robustness of the topologies, we compute the length of the MST, that shows the state of the system at a specific time. Survival ratios also indicate how the topology evolves over time. Finally, these survival ratios allow us a deeper investigation into the connections between markets in the 3-D analysis.

In what follows, we retain a rolling time window of $\Delta T = 480$ consecutive trading days.

### 2.5.1 The Nodes Strength in the Fully Connected Graphs

In order to examine the time evolution of our system, we investigate the nodes strength in the fully connected graph. This quantity, calculated for each node $i$, indicates the closeness of one node $i$ to all others. It is defined as follows:

$$S_i = \sum_{i \neq j} \frac{1}{d_{ij}}. \tag{2.5}$$

In our case, the node strength provides information on the intensity of the correlations linking a given node to the others. When $S_i$ is high, the node is close to all others. For the sake of simplicity, we use this measure in the spatial dimension only. As far as the maturity dimension is concerned, it was indeed not easy to represent the nodes strength for all futures contracts.[5] Figure 2.5 represents the time evolution of the nodes strength in the spatial dimension. The figure has been separated into four panels: the energy sector is at the top, with American products on the left and European ones on the right; the agricultural sector is at the bottom left and financial assets are at the bottom right.

Figure 2.5 shows that, at the end of the period, out of all the assets studied, the two crude oils and Heating oil show the greatest nodes strength. However, since 2010, the American node strength has decreased, which indicates a difference in the connectivity of the two crude oils. This is an interesting result, as there are, indeed, delivery problems on the American crude oil since that date. These problems raise the question of the relevance of the Light crude oil as a worldwide benchmark. The petroleum products are followed by soy oil, other agricultural assets, the S&P500 contract, gold, the exchange rate USD/EUR, and the gasoil. A remarkable evolution is the sharp rise in the equity connectivity in the post-Lehman period, as opposed to 2001–2007. This finding corroborates those of [3] and [18]. Finally, the more distant nodes are those representing the Eurodollar and the two natural gases.

As far as the time evolution of the node strength is concerned, the sectors exhibit different patterns: the integration movement, characterized by an increase in this measure, emerges earlier for the energy sector than for the agricultural sector. However, it decreases for energy at the end of the period (especially for the Light crude

[5] For more information on these results, please see [8].

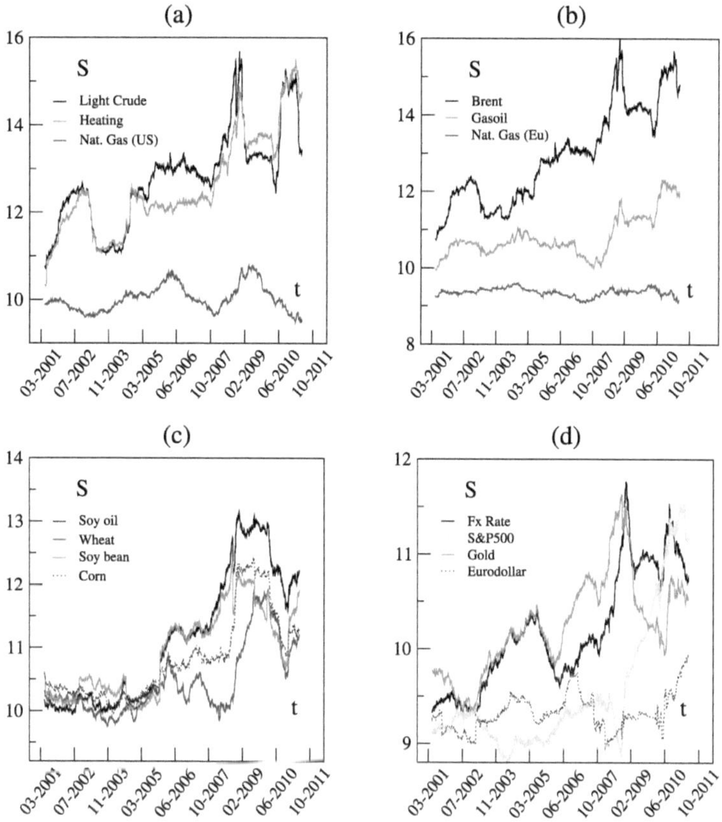


**Fig. 2.5** Nodes strength of the markets in the spatial dimension 2001–2011. (**a**) American energy products. (**b**) European energy products. (**c**) Agricultural products. (**d**) Financial assets

oil). The nodes strength of the agricultural products is characterized by a plateau from the middle of 2009 to the beginning of 2010, followed by a drawdown until the Fall 2010. Last but not least, most of the products exhibit a strong increase, except for natural gas and interest rate contracts. Thus, whereas the core of the graph becomes more and more integrated, the peripheral assets do not follow this movement.

## *2.5.2 The Length of the Minimum Spanning Trees*

The normalized tree's length can be defined as the sum of the lengths of the edges belonging to the MST:

$$\mathcal{L}(t) = \frac{1}{N-1} \sum_{(i,j)\in MST} d_{ij}, \tag{2.6}$$

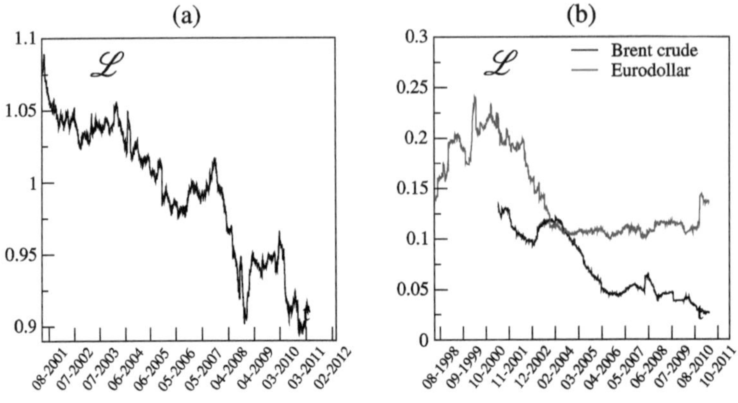


**Fig. 2.6** Time evolution of the normalized tree length. (**a**) Spatial dimension. (**b**) Maturity dimension for the Brent crude (*black line*) and the Eurodollar (*gray line*)

where $t$ denotes the date of the construction of the tree and $N-1$ is the number of edges. The length of a tree is longer as the distances increase, and consequently when correlations are low. Thus, the more the length shortens, the more integrated the system is. On the contrary, in the case of random co-movements, the length of the tree is equal to $\sqrt{2}$.

Figure 2.6 represents the dynamic behavior of the normalized length of the MSTs in the spatial and in the maturity dimensions. In the spatial dimension, the general pattern is that the length decreases, which reflects the integration of the system. This information confirms what was observed in the fully connected graph on the basis of the nodes strength. In addition, it shows that the most efficient transmission path for price fluctuations becomes shorter as times goes on. A more in-depth examination of the figure also shows a very important decrease between October 2006 and October 2008, as well as significant fluctuations in September and October 2008. We leave the analysis of such events for future studies.

In the maturity dimension, as integration increases, the normalized tree's length also diminishes. Figure 2.6 illustrates this phenomenon by representing the evolutions recorded for the Eurodollar contract and Brent crude. As far as the interest rate contract is concerned, the tree's length first increases, then in mid-2001 it drops sharply and remains fairly stable after that date. For crude oil, the decrease is constant and steady, except for a few surges.

### *2.5.3 Survival Ratios, the Stability of the Prices System and the Interconnections Between Markets*

The robustness of the MSTs over time is examined by computing the single-step survival ratio of the links, $S_R$. This quantity refers to the fraction of edges in the

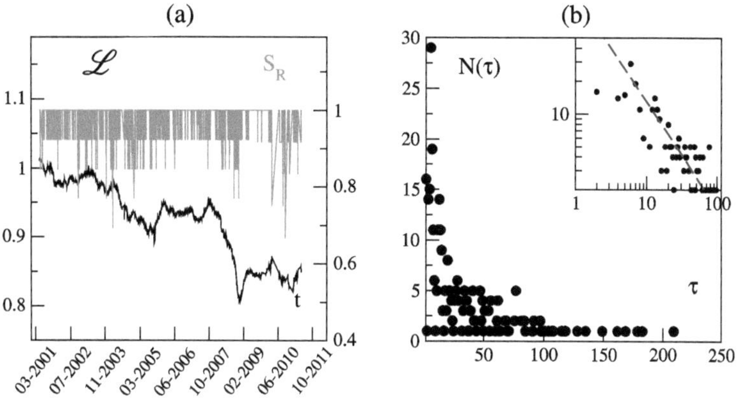


**Fig. 2.7** Properties of the pruned trees. (**a**) Survival ratios and pruned tree length. (**b**) Number of occurrences of stable periods of length $\tau$. *Inset*: same as in (**b**), but in log–log scale. The *dashed line* corresponds to $\tau^{-1}$

MST, that survives between two consecutive trading days [14]:

$$S_R(t) = \frac{1}{N-1}\big|E(t) \cap E(t-1)\big|. \tag{2.7}$$

In this equation, $E(t)$ refers to the set of the tree's edges at date $t$, $\cap$ is the intersection operator, and $|.|$ gives the number of elements contained in the set. The survival ratios are very important for our study. Under normal circumstances, the topology of the trees should be very stable and the value of the survival ratio around one.

Concerning the stability of the trees, especially in 3-D, when focusing on the whole system, it is interesting to distinguish between reorganizations occurring in a specific market (i.e., between different delivery dates of the same contract) and reorganizations that change the nature of the links between two markets or even between two sectors. However, (2.7) gives the same weight to every kind of reorganization, whatever its nature. The trouble is, a change in intra-maturity links does not have the same meaning, from an economic point of view, as a movement affecting the relation between two markets or sectors. Because we are interested in the strong events that affect the markets, inter-market and inter-sector reorganizations are more relevant.[6] Thus, in order to distinguish between these categories of displacements, we "prune" the 3-D trees, that is, we only consider the links between markets, whatever the maturity considered. This pruning does not mean that maturity is removed from the analysis, but that the information on the specific maturity that is responsible for the connection between markets is no longer identified.

Pruned trees enable us to compute the length and the survival ratios on the sole basis of market links. As shown by Fig. 2.7a, most of the time, the survival ratios remain constant, with a value greater than 0.9. Thus, the topology of the trees is

[6]One can find all the results concerning the survival ratios in [10].

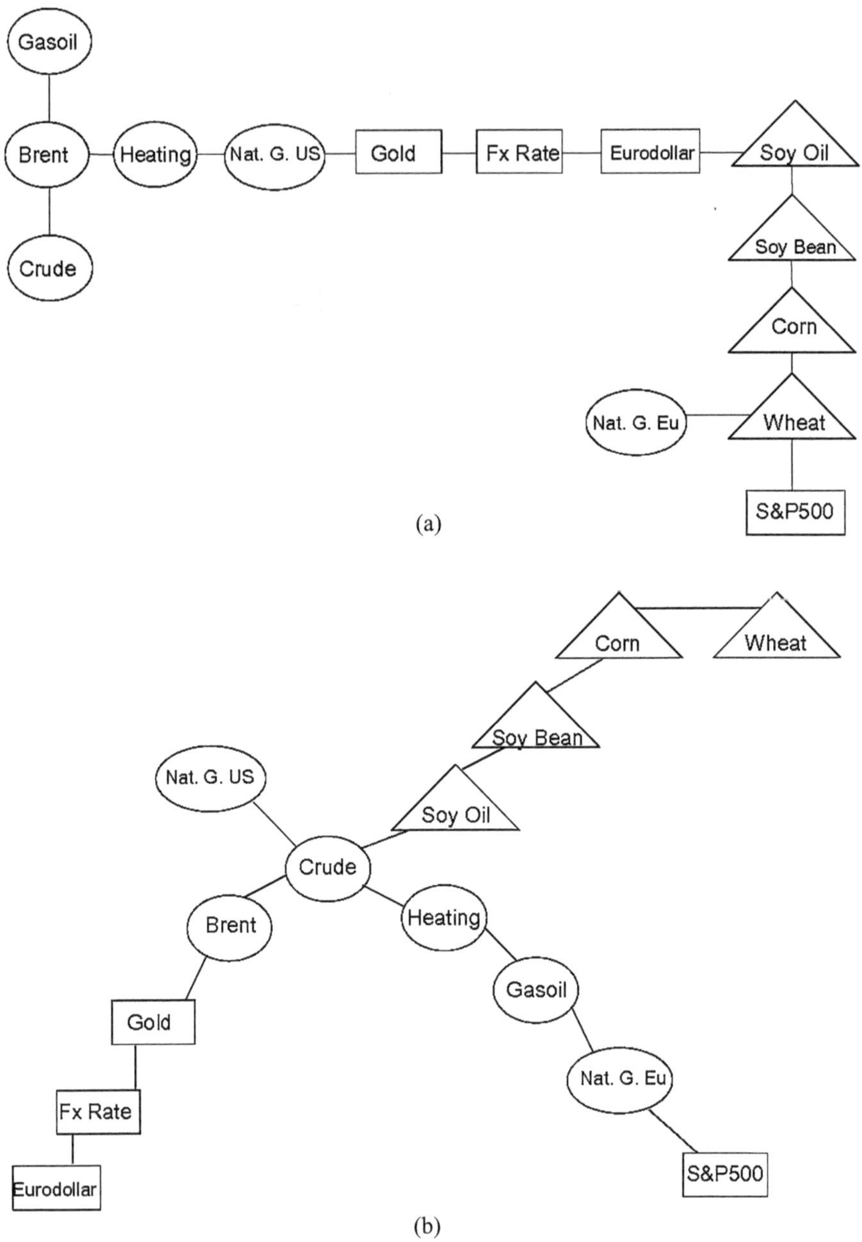


**Fig. 2.8** Pruned MST of the events 09/02/2004 (**a**) and 16/09/2008 (**b**)

very stable: the shape of the most efficient path for the transmission of price shocks does not change much over time.

Another interesting characteristic of the pruned survival ratios is that they provide information on the lifetime of a configuration of such trees. In what follows, we measure the length of time $\tau$ between two different consecutive configurations and compute the occurrences $N(\tau)$ of these periods. Figure 2.7b displays our results. It shows that $N(\tau)$ decreases quickly with $\tau$. The dashed line in the inset (in log–log scale) suggests that $N(\tau)$ is roughly proportional to $\tau^{-1}$. Such a scaling behavior indicates that there is neither a typical nor an average lifetime of a new configuration of the MST.

On Fig. 2.7a, it is possible to identify several events which caused a significant rearrangement of the tree. This is the case, for example, for two specific dates, namely 02/09/04 and 09/16/08, where 30 % of the edges has been shuffled. As illustrated by Fig. 2.8, a focus on these two dates shows that the trees are totally rearranged. In 2004, the MST becomes highly linear, the financial assets sector is at the center of the graph, and commodities appear mainly at the periphery of the system. Conversely, in 2008, the tree has a typical star-like shape showing an organization based on the different sectors studied.

## 2.6 Conclusions

We study systemic risk in energy derivative markets based on two choices. First, we focus on market integration, which is a favorable condition for the propagation of a price shock. Second, based on the fact that previous studies mainly focus on the spatio-temporal dimension of integration, we introduce the maturity dimension and perform a three-dimensional analysis.

In the context of empirical studies aiming to understand the organization and the dynamic behavior of a highly dimensional price system, our methodology, based on graph-theory, proves very useful. Moreover, Minimum Spanning Trees are particularly interesting in our framework, as they are filtered networks enabling us to identify the most probable and the shortest path for the transmission of a price shock.

We show that the topology of the MSTs tends towards a star-like organization in the spatial dimension, whereas the universal linear topology characterizes the maturity dimension. These two topologies merge in the 3-D analysis, and all of them are very stable. The star-like organization reproduces the three different sectors studied (energy, agriculture, and finance), and the chain-like structure reflects the presence of a Samuelson effect. The reasoning behind these findings is very important: the robustness of our methodology is embedded in these topologies.

Another contribution is to show that the American and European crude oils are both at the center of our large scale system; furthermore, they provide the links with the subsets of agricultural products and financial assets. Thus, crude oil is the best candidate for the transmission of price shocks. If such a shock appears at the periphery of the graph, it will necessarily pass through crude oil before spreading to other energy products and sectors. Moreover, a shock will have an impact on the whole system that will be all the greater the closer it is to the heart of the system.

Another important conclusion is that integration increases significantly on both the spatial and maturity dimensions. Such an increase can be observed in the whole price system. It is even more evident in the energy sector (with the exception of the natural gas markets), which becomes highly integrated at the end of our period. Thus, as time goes on, the heart of the price system becomes stronger whereas the peripheral assets do not change significantly. Moreover, the level of integration is higher in the maturity than in the spatial dimension: arbitrage operations are easier with standardized futures contracts written on the same underlying asset.

These results have very important consequences for regulatory as well as for diversification and hedging purposes.

Whereas the move towards integration started some time ago (and there is probably no way to refrain it), knowledge of its characteristics remains poor, especially from a holistic perspective. On the basis of this study, regulation authorities can see that their actions against systemic risk will not have the same impact depending on the market they are addressing. They should pay particular attention to the heart of the system: this is the place where price shocks spread more easily to other markets.

As far as diversification is concerned, portfolio managers must concentrate their positions on the most stable parts of the graph. More precisely, the benefits associated with diversification that rely on sub-indexes and focus on specific sectors of activity (agricultural products for example) should be more recurrent than those associated with large scale indexes.

Lastly, one important concern for hedging is the information conveyed by futures prices and its meaning. The increasing integration of derivative markets is probably not a problem for hedging purposes, until a price shock appears somewhere in the system. In such a case, the information related to the transmission path of the shock is important, as prices might temporarily become irrelevant.

These results call for further work. First, as previously exhibited, survival ratios make it possible to identify a few events leading to important reconfigurations of the trees. A thorough analysis of such phenomenon can provide the regulating authorities with a battery of stylized facts about the different possible manifestations of prices shocks and the signs announcing a future shock. Second, now that we have defined the paths for shock transmission, it is important to obtain directed graphs to determine the direction of the propagation of price movements. Third, a focus on the gas market, which exhibits a striking pattern of cross-maturity connections, can be of interest for energy specialists.

**Acknowledgements** The financial support of the French Energy Council and of the Finance and Sustainable Development chair is gratefully acknowledged.

## References

1. Banavar J, Maritan A, Rinaldo A (1999) Size and form in efficient transportation networks. Nature 399:130
2. Buyukşahin B, Haigh MS, Robe MA (2010) Commodities and equities: ever a "market of one"? J Altern Invest 76–95
3. Buyuksahin B, Robe MA (2010) Speculators, commodities and cross-market linkages. Working Paper, American University, Nov 2010

4. Chong J, Miffre J (2010) Conditional returns correlations between commodity futures and traditional assets. J Altern Invest 12:61–75
5. Garlaschelli D, Caldarelli G, Pietronero L (2003) Universal scaling relations in food webs. Nature 423:165–168
6. Gower JC (1966) Some distance properties of latent root and vector methods used in multivariate analysis. Biometrika 53(3/4)
7. Korniotis GM (2009) Does speculation affect spot prices levels? The case of metals with and without futures markets. Finance and Economics. Discussion Series Working Paper 2009-29. Board of Governors of the Federal Reserve
8. Lautier D, Raynaud F (2011) Energy derivative markets and systemic risk. Technical report, French Energy Council
9. Lautier D, Raynaud F (2011) Statistical properties of derivatives: a journey in term structures. Phys A, Stat Mech Appl 390(11):2009–2019
10. Lautier D, Raynaud F (2012) Systemic risk in energy derivative markets: a graph theory analysis. Energy J 33(3):217–242 (forthcoming)
11. Lautier D, Raynaud F (2012) High dimensionality in finance: the advantages of the graph-theory. In: Batten J, Wagner N (eds) Derivatives securities pricing and modelling, Emerald Publishing, June
12. Mantegna RN (1999) Hierarchical structure in financial markets. Eur Phys J B 11
13. Milonas N (1986) Price variability and the maturity effect. J Futures Mark 6:443
14. Onnela J-P, Chakraborti A, Kaski K, Kertész J (2003) Dynamic asset trees and black monday. Physica A 324
15. Onnela J-P, Chakraborti A, Kaski K, Kertész J, Kanto A (2003) Dynamics of market correlations: taxonomy and portfolio analysis. Phys Rev E 68(5):056110
16. Qian M-C, Jiang Z-Q, Zhou W-X (2010) Universal and nonuniversal allometric scaling behaviors in the visibility graphs of world stock market indices. J Phys A, Math Theor 43(33):335002
17. Samuelson PA (1965) Proof that properly anticipated prices fluctuate randomly. Ind Manage Rev 6
18. Tang K, Xiong W (2011) Index investing and the financialization of commodities. NBER, Working Paper No. 16325

# Chapter 3
# Omori Law After Exogenous Shocks on Supplier-Customer Network

**Yoshi Fujiwara**

**Abstract** We study the relaxation process of a supplier-customer network after mass destruction due to two giant earthquakes, Kobe 1995 and East Japan 2011, by investigating the number of chained failures. Firstly, a mass destruction and intervention of business activities in the damaged areas can be considered as a main shock. The exogenous shock was propagated on the supplier-customer network deteriorating financial states of other firms, even if they are not located in geographical neighbors. To quantify such aftershocks, we use chained failures on the network assuming that they indicate the trace of propagation of shocks. We show that the number of chained failures in its temporal change obeys an Omori-law, a power-law relaxation. This finding implies that the relaxation is much more sluggish than one would naively expect, and that it might be possible to estimate the extent and duration of aftershocks by using the empirical law. Several issues are discussed including the origin of the long-time relaxation.

## 3.1 Introduction

The Nobi Earthquake in 1891 is the largest inland earthquake in the recorded history of Japan. A century after the earthquake witnessed a long-time relaxation of aftershocks. Seismologist, Fusakichi Omori, in his pioneering work [1], found that the long-time relaxation obeys a power-law decay for the number of aftershocks by using the Nobi and two other earthquakes. This is known as *Omori-law* today. Surprisingly, after the centenary of his discovery, the aftershocks obey the Omori-law even until today for more than $10^4$ days, a long relaxation, indeed (see Fig. 1 in the review paper [2]).

Interestingly, social and economic systems as well as natural systems have been found to possess long-time relaxation similar to the Omori-law. Examples include financial market crashes [3], book sales ranking dynamics accompanying shocks [4], and so forth. See the references in [3, 4] for other such systems. These empirical

Y. Fujiwara (✉)
Graduate School of Simulation Studies, University of Hyogo, Kobe 650-0047, Japan
e-mail: yoshi.fujiwara@gmail.com

F. Abergel et al. (eds.), *Econophysics of Systemic Risk and Network Dynamics*, New Economic Windows,
DOI 10.1007/978-88-470-2553-0_3, 

findings, their models and theoretical explanations are addressing a same important question—how is the dynamics of such a complex system affected and reacting to endogenous and/or exogenous shocks? This is important for understanding the dynamics, predicting relaxation processes after shocks, and giving precursory signals before a main shock in some cases.

In the present paper, we consider two cases of "main shocks"; each case is a mass destruction of supplier-customer network at a nation-wide scale in Japan. They are large earthquakes of the East Japan Earthquake in 2011 and the Kobe Earthquake in 1995. And we shall study subsequent "aftershocks" in the economy by observing chained failures on the network.

According to the 2011 White Paper on Small and Medium Enterprises in Japan [5], a large number of firms were directly damaged by the East Japan Disaster due to the earthquake, tsunami, nuclear and other disasters, in the Tohoku area along the north-east coast. This primary and exogenous shock resulted in business failures at a considerable scale. Firstly, a mass destruction of firms and a set of intervention in many industrial sectors and geographical locations took place in those areas. Secondly, financially fragile firms were eventually forced to cease their business activities and went into bankruptcies or financially ill-conditions, even if they are not directly damaged. If these firms are irreplaceable in the supplier-customer links, other firms in the network were influenced. Thirdly, such influences were so serious that other firms in upstream or downstream side of the firms on the network eventually went into bankruptcies or financially ill-conditions as a kind of *Tsunami* effect.

These chained failures did not necessarily take place in geographical neighbors adjacent to the north-east coast, but actually took place at geographically remote areas as depicted in Fig. 3.1. The top figure shows the numbers of chained failures, which are carefully defined and investigated as we shall see, in all 47 prefectures. The bottom one shows the total amounts of debt of those failed firms. One can see that not only the Tokyo prefecture (at center) but also the northeast prefecture of Hokkaido (top) and east prefectures in Kyushu island (leftmost) were influenced seriously, even if they are all located far away from the Tohoku area.

The paper is organized as follows. In Sect. 3.2 we define aftershocks in terms of chained failures, where a chain refers to neighborhood on the supplier-customer network. We give a basic idea behind the method of counting the number of chained failures. In Sect. 3.3, we shall investigate the validity of Omori law by using two episodes of the Kobe and East Japan. We show that the aftershocks, quantitatively measured by the cumulative number of chained bankruptcies, satisfy a modified Omori formula. We shall discuss about several points for our preliminary results of this finding in Sect. 3.4. Section 3.5 summarizes our conclusion.

## 3.2 Aftershocks as Chained Failures

Although we do not employ the network structure directly as a dataset, it would be helpful to give a picture of underlying network data, which we shall briefly describe

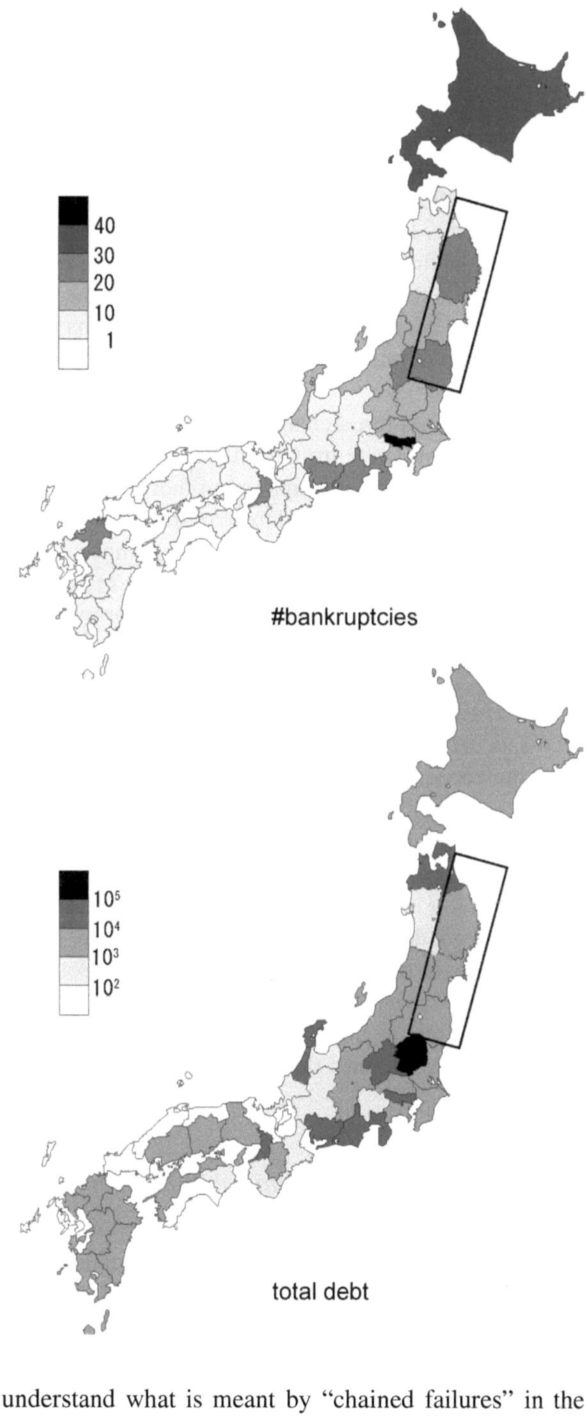


**Fig. 3.1** *Top*: the number of chained failures caused by the East Japan Earthquake in each prefecture is shown in gray levels. The central prefecture is Tokyo with the highest level, while the north island (Hokkaido) and the east island (Kyushu) were also influenced. *Bottom*: the total amount of debt of those failed firms in each prefecture is shown with the amount in million Yen. *Both*: the tilted rectangular boxes are the part of Tohoku area along the north-east coast, directly damaged by the natural disaster of earthquake and tsunami, and also by the nuclear disaster. *Data*: compiled on January 5 (2012) [6]

in Sect. 3.2.1 in order to understand what is meant by "chained failures" in the definition of aftershocks given in Sect. 3.2.2.

### 3.2.1 Supplier-Customer Network

Our underlying data for supplier-customer links at a nation-wide scale is based and compiled on the following idea. Let us say that a directional link is present as $u \to v$ in a supplier-customer network among firms, where firm $u$ is a supplier to firm $v$, or equivalently, $v$ is a customer of $u$. It would be difficult to know every transaction between firms. However, it is pointless to have a record that a firm buys a pencil from another firm. Necessary for understanding macroscopic economy are data of links such that the relation $u \to v$ is crucial for the business activity of either or both of $u$ and $v$. Therefore, if at least one of the firms at either end of a link nominates the other firm as one of most important suppliers or customers, then the link should be listed.

Leading credit research agencies in Tokyo [6, 7] regularly gather credit information on most of active firms from detailed survey at branch offices located across the nation. The credit information of individual firm includes its suppliers and customers, which are considered to be most crucial for its business activity. We assume that the gathered set of links are playing important roles in the supplier-customer network. In the compiled data, the number of firms is roughly a million, and the number of directional links is more than four million. The set of nodes in the network covers essentially most of the domestic firms that are active in the sense that their credit information is required.

In addition, the agencies compile every day exhaustive lists of bankrupted firms. The data are exhaustive in the sense that any bankrupted firm with a total amount of debt exceeding 10 million yen (roughly 70 thousand euro or 100 thousand US dollar) is listed therein. Each record includes the date of failure, the total amount of debt when bankrupted and categorized causes of bankruptcy. A bankruptcy or business failure is a critical financial state of a firm; its debt dominates its balance-sheet so that it has little equity, and the firm cannot no longer manage its business, although legal definitions differ for cases.

To put these two datasets together, one can define chained failures or bankruptcies as defined in what follows. See [8] for more details of the network and bankruptcies data, empirical analysis for network structure including hierarchical communities or modules, chained failures and corresponding avalanche-size distribution.

### 3.2.2 Aftershocks as Chained Failures

For the case of the East Japan Earthquake,because more than 0.7 million firms were present in the region (see [5]) among two million firms in the entire country,[1]the

[1]The total number of firms is based on a census taken by the National Tax Administration Agency. Other censuses seem to give overestimation including inactive firms or underestimation due to cursory survey.

**Table 3.1** Scheme for classification of bankruptcy causes. A bankruptcy can be caused by two or more combination of these classified effects; the most dominant cause is recorded in such a case

| Classification | Criterion for bankruptcy cause |
|---|---|
| solo failure | Poor performance in business, which includes business depression, excessive competition and extrinsic shocks |
| | Loose management, which includes failure of speculative investment, internal conflict and lack of efficient management |
| | Long-term accumulated deficit |
| | Insufficient working capital |
| | Accidental causes (disasters etc.) |
| | Deterioration of products in inventory |
| | Excessive investment in facilities and equipment |
| link effect | Secondary effect from bankruptcy of customer, subsidiary or collateral companies and failure of business-related firms |
| | Failure of accounts receivable |
| others | Refusal of credit by financial institutions |
| | Unclassified |

suppliers and customers, who depends their business on the damaged firms and the regional economy, were affected afterward. For example, suppliers had a delay, or often a loss, in the receipt of accounts receivable, causing an abrupt drop in sales which may have deteriorated their financial conditions subsequently.

We shall focus on bankruptcies in order to measure such secondary effects. A bankruptcy is a critical financial state of a firm in its business activity as described in the preceding section, so the data of bankruptcies are quite adequate for our purpose. Because the secondary effects propagate along the supplier-customer network, a bankruptcy may be regarded as a fracture under a stress strengthened in the neighbor of preceding increase of stress, namely a financial fragility. We assume that the process of stress propagation can be traced by observing such chained failures on the production network.

We employ two datasets by the leading credit research agencies, which carefully identified the causes of bankruptcies in exhaustive lists of domestic failures [6, 7]. The details of identification and classification are given in [9]. We can summarize them in Table 3.1.

Furthermore, they carefully investigated the causes of bankruptcies after the Kobe Earthquake in 1995 and the East Japan Earthquake in 2011. The solo failures and link effects may have been originated from business depression and other bankruptcies in the primary shock. A typical case of solo failures is an abrupt drop of sales due to an exogenous shock in a damaged region, when the firm depends crucially on the region for its sales. Typical link effects are what we already mentioned above. We define that these cases correspond to *chained failures* in secondary effects due to the main shock. Explicitly, the cases are:

- Direct type: failure due to severe damage of facilities, infra-structures, factories, machines and so forth due to the disaster;

- Indirect type: failures of financial fragile firms due to afterward effect, such as depression, drop in sales, loss of receivables and so forth;

while the cases of latent fragility prior to the main shock, but later manifested as bankruptcies, are carefully excluded. Figure 3.1 depicted the number and total assets of chained failures due to the East Japan Earthquake. We assume that they indicate the trace of propagation of shocks.

## 3.3 Omori-Law for Aftershocks on Supplier-Customer Network

The Omori law is a power-law decay of aftershock activity with time after a large earthquake. It states that the frequency of aftershocks per unit time, $n(t)$ at time $t$ decays as

$$n(t) = K(t+c)^{-p} \tag{3.1}$$

where $K$ is a positive constant which determines the magnitude of $n(t)$ and $c$ is a positive constant to avoid divergence at the origin, $t = 0$. $p > 0$ is the exponent of the power-law decay. The original proposal corresponds to the case $p = 1$, and was later modified into the above form, a modified Omori formula (see [2]).

Equivalently, but more suitable for comparison with data, the cumulative number $N(t)$ can be written by the following:

$$N(t) = \int_0^t dt' n(t') = \frac{K}{1-p}\left[(t+c)^{1-p} - c^{1-p}\right] \tag{3.2}$$

for $p \neq 1$.

Because $N(t)$ is accumulation of $n(t)$, the statistical fluctuation and measurement errors in $N(t)$ is much reduced compared to them in $n(t)$. It is customary to measure $N(t)$ to estimate the parameters of $p$ and $K, c$ (see [3]).

Let us put the monthly data of chained failures described in Sect. 3.2 into the cumulative number $N(t)$, where $t$ is the time of occurrence of the extreme shocks. The result is shown in Fig. 3.2.[2]

For the Kobe Earthquake in 1995, the chained failures were recorded for three years by Teikoku Data Bank, Inc. (TDB) [7] (data 1); for one year by Tokyo Shoko Research, Inc. (TSR) [6] (data 2). We fit the data of TDB by (3.2) in a nonlinear least-square fitting. The parameter $p$ was estimated as $p = 0.68 \pm 0.18$ with the standard error. Although the two data have different magnitude of $N(t)$ based on independent investigation, they simply gives different scales of the parameter $K$ if they obey a same power-law. The data 2 is rescaled vertically in Fig. 3.2, which can be fitted well by the power-law with the same exponent $p$.

[2]It should be remarked that earlier result based on a fewer number of data for the East Japan Earthquake is written in [10]. The present paper gives extended results and detailed description of the network.

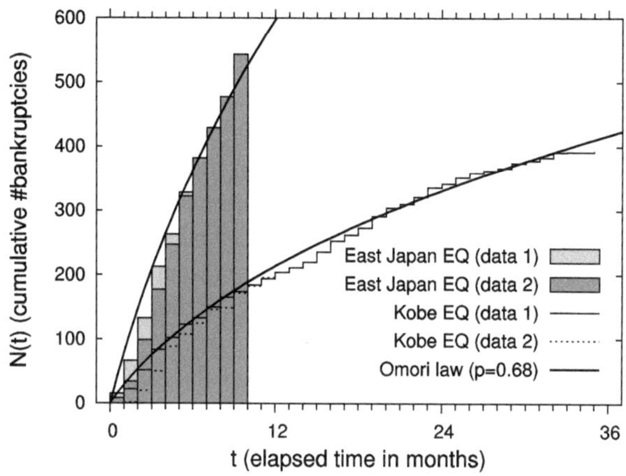


**Fig. 3.2** Cumulative number of bankruptcies $N(t)$ and elapsed time $t$ since the events of two earthquakes, Kobe 1995 and East Japan 2011. *Step lines* are two data $N(t)$ of the Kobe case for three years by two credit research agencies, TDB [7] (*data 1*) and TSR [6] (*data 2*) (the latter is scaled with its number for comparison). *Gray bars* are monthly data of the East Japan case for ten months by the two agencies. The *two curves* are modified Omori formula with the same parameter $p = 0.68$, which is estimated by least-square fit for the Kobe data

Similarly, although the data for the East Japan Earthquake has an elapsed time of ten months, it is observed that the fitting is satisfactory with respect to the same power-law (see gray bars for the actual data and a line for the Omori law). One can also see that the East Japan case has a much larger impact in terms of the overall scale in the number of failures. The parameters $p$ and $K$ are to be determined by new data available in coming months.

Note that while the data 1 was for three years for the Kobe case, longer investigation might reveal that the effect is longer than the first three years. Actually, if one naively extrapolates the Omori law, there is a possibility that chained failures had been still going on beyond into the fourth or even longer years.

We have shown that the Omori law holds for the number of aftershocks measured by chained failures after the exogenous shocks of the two large earthquakes. Let us discuss about the implication of this finding and also necessary verification to be done in our future work.

## 3.4 Discussion

While our result in the preceding section is encouraging, there are several issues to be investigated further. We shall discuss about them in this section.

First, one could employ other methods of measuring aftershocks, especially with "magnitudes". We used the number of chained failures on the production network in order to measure aftershocks, because a bankruptcy is the most clear evidence that a firm is in a so critically deteriorated state that it fails under a stress caused by the chained failures. In a more mild case, however, a firm can be affected seriously in its financial state but may not go into bankruptcy yet. If one could measure the magnitude of stress by an abrupt drop of sales, a sudden increase of debt and so forth, for example, one would be able to quantify the extent of aftershocks including their magnitude. This method will be pursued in future.

Second, if our picture of long-time relaxation captures the dynamics on the supplier-customer network, one should be able to find similar phenomena after other economic shocks such as the Lehman shock, the present and past financial crises. We believe the above mentioned method might help to analyze other economic shocks as well.

Third, let us consider possible implication of the Omori law. The fact that the relaxation process obeys a power-law implies that the influence of secondary effects after the primary damage in the network is extremely sluggish. Figure 3.2 indicates two things—duration of relaxation and its extent. Suppose that one can observe the first half-year, for instance, then by fitting the data by a modified Omori formula with a set of parameters, one can predict how long the relaxation will take and how large effects it will bring about under the hypothesis of power-law relaxation. What is more important than such a prediction is the possibility that one can identify industrial sectors and geographical regions that are fragile or robust under such a propagation of failures. Such information would be useful to plan a recovery of and investment into sectors and regions.

Fourth, let us mention about the origin of Omori law. The presence of a power-law relaxation is often found in various complex systems (see [3, 4] for example, and references therein). The origin of Omori law even in the case of earthquake is not fully understood. Nevertheless, many proposed mechanisms are based on the basic idea that the entire region of aftershocks is composed of a number of small and heterogeneous regions containing faults, among which the stress of main earthquake is released subsequently with delayed fractures of the faults. The idea of delayed fractures is quite analogous to our case. The network has a link effect [11, Chap. 6.2] in which each past failed firm may quite possibly give a stress onto other firms in its neighbor with the risk of failure. This influence of one failure onto others is not instantaneous as firms would react at different time-scale depending on financial conditions. This latency causes aftershocks of aftershocks superimposing a number of Omori-type relaxation. This approach similar to [4] seems promising to understand the value of $0 < p < 1$, being different from the earthquake aftershocks (usually $p > 1$ [2]).

## 3.5 Summary

We studied the relaxation process of a supplier-customer network after mass destruction due to two giant earthquakes, Kobe 1995 and East Japan 2011. Specifically, we investigated the number of chained failures as "aftershocks", which is defined and measured by observed chained failures on the production network. In such episodes of disasters, a mass destruction and intervention of business activities is considered as an exogenous main-shock. The exogenous shock was propagated on the supplier-customer network deteriorating financial states of other firms, even if they are not located in geographical neighbors. Assuming that the chained failures indicate the trace of propagation of aftershocks, we show that the number of chained failures in its temporal change obeys an Omori-law, a power-law relaxation.

Our finding in this paper, at this preliminary stage, implies that the relaxation is much more sluggish than one would naively expect, and that it might be possible to estimate the extent and duration of aftershocks by using the empirical law. There are several issues including other methods of quantifying aftershocks, consideration of magnitudes of aftershocks, study of economic disasters rather than natural ones, the origin of Omori law, and so forth.

**Acknowledgements** The author would like to thank the organizers, F. Abergel, B.K. Chakrabarti, A. Chakraborti, A.K. Ghosh of the Econophys-Kolkata VI "Econophysics of systemic risk and network dynamics" for invitation and hospitality. This work is supported in part by the Program for Promoting Methodological Innovation in Humanities and Social Sciences by Cross-Disciplinary Fusing of the Japan Society for the Promotion of Science.

## References

1. Omori F (1894) On aftershocks of earthquakes. J Coll Sci, Imp Univ Tokyo 7:111–200
2. Utsu T, Ogata Y, Matsu'ura RS (1995) The centenary of the Omori formula for a decay law of aftershock activity. J Phys Earth 43:1–33
3. Lilo F, Mantegna RN (2003) Power-law relaxation in a complex system: Omori law after a financial market crash. Phys Rev D 68(1):016119
4. Sornette D, Deschatres F, Gilbert T, Ageon Y (2004) Endogenous versus exogenous shocks in complex networks: an empirical test using book sale ranking. Phys Rev Lett 93(22):228701
5. The Small and Medium Enterprise Agency, Japanese Government. http://www.chusho.meti.go.jp
6. Tokyo Shoko Research, Ltd. http://www.tsr-net.co.jp
7. Teikoku Databank, Ltd. http://www.tdb.co.jp
8. Fujiwara Y, Aoyama H (2010) Large-scale structure of a nation-wide production network. Eur Phys J B 77(4):565–580
9. Fujiwara Y (2008) Chain of firms bankruptcy: a macroscopic study of link effect in a production network. Adv Complex Syst 11(5):703–717
10. Fujiwara Y (2012) Omori law after large-scale destruction of production network. Prog Theor Phys Suppl 194:158–164
11. Aoyama H, Fujiwara Y, Ikeda Y, Iyetomi H, Souma W (2010) Econophysics and companies—statistical life and death in complex business networks. Cambridge University Press, Cambridge

# Chapter 4
# Aftershock Prediction for High-Frequency Financial Markets' Dynamics

**Fulvio Baldovin, Francesco Camana, Michele Caraglio, Attilio L. Stella, and Marco Zamparo**

**Abstract** The occurrence of aftershocks following a major financial crash manifests the critical dynamical response of financial markets. Aftershocks put additional stress on markets, with conceivable dramatic consequences. Such a phenomenon has been shown to be common to most financial assets, both at high and low frequency. Its present-day description relies on an empirical characterization proposed by Omori at the end of 1800 for seismic earthquakes. We point out the limited predictive power in this phenomenological approach and present a stochastic model, based on the scaling symmetry of financial assets, which is potentially capable to predict aftershocks occurrence, given the main shock magnitude. Comparisons with S&P high-frequency data confirm this predictive potential.

## 4.1 Introduction

It is not uncommon for financial indexes or asset prices to experience exceptionally large negative or positive returns which trigger periods of high volatility, the case

F. Baldovin (✉) · A.L. Stella · M. Zamparo
Dipartimento di Fisica, Sezione INFN and Sezione CNISM Università di Padova, Via Marzolo 8, I-35131 Padova, Italy
e-mail: baldovin@pd.infn.it

A.L. Stella
e-mail: stella@pd.infn.it

M. Zamparo
HuGeF, Via Nizza 52, 10126 Torino, Italy
e-mail: marco.zamparo@hugef-torino.org

F. Camana · M. Caraglio
Dipartimento di Fisica, Università di Padova, Via Marzolo 8, I-35131 Padova, Italy

F. Camana
e-mail: camana@pd.infn.it

M. Caraglio
e-mail: caraglio@pd.infn.it

F. Abergel et al. (eds.), *Econophysics of Systemic Risk and Network Dynamics*, New Economic Windows,
DOI 10.1007/978-88-470-2553-0_4, © Springer-Verlag Italia 2013

of abnormal negative returns corresponding to market crashes. An understanding of the dynamical response of the market to a *main shock* is of great interest because it may help, e.g., in the definition of emergency plans for financial crises, or for risk management.

There is a clear analogy between the behavior of volatility after a main financial shock and that of the seismic activity after an earthquake of exceptional magnitude in geophysics [14]. Omori [11], with a subsequent modification by Utsu [18], established an important empirical law describing the frequency of occurrence of seismic events above a given threshold after a main earthquake. The characterizing feature of this law is the decay as a power of time, $t$, of the rate of occurrence of aftershocks above the threshold, indicating the absence of a characteristic time scale in the manifestly non-stationary Omori regime. More precisely, according to Omori the number, $n(t)$, of aftershocks per unit time above a given threshold $\sigma_a$ is given by

$$n(t) = K(t+\tau)^{-p}, \tag{4.1}$$

where $K$, $\tau$, $p$ depend on the aftershock threshold $\sigma_a$, and also on the specific magnitude of the main shock earthquake. Equivalently, the Omori law can be expressed in an integral form as

$$N(t) = \frac{K}{1-p}\left[(t+\tau)^{1-p} - \tau^{1-p}\right] \tag{4.2}$$

if $p \neq 1$, or $N(t) = K \ln(t/\tau + 1)$ if $p = 1$, where $N(t)$ is the cumulative number of aftershocks up to time $t$ after the main shock. Lillo and Mantegna [8] were the first to verify the validity of an analog of the Omori law for the volatility in Finance after a main crash. They also showed [9] that standard dynamical models of index evolution, like GARCH, are not adequate to reproduce financial Omori-like regimes. Several studies [8–10, 12, 15, 16, 20] verified the presence of Omori regimes under various market conditions, triggered by financial crashes [8, 9, 15, 16, 20], by volatility shocks [10], and even by U.S. Federal Open Market Commission meetings [12]. In particular, the Omori law in finance has been upgraded to a more general characterization of market dynamics by Weber et al. [20], who pointed out that this law holds on a wide range of time scales, with aftercrashes of a main shock playing the role of main crashes for even smaller aftercrashes, etc.

The above mentioned studies make clear the connection between financial Omori processes and long-range dependence of the volatility. They also show that a modulating, time dependent scale for the returns must be considered in order to account for the manifest non-stationarity of the Omori process. At the same time, they emphasize the limits in the predictive value of the Omori law. For example, the parameters $K$ and $\tau$ need to be adjusted for each aftershock threshold considered (see below). This holds also for the exponent $p$ of the power law decay, which should be expected to be the most robust parameter. In addition, there is no idea of how the parameters could be linked to the magnitude of the main shock. These limits reflect a lack of adequate modeling for the dynamics of financial indexes, especially in regimes like those covered by the Omori law. In recent contributions [2, 3], some of the present authors have proposed a model for the dynamics at high frequency of exchange rates or stock market indexes, which takes into account most of the relevant stylized facts. Among them, the martingale character of index evolution, the manifest non-stationarity of volatility detected in well defined daily windows of trading activity, the anomalous scaling properties of the aggregate return probability density function (PDF) in the same windows, and the strong time autocorrelation of the elementary absolute return. This model for high-frequency data, which applies more general ideas about the time evolution of financial indexes [4, 7, 17], has also been tested [2] by comparing its predictions with the statistics of ensembles of daily histories all supposed to reproduce the same underlying stochastic process. It has been also shown [3] that some arbitrage opportunities revealed by the model could be successfully exploited by appropriate trading strategies.

In the present contribution we address the problem of describing with such a model the Omori processes which may be detected within these daily windows. Our goal is to show that, after proper calibration, this model allows the *prediction* of the aftershock rate within an Omori regime, given the value of the main shock magnitude. Indeed, we provide analytical expressions for the rate of financial aftershocks with explicit dependence on the magnitude of the main shock and on the aftershocks threshold. By comparing our predictions with high frequency data from the S&P 500 index we show that these quantities are sufficient to determine the Omori response without further fitting parameters. Our success is partly due to the fact that we are able to identify the Omori processes within a context for which non-stationarity is well established [6] and amenable to modeling [2, 3]. In an interday context, the question of the applicability of the models of Ref. [4] to Omori regimes has already been raised in Ref. [7].

This note is organized as follows. In Sect. 4.2 we briefly recall the model of Refs. [2, 3] and present the procedure of calibration. In Sect. 4.3 we discuss the selection of Omori-like processes from our database and show how our model can be used to analytically describe these processes. In Sect. 4.4 we compare the results of the properly calibrated model with the statistical records at our disposal for the S&P 500 index. Section 4.5 is devoted to general discussion and conclusions.

## 4.2 Model Calibration

Let us consider the successive (log-)returns over ten minutes intervals of the S&P 500 index $S(t)$ for daily windows from 9.40 a.m., Chicago time, to 1.00 p.m.:

$$R_t \equiv \ln S(t+1) - \ln S(t), \quad t = 0, 1, \ldots, 19, \tag{4.3}$$

where the time is measured in ten-minute units and we have set $t = 0$ at 9.40 a.m.[1] A statistics made over the ensemble of 6283 available daily histories from 1985 to

[1] In order to keep contact with ordinary notations for the Omori law, in this paper we change slightly our usual conventions by shifting the origin of time by one unit with respect to, e.g., Refs. [2, 3].

2010 shows [3] that a stochastic process supposed to generate the successive returns $R_t$ in a generic history of the ensemble is consistent with the following joint PDF:

$$p_{R_0,R_1,\ldots,R_t}(r_0,r_1,\ldots,r_t)=\int_0^\infty d\sigma\,\rho(\sigma)\prod_{i=0}^{t}\frac{\exp\left(-\frac{r_i^2}{2\sigma^2 a_i^2}\right)}{\sqrt{2\pi\sigma^2 a_i^2}},\tag{4.4}$$

where

$$a_i=\left[(i+1)^{2D}-i^{2D}\right]^{1/2}\tag{4.5}$$

with $D\geq 0$, $i=0,1,\ldots,19$, and $\rho(\sigma)\geq 0$ with

$$\int_0^\infty d\sigma\,\rho(\sigma)=1.\tag{4.6}$$

This joint PDF is a convex combination of products of Gaussian PDF's for each individual return. The PDF $\rho(\sigma)$ weights this combination and introduces a nontrivial dependence of the returns from the preceding ones. For $D\neq 1/2$, the coefficients $a_i$ make the process increments non-stationary, and modulated by the exponent $D$.

The calibration of the model can be done by direct comparison of its predictions with the main features of the PDF's of the 10-minute returns $R_i$'s, or, alternatively, with those of the aggregate returns $\sum_{i=0}^t R_i$ [2, 3]. Here we follow the second option. Since the model predicts for the PDF of the aggregate return $\sum_{i=0}^t R_i$ satisfaction of an anomalous scaling of the form

$$p_{\sum_{i=0}^t R_i}(r)=\frac{1}{(t+1)^D}g\left(\frac{r}{(t+1)^D}\right),\tag{4.7}$$

where the scaling function $g$ is expressed as

$$g(r)=\int_0^\infty d\sigma\,\rho(\sigma)\frac{\exp(-\frac{r^2}{2\sigma^2})}{\sqrt{2\pi\sigma^2}},\tag{4.8}$$

one can determine $D$ through a fitting of the power law $t$-dependence of the moments of $p_{\sum_{i=0}^t R_i}$. Indeed, for $q\in\mathbb{R}$, according to (4.7)

$$\mathbb{E}\left[\left|\sum_{i=0}^t R_i\right|^q\right]=\mathbb{E}\left[|R_0|^q\right]t^{\gamma(q)}\tag{4.9}$$

with $\gamma(q)=qD$, and provided that the moment $\mathbb{E}[|R_0|^q]$ exists. In Fig. 4.1 we report the empirical values for $\gamma(q)$, using (4.9) as an ansatz. To calibrate $D$, we make a linear data regression for $q\leq 2$, since for higher moments a multiscaling behavior [19] is detected (see Fig. 4.1). The result is $D\simeq 0.35$.

A particularly simple expression for the joint PDF $p_{R_0,R_1,\ldots,R_t}$ is achieved if the integration on $\sigma$ can be worked out explicitly in (4.4). This is indeed the case if we

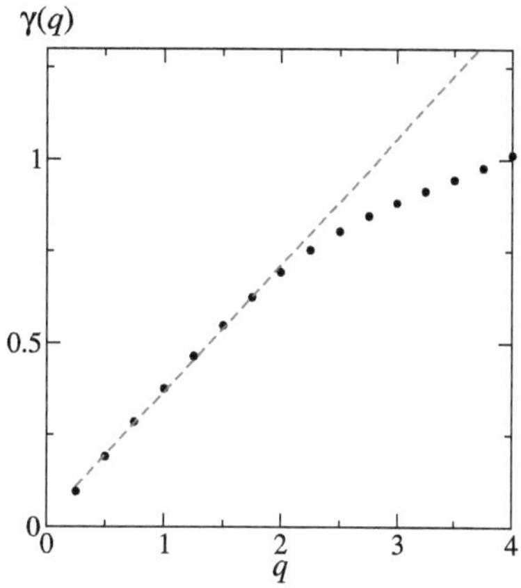


**Fig. 4.1** Calibration of the scaling exponent $D$. The empirical values for $\gamma(q)$ are reported using (4.9) as an ansatz (*points*). A linear regression for $0 < q \le 2$ gives $\gamma(q) = qD$ with $D \simeq 0.35$ (*dashed line*)

choose an inverse-gamma distribution for $\sigma^2$ [7]. Equivalently, we may set

$$\rho(\sigma) = \frac{2^{1-\frac{\alpha}{2}}\beta^\alpha}{\Gamma(\frac{\alpha}{2})\sigma^{\alpha+1}} \exp\left(-\frac{\beta^2}{2\sigma^2}\right), \tag{4.10}$$

where the exponent $\alpha$ determines the long-range behavior of $g$ according to $g(r) \sim 1/r^{\alpha+1}$ for $|r| \gg 1$, and $\beta$ is a scale parameter determining the distribution width. Performing the integration on $\sigma$ in (4.4) we obtain a multi-variate Student PDF:

$$p_{R_0,R_1,\ldots,R_t}(r_0, r_1, \ldots, r_t) = \frac{\beta^\alpha \Gamma(\frac{\alpha+t+1}{2})}{\pi^{\frac{t+1}{2}}\Gamma(\frac{\alpha}{2})}\left(\beta^2 + \frac{r_0^2}{a_0^2} + \frac{r_1^2}{a_1^2} + \cdots + \frac{r_t^2}{a_t^2}\right)^{-\frac{\alpha+t+1}{2}}. \tag{4.11}$$

As we will show in the following, an explicit form for $p_{R_0,R_1,\ldots,R_t}$ enables us to obtain a simple analytic expression for $N(t)$. Unlike in previous papers [2, 3], we thus choose here the functional form in (4.10) for $\rho$. Besides $D$, the other parameters of the model, $\alpha$ and $\beta$, are calibrated by first data-collapsing the empirical PDF's for $\sum_{i=0}^t R_i$ according to (4.7) with $D = 0.35$, and then by fitting $\alpha$ and $\beta$ on this data-collapse using (4.11) with $t = 0$. The result is given in Fig. 4.2. In summary, the result of the calibration procedure is the triple $(\alpha, \beta, D) = (3.5, 2.9 \cdot 10^{-3}, 0.35)$.

The ensemble of histories at our disposal is relatively poor. This implies, as can be appreciated in Fig. 4.2, that some rare events fall significantly out of the scaling function, since a much larger number of histories would be needed to correctly characterize their frequency of occurrence. The multiscaling behavior shown in Fig. 4.1 could be at least partly related to this effect. The Omori events are precisely related to extreme events. In order to obtain a reliable statistics of the aftershocks, we impose thus an upper bound $\sigma_{max}$ to the absolute value of the returns $R_i$'s included in our empirical analysis (see Fig. 4.2). Once done this, the overall agreement of the empirical data with the various model predictions gives a convincing validation of the model itself (see also [3]). Still, the agreement shown in what follows with respect to the Omori processes must be intended as a first important result, which calls for more extensive analysis also in terms of the calibration procedure.

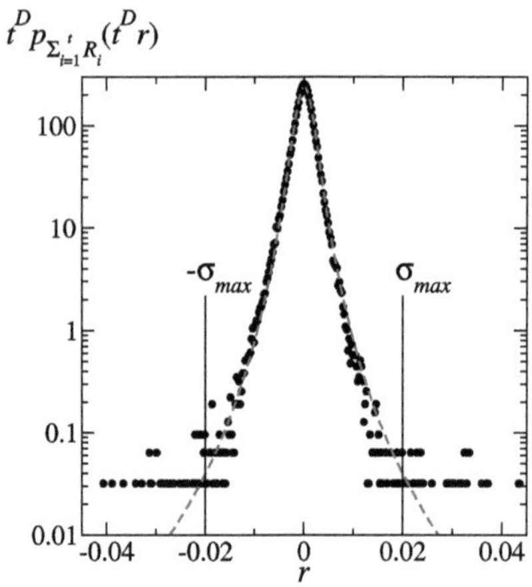


**Fig. 4.2** Calibration of the parameters $\alpha$ and $\beta$. The empirical PDF's for $\sum_{i=0}^{t} R_i$ at various $t$ are rescaled according to (4.7) with the previously calibrated $D = 0.35$ (*points*). The parameters $\alpha$ and $\beta$ are then fitted using (4.11) with $t = 0$ (*dashed line*), yielding the values $\alpha = 3.5$ and $\beta = 2.9 \cdot 10^{-3}$. An upper bound to the empirical analysis is posed at the $\sigma_{max} = 0.02$ for the 10-minute volatility

## 4.3 Aftershock Prediction

As already mentioned above, in the present analysis we are going to identify and select Omori processes, which are manifestations of non-stationarity, within a process which manifestly turns out to be with non-stationary returns in its ensemble of daily realizations. This is a simplification which marks an important difference with respect to the problem of modeling the Omori regimes revealed in Refs. [8–10, 12, 15, 16, 20], where they were extracted from single time series expected to be globally stationary on long time scales. In the perspective of our approach here, dealing with a process which is by itself time-inhomogeneous offers the advantage that the selection of Omori processes does not imply the need of identifying how their non-stationarity emerges from an otherwise stationary global behavior. In a version of our model suited for describing single, long time series of returns [5, 17], the necessity to consider random exogenous factors influencing the market, leads us to switch-on at random times some time-inhomogeneities formally similar to those characterizing the model of the previous section. This is achieved by setting $a_t = 1$ concomitantly with these random events (see also [1, 2, 4, 7]). In such a context it is not *a priori* clear whether or not the start of an Omori process should imply putting $a_t = 1$ in correspondence with the time $t$ of the main shock. This difficulty is also accompanied by the need of implementation of a more complicated calibration procedure [5] with respect to the one presented here.

As shown below, remarkable results of our analysis in this note are:

(i) that the selected processes are legitimately classified as Omori-like in the sense that they can all be fitted by the Omori law;
(ii) that the description one obtains for them based on the model presented in the previous section contains explicit dependencies on the intensities of the main shock and on the aftershocks thresholds.

This endows our approach to the Omori regimes of a predictive potential which, if confirmed by further analysis, could be exploited by decision-makers under crisis conditions.

We select as Omori processes all those histories in the above S&P 500 ensemble for which the initial absolute return, $|r_0|$, besides being smaller than $\sigma_{max}$, also exceeds a main shock threshold $\sigma_m$. At variance with the analysis in Refs. [8–10, 12, 15, 16, 20], we consider, in place of a single time series, groups of histories for which $\sigma_m \le |r_0| \le \sigma_{max}$. As far as the aftershocks are concerned, we record for each of these histories the elementary returns which exceed in absolute value an aftershock threshold $\sigma_a$ and are below the main shock value $|r_0|$: $\sigma_a \le |r_i| \le |r_0|$, for $i \ge 1$. The parameter $\sigma_a$ is an important one to be fixed in any analysis of the Omori law. Again, by imposing the aftershock magnitude to be smaller than that of the main shock we reduce the influence of extreme events in our limited dataset. We decided to search for main shocks occurring right at the beginning of the daily time window described by our model for two main reasons. In first place the ensemble average volatility on 10 minutes intervals is maximal in the first interval. Secondly, a main shock occurring right at the beginning of the time window leaves the maximum possible time for the development of the subsequent Omori process. While we will limit ourselves below to discuss such optimal case, different choices are of course possible.

According to the above selection procedure of the Omori processes, the cumulative number of aftershocks $N_{|r_0|}(t)$ after a main shock of magnitude $|r_0|$ is given by

$$N_{|r_0|}(t) = \mathbb{E}\left[\sum_{i=1}^{t} \mathbb{1}_{(\sigma_a \le |R_i| \le |R_0|)} \mid |R_0| = |r_0|\right], \tag{4.12}$$

where $\mathbb{1}_{(\sigma_a \le |R_i| \le |R_0|)}$ is the indicator function, yielding 1 if $\sigma_a \le |R_i| \le |R_0|$ and zero otherwise. Using (4.11), through a change of variable it is straightforward to show

$$\begin{aligned} N_{|r_0|}(t) &= \sum_{i=1}^{t} 2 \int_{\sigma_a}^{|r_0|} dr_i \frac{p_{R_0,R_i}(r_0, r_i)}{p_{R_0}(r_0)} \\ &= \frac{2}{\sqrt{\pi}} \frac{\Gamma(\frac{\alpha+2}{2})}{\Gamma(\frac{\alpha+1}{2})} \sum_{i=1}^{t} \int_{\frac{\sigma_a}{a_i\sqrt{\beta^2+r_0^2}}}^{\frac{|r_0|}{a_i\sqrt{\beta^2+r_0^2}}} dx \left(1+x^2\right)^{-\frac{\alpha+2}{2}}. \end{aligned} \tag{4.13}$$

If in the considered ensemble of histories there are $M$ realizations $\{r_i^{(m)}\}_{m=1,2,\ldots,M}$ in which we register a main shock, i.e., $\sigma_m \le |r_0^{(m)}| \le \sigma_{max}$, then the cumulative number of aftershocks $N(t)$ is obtained through the sample average

$$N(t) = \frac{1}{M} \sum_{m=1}^{M} N_{|r_0^{(m)}|}(t), \tag{4.14}$$

where we stress the fact that each $N_{|r_0^{(m)}|}(t)$ is conditioned to the main shock magnitude $|r_0^{(m)}|$. Notice that since with the available dataset the selected main shocks constitute a small sample (see next section), we use here the sample average rather than the ensemble one to get the number of aftershocks conditioned to $\sigma_m \le |R_0| \le \sigma_{max}$.

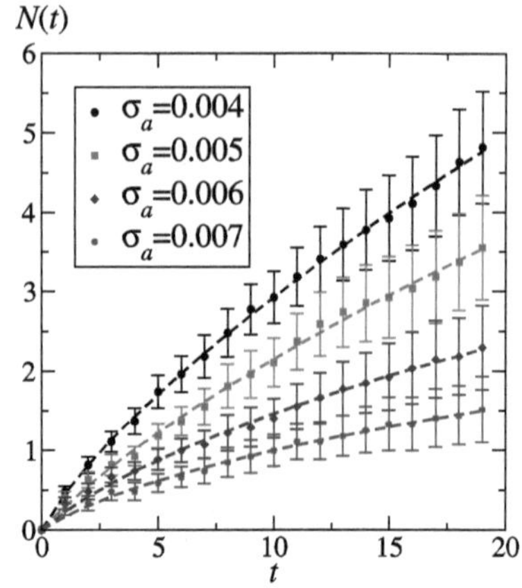


**Fig. 4.3** Fitting of the empirical aftershock at different thresholds $\sigma_a$ (*points with error bars*), with the Omori law in (4.2) (*dashed lines*). Fitted parameters are reported in Table 4.1

**Table 4.1** Omori parameters in (4.2) fitted from the empirical data

| $\sigma_a$ | $K$ | $p$ | $\tau$ |
|---|---|---|---|
| $4 \cdot 10^{-3}$ | 0.44 | 0.29 | 0.52 |
| $5 \cdot 10^{-3}$ | 0.40 | 0.34 | 2.00 |
| $6 \cdot 10^{-3}$ | 0.35 | 0.49 | 2.00 |
| $7 \cdot 10^{-3}$ | 0.28 | 0.59 | 2.07 |

## 4.4 Comparison of the Model Predictions with the Statistics of Aftershocks

Our choice for the thresholds $\sigma_m$ and $\sigma_{max}$ is such that the absolute first returns for which $\sigma_m \leq |r_0^{(m)}| \leq \sigma_{max}$ are quite exceptional. They occur with $27/6283 \simeq 0.4\ \%$ frequency in our ensemble; Only 3 realizations have $|r_0| > \sigma_{max}$ and are thus excluded. Accordingly, we analyze the averaged $N(t)$ of aftershocks for these $M = 27$ main shocks. A first point to clarify is whether the recorded rates are well fitted by the Omori law in (4.2). This is shown in Fig. 4.3, where many sets of data for $N(t)$, obtained with different aftershocks thresholds $\sigma_a$, are indeed fitted by the Omori equation (4.2). In Table 4.1 one also realizes that $K$, $\tau$ and $p$ need to be varied for each $\sigma_a$ in order to reach a satisfactory fit. In particular, by changing $K$ and $\tau$ it is even possible to obtain reasonable fittings also with $p > 1$ (see also [10]). This parameters variability makes it very difficult to use the Omori law to predict the aftershock occurrence for a given main shock magnitude and aftershock threshold.

Model predictions on the same set of data fitted in Fig. 4.3 are instead given in Fig. 4.4. Dashed lines in Fig. 4.4 are obtained on the basis of (4.13, 4.14) with the parameters $(\alpha, \beta, D)$ resulting from the calibration discussed in Sect. 4.2. The only difference among the curves is the value of the aftershock threshold $\sigma_a$. The agreement of the analytical predictions with the data and the sensitivity of the curves to the variation of the aftershock threshold are remarkable. This shows that our

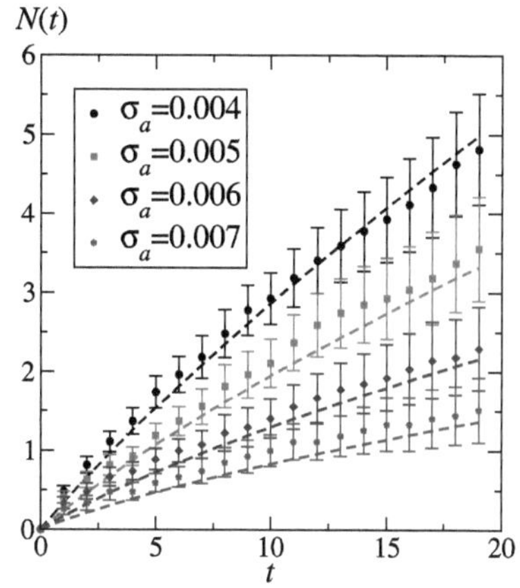


**Fig. 4.4** Comparison between the analytical model predictions for different aftershock thresholds $\sigma_a$ (*dashed lines*) with the same empirical S&P data reported in Fig. 4.3 (*points with error bars*)

model potentially provides a satisfactory and parameter-free description of Omori processes.

## 4.5 Conclusions

We have shown here in the case of the S&P 500 index, that a model suited for the description of the high frequency market dynamics allows also to predict Omori regimes following exceptional extreme events. Within the class of events considered, the model specifies the dependence on the main shocks intensities and on the aftershocks threshold. As such, its description goes far beyond the limits of the Omori phenomenological law.

Besides providing a further validation of the model of Refs. [2, 3], the results presented here encourage to extend similar analysis to cases in which the Omori processes are to be selected within a process which is globally stationary. For the modeling of these processes, our recipe [5] is that of switching-on at random some non-stationarities ascribable to coefficients like the $a_t$ defined above. Global stationarity of the process on long time scales is then guaranteed by the fact that empirical averages are in this case made by considering time intervals sliding along the single long history [5]. While it is conceivable that in many cases main shocks are localized close to resets of the time inhomogeneity, this is not true in general. Some attempts to strictly identify main shocks with restarts of them inhomogeneity in the model ($a_t = 1$) already gave some preliminary agreement with the data. A more general discussion is however needed [5].

**Acknowledgements** This work is supported by "Fondazione Cassa di Risparmio di Padova e Rovigo" within the 2008-2009 "Progetti di Eccellenza" program.

## References


1. Andreoli A, Caravenna F, Dai Pra P, Posta G (2012) Scaling and multiscaling in financial indexes: a simple model. Adv Appl Probab (in press, 2012). arXiv:1006.0155
2. Baldovin F, Bovina D, Camana F, Stella AL (2011) Modeling the non-Markovian, non-stationary scaling dynamics of financial markets. In: Abergel F, Chakrabarti BK, Chakraborti A, Mitra M (eds.) Econophysics of order-driven markets, 1st edn. New Economic Windows, Springer, New York, pp 239–252
3. Baldovin F, Camana F, Caporin M, Stella AL (2012) Ensemble properties of high frequency data and intraday trading rules (submitted). arXiv:1202.2447
4. Baldovin F, Stella AL (2007) Proc Natl Acad Sci USA 104:19741
5. Baldovin F, Stella AL, Zamparo M (in preparation)
6. Bassler KE, McCauley JL, Gunaratne GH (2007) PNAS 104:17287
7. Challet D, Peirano PP (2012) The ups and downs of modeling financial time series with Wiener process mixtures (submitted). arXiv:0807.4163
8. Lillo F, Mantegna RN (2003) Phys Rev E 68:016119
9. Lillo F, Mantegna RN (2004) Physica A 338:125
10. Mu G-H, Zhou W-X (2008) Physica A 387:5211
11. Omori F (1984) J Coll Sci Imp Univ Tokyo 7:111
12. Petersen AM, Wang F, Havlin S, Stanley HE (2010) Phys Rev E 81:066121
13. Petersen AM, Wang F, Havlin S, Stanley HE (2010) Phys Rev E 82:036114
14. Scholz CH (2002) The mechanics of earthquakes and faulting. Cambridge University Press, Cambridge
15. Selçuk F (2004) Physica A 333:306
16. Selçuk F, Gençay R (2006) Physica A 367:375
17. Stella AL, Baldovin F (2010) J Stat Mech P02018
18. Utsu T (1961) Geophys Mag 30:521
19. Wang F, Yamasaki K, Havlin S, Stanley HE (2008) Phys Rev E 77:016109
20. Weber P, Wang F, Vodenska-Chitkushev I, Havlin S, Stanley HE (2007) Phys Rev E 76:016109

# Chapter 5
# How Unstable Are Complex Financial Systems? Analyzing an Inter-bank Network of Credit Relations

**Sitabhra Sinha, Maximilian Thess, and Sheri Markose**

**Abstract** The recent worldwide economic crisis of 2007–09 has focused attention on the need to analyze systemic risk in complex financial networks. We investigate the problem of robustness of such systems in the context of the general theory of dynamical stability in complex networks and, in particular, how the topology of connections influence the risk of the failure of a single institution triggering a cascade of successive collapses propagating through the network. We use data on bilateral liabilities (or exposure) in the derivatives market between 202 financial intermediaries based in USA and Europe in the last quarter of 2009 to empirically investigate the network structure of the over-the-counter (OTC) derivatives market. We observe that the network exhibits both heterogeneity in node properties and the existence of communities. It also has a prominent core-periphery organization and can resist large-scale collapse when subjected to individual bank defaults (however, failure of any bank in the core may result in localized collapse of the innermost core with substantial loss of capital) but is vulnerable to system-wide breakdown as a result of an accompanying liquidity crisis.

## 5.1 Introduction

Isaac Newton, possibly the greatest physicist of all time, is believed to have once said that while he could calculate the motions of cosmic bodies, his theories are useless for understanding the madness of crowds [1]. This statement was supposedly made in the context of the mass frenzy that was seen among the general public during the height of the South Sea Bubble of 1720, one of the most famous episodes

S. Sinha (✉)
The Institute of Mathematical Sciences, CIT Campus, Taramani, Chennai 600113, India
e-mail: sitabhra@imsc.res.in

M. Thess
Institut für Theoretische Physik, Technische Universität Berlin, Hardenbergstrasse 36, D-10623 Berlin-Charlottenburg, Germany

S. Markose
Department of Economics, University of Essex, Wivenhoe Park, Colchester C04 3SQ, Essex, UK

F. Abergel et al. (eds.), *Econophysics of Systemic Risk and Network Dynamics*, New Economic Windows,
DOI 10.1007/978-88-470-2553-0_5, © Springer-Verlag Italia 2013 

of financial speculation and the panic triggered by its subsequent collapse [2]. Indeed, until about the 1990s, the methods of physics were largely thought to be inapplicable for the study of society, of which economic activity is an integral part. However, in recent days, this perception has begun to change and it is instructive to learn that Paul Samuelson, one of the leading figures of contemporary mainstream economics, mentioned during an interview in 1998 that the physics of avalanches was a better guide for understanding the "irrational exuberance" of the overvalued markets of this period than what standard economics textbooks teach [3]. One might be tempted to think that physics has at last come of age to be applied fruitfully for understanding economic phenomena, even though we might be at the same stage in our search for a physical theory of Economics (a discipline that has often been referred to as *Econophysics* [4]) as natural philosophers were at the time of Newton in their quest for a theory of the physical universe.

The late 2000s financial crisis, possibly the worst economic disaster since the Great Depression of the 1930s, has brought to fore once again the poverty of mainstream economics when forced to explain the real world rather than idealized systems using elegant but unrealistic assumptions of perfect competition and complete knowledge. The inability of standard theories to understand the mechanisms that result in such system-wide failures of financial markets is deeply worrying, as these crises are damaging not just on their own account—often involving collapse of large financial institutions, extensive intervention in the financial sector by the government and sometimes involving bailouts costing the taxpayer enormous sums of money—but by affecting the stock market (and in the 2007–09 crisis, also the housing market) and constricting liquidity, they can depress the rest of the economy. The potential of financial crashes to drive the economy into severe recession is not a recent phenomenon but have been seen in earlier instances of market collapses [5]. The lack of availability of credit in the aftermath of such crises can result in failure of businesses causing large-scale unemployment. This in turn reduces the overall income and leads to a significant drop in consumer spending, which slows down the economy further. It is therefore of critical importance to come up with an alternative theoretical framework to understand the genesis of financial crisis, with the aim of averting disaster before it strikes by learning to recognize warning signs of an impending collapse. An even more desirable outcome will be to arrive at principles for designing robust financial structures that are much less likely to suffer system-wide failure than at present.

A theoretical approach to understand financial crisis has to consider what kind of conditions can make a large-scale collapse of financial institutions likely. This is in fact related to the general question of why and how do economic institutions fail, one of the most fascinating topics of modern economics [6]. As we know from our everyday experience that large events need not have been triggered by an extraordinary stimulus, an important related question in this context is whether the failure of a single economic entity can drive events so as to result in a cascading chain of successive collapses among several inter-connected institutions— eventually leading to a large-scale breakdown of the financial system [7]. Indeed, it was the specter of such a catastrophe that prompted governments across the world to spend enormous sums of money to try and save banks and financial institutions that were considered to be "too big to fail" in the sense of being so intricately and intimately connected to a large number of other institutions that they would bring down a significant fraction of them if allowed to fail. At an even larger scale, one can imagine that, if unchecked, such a series of failures can spread world-wide, aided by the existence of a densely connected global financial network that has been made possible by the communication revolution [8]. It is conceivable that the general collapse of the financial infrastructure coupled with the economic chaos that might ensue may well be enough to trigger the collapse of our civilization and usher in a new "dark age", similar to the what has been repeatedly seen in history [9].

As collapse is usually manifested by a drastic reduction in the *complexity* of the system, be it in terms of the diversity of entities it is able to support or the types or nature of interactions between such entities which are allowed, it is natural to ask whether increasing complexity can itself make a system more prone to failure. This question has attracted a great deal of scientific interest for the past four or five decades, especially among ecologists [10]. At first it may appear counter-intuitive that a greater variety of elements and a strong degree of interactions among them can lead to instabilities. Indeed, the risk incurred through lending by financial institutions is sought to be reduced by securitization and selling debt instruments to other institutions, thereby connecting them together in a large web of mutual liabilities. The principle behind this practice appears to be that by sharing a large sum among many agents, the risk of default to each individual entity is reduced. In other words, as in selling insurance, if one increases and diversifies as much as possible the population which is insured, it reduces the risk that at any given time a significantly large fraction of the insured individuals will fail and that the insurer has to pay out large sums simultaneously. Arguments along these lines have been forwarded previously in other areas, for example in ecology, to contend that larger complexity actually makes a system more stable. However, in the early 1970s it was shown conclusively by Robert May [11] through linear stability analysis of large randomly connected networks that increasing the number of elements and/or the number of connections between them, as well as, increasing the strength of interactions, makes the system more unstable. In other words, a complex system is more likely to be knocked out of its equilibrium state by a small perturbation at any of its constituent elements, as compared to a simpler system.

Over the past four decades, the pioneering result of May has been debated intensely by scientists from various disciplines (dubbed as the *diversity-stability* debate [10]) and the exact conditions under which these results apply have been sought to be determined. A significant challenge to the general validity of these results had been that (a) the analysis was based on *linearization* of the system about a "fixed point" (or static) equilibrium, and (b) the system considered comprised *randomly connected* elements. However, subsequent studies of the global stability (e.g., measured in terms of the persistence of the constituent elements) of dynamical systems in various regimes, viz., exhibiting periodic and chaotic behavior apart from fixed point dynamics, has shown the original results to be valid even in this more general setting [12]. Similarly, the advent of new models of networks in the late 1990s, e.g., those exhibiting the "small-world" property [13] and those having "scale-free" distribution of degree[1] [14, 15], which arguably better represent the connection topology of complex systems seen in reality, has resulted in a series of studies of the stability when the complexity of such networks is increased. Again, it appears that increasing the size and connection density (as well as, strength of the connection weights) of these networks make them more, rather than less, unstable [16, 17]. Thus, it appears that despite the appealing intuition of the insurance hypothesis, increased connectance between a large number of dynamically evolving entities does increase the risk of overall system failure, a result whose implications for economic systems is obvious [18].

The recent crisis of 2007–09 has, therefore, brought forth calls by scientists (including from Robert May himself) to apply the lessons learnt in ecology through analyzing the stability of complex food webs to the problem of robustness in large, strongly connected networks of financial institutions [19, 20]. For example, structural properties of robust networks that can be identified as contributing to the dynamical stability of the system can be implemented in designing artificial entities such as the financial network to decrease their likelihood of failure when subjected to episodes of stress. It has been pointed out that "ecosystems are robust by virtue of their continued existence" [19], i.e., only those networks have survived (and are therefore seen today), whose structure enabled them to withstand the high degree of fluctuation in their environment and in the dynamics of their constituent species. On the other hand, financial networks have emerged very recently through the uncoordinated decisions of a large number of agents, often having divergent aims and interests. The connection topology of the network has not been developed based on robust design principles nor has the system been subjected to evolution through a series of successive failures and regrowth to attain a relatively stable configuration. In order to assess the fragility of the existing system (prior to redesigning it to make it more stable), we have to first reconstruct the network of interactions between financial institutions and study the dynamical stability implications of such a structure through simulations. Such an analysis can alert us to either "keystone" nodes in the network whose removal through failure can result in a significant number of other nodes failing in rapid succession.

Several such empirical studies of the inter-dependency networks of financial institutions have recently appeared in the literature. In particular, a very large network of over 7500 banks in USA connected through the Fedwire interbank payment network operated by the Federal Reserve System has been analyzed to reveal a sparsely connected system (only 0.3 % of the potential number of connections are actually observed) which nevertheless has relatively low average path length—a signature of the "small-world" phenomenon seen in many other networks—thus, indicating the existence of an extremely compact structure [21]. More importantly, the majority of the links correspond to weak flows, and focusing on the small set of high-value transactions reveals the existence of a core—a small set of 25 banks which are densely inter-connected—to which other banks (constituting the periphery) con-

[1] The degree of a node is the number of links it possesses.

nect. Such core-periphery organization, seen across many other complex systems, ranging from networks of neurons [22] to language [23], has also been reported for other networks of financial institutions such as the web of credit relations among Austrian banks [24]. A recent study on financial contagion propagation has claimed that applying stress in the core and in the periphery will have very different consequences for the system [25]. Another study has found a distinct bimodal capacity of nodes to propagate contagion, with those in the periphery having little or no effect while failure of nodes in the core can destroy the system [43]. The Fedwire network also exhibits scale-free degree distribution [21], another property it shares with many other inter-bank networks including those of Austria [24] and Japan [26]. Similar studies of the relation between properties of the connection topology of the credit network among financial institutions and its stability have also been carried out for other systems, including the European money market for overnight loans (required for maintaining liquidity) between Italian banks [27] and the Japanese credit network between banks and large firms [28, 29].

It is in this context that we report our analysis of a network of bilateral liabilities (or exposures) between 202 financial intermediaries (FI)[2] based in USA and Europe reconstructed from data for the last quarter of 2009 (the period during which the financial crisis of 2007–09 reached its denouement). Systemic risk in this system can be quantified as the probability that the failure of an individual entity results in a cascading series of defaults propagating through the network of mutual liabilities, with an institution failing when it is unable to honor its commitments to creditors as a result of its debtors failing and thereby defaulting on their commitments to it. Analysis of topological properties of the network reveal many of the same features seen in other financial networks, such as core-periphery organization and long-tailed distributions of degree and strength.[3] However, more important than such static properties is the dynamical response of the network to local perturbations (specifically, the failure of a particular FI). We have used a simple and intuitively appealing model of failure propagation in the network that takes into account the Tier-I core capital of each institution in addition to the information about bilateral liabilities, to study the impact of the collapse of each constituent FI on the rest of the network. This allows us to identify "super-spreader" nodes in the system whose collapse can trigger failure of a large fraction of elements in the network. A crucial parameter that affects this process is the critical fraction ($q$) of core capital of an FI that its net loss (as a result of failure of FIs connected to it via mutual liabilities) must exceed in order for it to collapse. Although the actual value of this critical fraction cannot be reliably determined from the empirical data, by studying the behavior of the system over a large range of $q$, it appears that a global or system-wide

[2]A financial intermediary is an institution, such as a bank, a credit union or a mortgage loan company, that transfers funds from investors (lenders) to those requiring capital (borrowers). For instance, a bank uses its deposits to provide loans or mortgages thereby mediating transactions between surplus and deficit agents [30].

[3]Strength of a node is the sum of weights of all links belonging to it.

collapse is unlikely except at very small (and possibly unrealistic) values of this parameter. Our observation that the propagation of disturbances along the network of explicit financial linkages (also referred to as *contagion* in the economics literature) is unlikely to cause system-wide collapse is good news and agrees with an earlier study [31]. The bad news is that the members of the highly clustered inner core who constitute the leading broker-dealers are potentially in need of tax-payer bailouts as failure of any member in the core may trigger failure of the entire core. We also see that an accompanying liquidity crisis can simultaneously decrease $q$ as more FIs fail with time, thereby triggering even more failures and decreasing $q$ even further. This coupling between the collapse of financial institutions and the reduction in availability of capital can thus drive a chain reaction of failures that can eventually cause the entire system to breakdown. Our results thus paint a nuanced picture of inter-bank networks, which can be viewed as "robust-yet-fragile" [32], and points out the importance of liquidity crisis that may accompany a cascading series of bank failures in triggering system-wide crisis in complex financial systems.

## 5.2 The Network of Financial Intermediaries

As already mentioned above, banks function as financial intermediaries between lenders who deposit money in the bank and borrowers who take out loans or mortgages. On the assumption that at any given time at most a small number of depositors will be withdrawing a substantial portion of their money, a bank holds only a small fraction of the total amount deposited in reserve to cover regular transactions with their customers and invests the rest in profit-earning enterprises [33]. While lending out their money provides banks income through interest payments, this also exposes them to credit risk of the borrowers defaulting on their promised payments. Another source of risk for banks can be a sudden devaluation in some of their external assets, such as, the drop in real estate prices following the end of a housing bubble. If such losses are a substantial fraction of its capital, a bank may find it difficult to honor its commitments to its lenders. Under such circumstances, if the bank faces a liquidity crisis and is unable to raise a loan to cover its liabilities, it can fail. Thus, banks need to have an optimized operational procedure in order to maximize their return (by lending out as large a fraction of their total deposits as possible) while at the same time minimizing the resulting risk.

A widely used method of risk reduction in modern finance is through the use of risk management instruments known as *derivatives* which are contracts between two parties specifying payoffs that will be made between them at some future date based on the value of an underlying asset such as foreign exchange rates, bonds/interest rates, commodities and equities [34]. As under normal circumstances a derivative and its corresponding underlying asset are expected to change their value in the same direction and by roughly the same amount, one can protect against loss by hedging, i.e., holding opposite positions in the underlying asset and derivative markets at the same time, so that losses in one market can be offset by gains in the other. In the specific context of credit risk, a credit default swap (CDS) is a type of derivative that can act as insurance for the lending agency against non-payment of debt. Thus, by purchasing CDS, a bank can transfer its credit risk (incurred by lending to a third party) to the seller of the swap, as in return for a series of payments (equivalent to an insurance premium) the seller agrees to compensate the buyer in the event of the default by paying off the debt. While for any individual institution such risk sharing by purchasing and selling derivatives may appear appealing, at the level of the overall system such practices bind together the different entities into a strongly interconnected entity where the failure of any one bank does not remain localized in its effects but spreads through the system. The systemic risk inherent in such a situation is worsened by a limited number of counterparties dominating the market in selling risk management instruments.

The subject of our study is the network of bilateral assets and liabilities of 202 financial intermediaries (listed in Table 5.1) aggregated over all categories of derivative products (including foreign exchange contracts, interest rate swaps, equities, CDS and commodities). In order to measure credit exposure of a FI, one first needs to identify the derivatives contracts which would result in loss of value to the institution if the counterparty defaults [35]. In the absence of bilateral netting[4] and any collateral from counterparties, the Gross Positive Fair Value (GPFV) is the aggregate fair value of all contracts where the FI is owed money by its counterparties. Thus, GPFV is the maximum credit exposure or losses which the FI can incur if its counter-parties default. Conversely, the sum total of values of all contracts where a FI owes money to its counterparties is referred to as Gross Negative Fair Value (GNFV), and it is the maximum loss incurred by the counterparties in the absence of netting agreement or bank collateral. Derivatives liabilities and assets are estimated by adjusting the gross payables and receivables (respectively) for collateral, bilaterally netting where agreements exist and summing over all counterparties.

The firm level data on derivative assets and liabilities used in our study were obtained from FDIC Call Reports for the fourth quarter of 2009 for US banks that operate solely as national associations, and from individual Annual Financial Statements for the global US banks and Europeans FIs. The firm level derivative liability (asset) is the positively (negatively) signed sum over all counterparties and products of the bilaterally netted market value of derivatives receivables and payables. An algorithm described in Ref. [37] is used to reconstruct a bilateral matrix for derivatives liability or asset between FIs from the firm-level data upto some margin of error. The starting point for the network reconstruction is this bilateral gross flow matrix between the FIs, **B**, where $B_{ij}$ represents the flow of financial obligation from the seller (row FI $i$) of the derivative to the buyer (column FI $j$). Thus, $N_i = \sum_j B_{ij}$ is the GNFV of bank $i$, representing the total derivatives obligations owed by it to other FIs, while $P_j = \sum_i B_{ij}$ is the GPFV of bank $j$, i.e., the total sum owed to it by all other FIs. The matrix will in general be asymmetric ($B_{ij} \neq B_{ji}$) and will have zeros along the diagonal ($B_{ii} = 0$) as banks do not lend to/borrow from themselves.

[4] Bilateral netting, whose primary purpose is to reduce exposure to credit risk, is an arrangement between two parties to exchange only the net difference in their obligations to each other [36].

**Table 5.1** The list of 202 banks analyzed in this article arranged in decreasing order according to their Tier-I core capital

| *i* | Financial intermediary | Core capital (billions USD) | *i* | Financial intermediary | Core capital (billions USD) |
|---|---|---|---|---|---|
| 1 | Bank of America | 111.92 | 40 | Marshall & Ilsley Bank | 3.95 |
| 2 | Royal Bank of Scotland | 98.28 | 41 | Harris Natl Asso | 3.52 |
| 3 | Citibank | 96.83 | 42 | First Tennessee Bank | 3.36 |
| 4 | JP Morgan Chase | 96.37 | 43 | Huntington Natl Bank | 2.87 |
| 5 | BNP Paribas | 90.37 | 44 | UBS Bank USA | 2.52 |
| 6 | Barclays | 77.56 | 45 | Citizens Bank of Pennsylvania | 2.43 |
| 7 | Lloyds | 74.27 | 46 | RBC Bank (USA) | 2.43 |
| 8 | UniCredit | 56.07 | 47 | Zions First Natl Bank | 1.81 |
| 9 | Deutsche Bank | 49.42 | 48 | Associated Bank | 1.78 |
| 10 | Morgan Stanley | 46.67 | 49 | City Natl Bank | 1.60 |
| 11 | Credit Agricole | 44.53 | 50 | Frost Natl Bank | 1.32 |
| 12 | Wells Fargo Bank | 43.77 | 51 | Amegy Bank | 1.27 |
| 13 | UBS | 42.32 | 52 | Webster Bank | 1.27 |
| 14 | Wachovia Bank | 39.79 | 53 | BanCorpSouth Bank | 1.14 |
| 15 | Credit Suisse | 39.49 | 54 | Bank of Oklahoma | 1.08 |
| 16 | HSBC | 35.48 | 55 | PrivateBank and Trust Co | 1.06 |
| 17 | Societe Generale | 34.69 | 56 | Mizuho Corp Bank (USA) | 1.05 |
| 18 | Dexia | 25.24 | 57 | Whitney Natl Bank | 1.00 |
| 19 | Standard Chartered | 24.58 | 58 | Susquehanna Bank | 0.99 |
| 20 | PNC Bank | 24.49 | 59 | RaboBank | 0.97 |
| 21 | Citibank (South Dakota) | 19.71 | 60 | California Bank & Trust | 0.96 |
| 22 | Goldman Sachs | 17.15 | 61 | Northwest Savings Bank | 0.92 |
| 23 | US Bank Natl Asso | 16.25 | 62 | Arvest Bank | 0.88 |
| 24 | Fifth Third Bank | 13.57 | 63 | WesternBank Puerto Rico | 0.84 |
| 25 | Branch Banking & Trust Co | 13.54 | 64 | Trustmark Natl Bank | 0.84 |
| 26 | Suntrust Bank | 11.97 | 65 | Signature Bank | 0.84 |
| 27 | State Street | 11.38 | 66 | Firstmerit Bank | 0.83 |
| 28 | Regions Bank | 10.58 | 67 | MB Financial Bank | 0.82 |
| 29 | New York Mellon | 10.15 | 68 | Woodlands Commercial Bank | 0.75 |
| 30 | TD Bank | 9.27 | 69 | Bank of Hawaii | 0.75 |
| 31 | Capital One | 8.42 | 70 | Investors Savings Bank | 0.75 |
| 32 | RBS Citizens | 8.24 | 71 | Israel Discount Bank of New York | 0.72 |
| 33 | KeyBank Natl Asso | 8.0 | 72 | United Community Bank | 0.72 |
| 34 | Union Bank | 7.21 | 73 | National Penn Bank | 0.7 |
| 35 | Comerica Bank | 5.76 | 74 | Doral Bank | 0.69 |
| 36 | Manufacturers and Traders Trust Co | 4.99 | 75 | Columbus Bank & Trust Co | 0.67 |
| 37 | Bank of the West | 4.80 | 76 | Apple Bank for Savings | 0.64 |
| 38 | Northern Trust Co | 4.76 | | | |
| 39 | Compass Bank | 4.58 | | | |

**Table 5.1** (Continued)

| *i* | Financial intermediary | Core capital (billions USD) | *i* | Financial intermediary | Core capital (billions USD) |
|---|---|---|---|---|---|
| 77 | Banco Santander Puerto Rico | 0.57 | 115 | American Chartered Bank | 0.19 |
| 78 | IberiaBank | 0.55 | 116 | Bank of Nevada | 0.18 |
| 79 | Nevada State Bank | 0.54 | 117 | American Natl Bank | 0.17 |
| 80 | 1st Source Bank | 0.53 | 118 | Stockman Bank of Montana | 0.16 |
| 81 | Natl Bank of Arizona | 0.53 | 119 | American Natl Bank of Texas | 0.15 |
| 82 | UMB Bank | 0.53 | 120 | First United Bank & Trust Co | 0.15 |
| 83 | Sterling Savings Bank | 0.5 | 121 | Bank of Kentucky | 0.13 |
| 84 | Texas Capital Bank | 0.49 | 122 | StockYards Bank & Trust Co | 0.13 |
| 85 | Southwest Bank | 0.48 | 123 | Wilson Bank & Trust | 0.13 |
| 86 | Safra Natl Bank of New York | 0.48 | 124 | Bank of North Carolina | 0.13 |
| 87 | Bank Leumi USA | 0.45 | 125 | Bank Rhode Island | 0.12 |
| 88 | Bank of North Georgia | 0.43 | 126 | Community Bank of Texas | 0.12 |
| 89 | Pinnacle Natl Bank | 0.42 | 127 | FSG Bank | 0.12 |
| 90 | Natl Bank of South Carolina | 0.4 | 128 | Community Trust Bank | 0.1 |
| 91 | Chemical Bank | 0.4 | 129 | Commerce Bank of Washington | 0.09 |
| 92 | Hancock Bank | 0.38 | 130 | Paragon Commercial Bank | 0.09 |
| 93 | Banco Bilbao Vizcaya Argentaria PR | 0.36 | 131 | ICE Trust US LLC | 0.08 |
| 94 | Columbia State Bank | 0.36 | 132 | Bryant Bank | 0.08 |
| 95 | R-G Premier Bank of Puerto Rico | 0.34 | 133 | Colorado Capital Bank | 0.08 |
| 96 | Rockland Trust Co | 0.33 | 134 | South Shore Savings Bank | 0.08 |
| 97 | Sun Natl Bank | 0.3 | 135 | D L Evans Bank | 0.08 |
| 98 | Hancock Bank of Louisiana | 0.3 | 136 | Commercial Bank | 0.07 |
| 99 | Sandy Spring Bank | 0.29 | 137 | Capstar Bank | 0.07 |
| 100 | Stellarone Bank | 0.28 | 138 | Northwestern Bank | 0.07 |
| 101 | S & T Bank | 0.27 | 139 | Gulf Coast Bank & Trust Co | 0.07 |
| 102 | Vectra Bank of Colorado | 0.27 | 140 | Business First Bank | 0.06 |
| 103 | Centennial Bank | 0.27 | 141 | Guaranty Bank | 0.06 |
| 104 | Wells Fargo HSBC Trade Bank | 0.26 | 142 | Guaranty Bond Bank | 0.06 |
| 105 | First American Bank | 0.26 | 143 | Avenue Bank | 0.05 |
| 106 | Mainsource Bank | 0.23 | 144 | State Bank & Trust Co | 0.05 |
| 107 | Boston Pvt Bank & Trust Co | 0.22 | 145 | Marine Bank | 0.05 |
| 108 | Bangor Savings Bank | 0.21 | 146 | Northeast Bank | 0.05 |
| 109 | First Security Bank | 0.21 | 147 | Horicon Bank | 0.05 |
| 110 | First Commercial Bank | 0.2 | 148 | Citizens Natl Bank | 0.05 |
| 111 | Integra Bank Natl Asso | 0.2 | 149 | Town North Bank | 0.05 |
| 112 | Berkshire Bank | 0.2 | 150 | American State Bank | 0.05 |
| 113 | Enterprise Bank & Trust | 0.2 | 151 | Community Natl Bank of Texas | 0.05 |
| 114 | Frontier Bank | 0.19 | 152 | First State Bank of East Detroit | 0.05 |

**Table 5.1** (Continued)

| $i$ | Financial intermediary | Core capital (billions USD) | $i$ | Financial intermediary | Core capital (billions USD) |
|---|---|---|---|---|---|
| 153 | Clinton Savings Bank | 0.05 | 179 | Providence Bank | 0.02 |
| 154 | Jersey Shore State Bank | 0.05 | 180 | Carroll County State Bank | 0.02 |
| 155 | Passumpsic Savings Bank | 0.04 | 181 | State Bank of Faribault | 0.02 |
| 156 | Coastal States Bank | 0.04 | 182 | Summit Bank | 0.02 |
| 157 | Southern Bank | 0.04 | 183 | FirstBank | 0.02 |
| 158 | Lincoln Savings Bank | 0.04 | 184 | Touchmark Natl Bank | 0.02 |
| 159 | United Bank & Trust | 0.04 | 185 | State Bank & Trust Co | 0.02 |
| 160 | Oakworth Capital Bank | 0.03 | 186 | First Natl Bank | 0.02 |
| 161 | Central Bank | 0.03 | 187 | Commerce Bank of Oregon | 0.01 |
| 162 | Hometown Bank | 0.03 | 188 | Canyon Community Bank | 0.01 |
| 163 | Security Financial Bank | 0.03 | 189 | Nebraska Natl Bank | 0.01 |
| 164 | Progress Bank & Trust | 0.03 | 190 | First Natl Bank of Junction City | 0.01 |
| 165 | State Bank Financial | 0.03 | 191 | First Vision Bank of Tennessee | 0.01 |
| 166 | Cornerstone Bank | 0.03 | 192 | New Frontier Bank | 0.01 |
| 167 | Bank of South Carolina | 0.03 | 193 | Citizens State Bank | 0.01 |
| 168 | C US Bank | 0.03 | 194 | Keokuk County State Bank | 0.01 |
| 169 | Texas Bank | 0.03 | 195 | Boone Bank & Trust Co | 0.01 |
| 170 | Biddeford Savings Bank | 0.03 | 196 | Northwoods State Bank | 0.01 |
| 171 | Paragon Natl Bank | 0.03 | 197 | Cleveland State Bank | 0.01 |
| 172 | Cornerstone Community Bank | 0.03 | 198 | Farmers Bank | 0.01 |
| 173 | South Central Bank of Barren County | 0.03 | 199 | Farmers Savings Bank & Trust | 0.01 |
| 174 | Somerset Hills Bank | 0.03 | 200 | Business Bank | 0.01 |
| 175 | Platte Valley Bank | 0.03 | 201 | Mount Vernon Bank & Trust Co | 0.01 |
| 176 | Keysource Commercial Bank | 0.02 | 202 | West Town Savings Bank | 0.01 |
| 177 | First State Bank | 0.02 | | | |
| 178 | Premier Commercial Bank | 0.02 | | | |

For simplicity we have then constructed an antisymmetric matrix **M** of netted positions between FIs, i.e., $M_{ij} = B_{ij} - B_{ji} = -M_{ji}$. For each FI $i$, a positive (negative) entry $M_{ij}$ along the $i$-th row gives the net sum payable to (receivable from) the counterparty FI $j$. To analyze a chain of cascading failures following the collapse of bank $i$, only the positive entries of **M** are relevant—as the contagion flows from the failed FI to its net creditor FIs (i.e., those counterparties to which it owes more than what they have borrowed from it). Thus, the matrix **J** we use to construct the network of bilateral exposures among the FIs is obtained from **M**, by replacing all negative matrix entries with zeros, i.e., $J_{ij} = M_{ij}$ if $M_{ij} \geq 0$ and $J_{ij} = 0$ otherwise. This represents a weighted, directed network of financial institutions, with a link being directed from a bank to its net creditors and the link weight being the net liability (in units of billions of US Dollars).

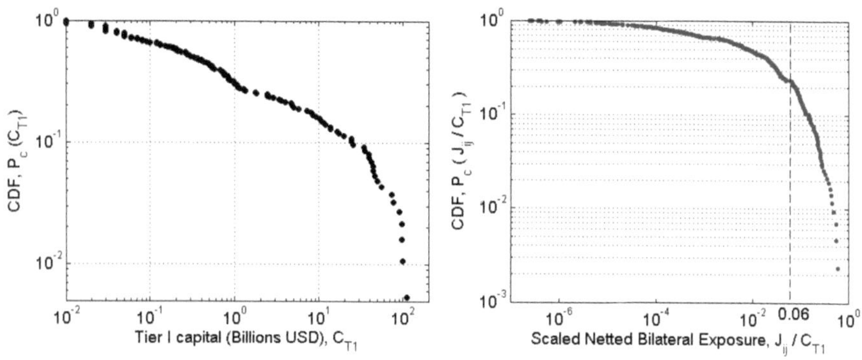


**Fig. 5.1** The cumulative distribution function for (*left*) the core capital ($C_{T1}$) of the 202 FIs considered and (*right*) the netted bilateral exposure $J_{ij}$ scaled by the Tier-I capital of the creditor bank. Assuming a value of 0.06 for the critical fraction $q$, all links to the right of the *broken vertical line* will spread contagion in the network

While most FIs in the network either send or receive at least one link, there are 15 nodes which have neither incoming nor outgoing links. In addition there is one other node for which the sum borrowed from another FI exactly equals the sum it has lent, so that on netting it does not have any net liability with respect to other FIs. The sixteen isolated FIs are City National Bank (node 49), Northwest Savings Bank (61), Apple Bank for Savings (76), Bangor Savings Bank (108), American National Bank of Texas (119), D L Evans Bank (135), Northeast Bank (146), Lincoln Savings Bank (158), Progress Bank & Trust (164), Providence Bank (179), Carroll County State Bank (180), Commerce Bank of Oregon (187), Canyon Community Bank (188), New Frontier Bank (192), Keokuk County State Bank (194) and Cleveland State Bank (197).[5] The largest connected component (LCC) of the network of netted bilateral obligations between FIs comprises $N_{LCC} = 186$ nodes, which have only 424 connections (out of the $N_{LCC}(N_{LCC} - 1)/2 = 17205$ total number of possible links) between them and is therefore very sparse.

In addition to the data on bilateral exposure, we also have information about the Tier-I capital, $C_{T1}$ (in units of billions of USD) of each FI (Fig. 5.1(left)), which measures the financial strength of a bank and comprises the core capital consisting primarily of common stock and disclosed reserves (or retained earnings) [38]. Internationally set standards (the Basel agreements) specify the desired minimum ratio of the core capital of a bank to the total risk-weighted assets held by it in order to provide protection against defaults or sudden loss in value. In our model for failure propagation in the inter-bank network, we specify a critical fraction $q$ of the Tier-I capital of an FI, which, if exceeded by the total net loss of the bank resulting from failures of one or more of its debtor counterparties, will cause its

[5]Except for the D L Evans Bank, for which the GNFV exactly equals the GNPV so that the total netted exposure is zero, all the other banks have no bilateral exposure at all with respect to any other bank in the network.

own failure. Figure 5.1(right) shows the distribution of the netted liabilities of the FIs scaled by the core capital of the creditor counterparty, i.e., $J_{ij}/C_{T1}$. If a node $j$ defaults, then the resulting perturbation will bring down its neighboring node $i$ only if $J_{ij}/C_{T1}(i) > q$ (a value of $q = 0.06$ is shown as a broken vertical line in the figure). Thus, the distribution of $J_{ij}/C_{T1}$ determines the stability of nodes with respect to local perturbations (failure of a single FI).

## 5.3 Topological Properties

Calculating the standard topological properties from the directed network represented by $J$ shows us that it shares many of the features of other inter-bank networks which have been reported in earlier studies. For example, it exhibits the characteristics of a "small-world" network [13] having both low average path length ($\langle l \rangle = 3.6$) and high clustering coefficient ($C = 0.24$). The undirected and non-weighted network shows disassortative mixing by degree (assortative coefficient $r = -0.28$), i.e., nodes with low degree preferentially connect to those having high degree. This may be related to the strong core-periphery structure seen in the network, where a small number of highly (and strongly) interconnected banks form the central nucleus to which most of the other banks of the network connect.

We use a generalization of the core decomposition technique applied to directed networks described in Ref. [22] to obtain the in-degree and out-degree $k$-core—a subnetwork containing only those nodes which have at least $k$ incoming and outgoing links (respectively)—for the unweighted network corresponding to $J$. The cores corresponding to in-degree and out-degree need not be identical although they may have nodes in common, and this is indeed what is observed. We observe that 19 banks belong to both the innermost in-degree and out-degree cores (Nodes 1–7, 9–10, 12, 14–17, 20, 22, 26, 27, 29 and 33—see Table 5.1 for the identity of these FIs), while 4 banks belong only to the out-degree innermost core (Nodes 8, 11, 13 and 33) and only 1 bank (Node 19) belong only to the in-degree innermost core. Thus a set of 24 banks, all having relatively high core capital, form the highly interconnected central nucleus of the network to which the other banks connect.

While the in-degree and out-degree of an FI can give a sense of its "centrality" (i.e., importance) in the network, an even better measure is to use eigenvector centrality, which not only considers how many connections a node has, but also weighs this with the importance (or centrality score) of each neighbor. It is measured by simply considering the eigenvector corresponding to the largest eigenvalue of the adjacency matrix for the network, with the vector components corresponding to each node being their eigenvector centrality score. When a node has high eigenvector centrality, this could be either because it has many neighbors or it has relatively large number of important neighbors or both [39]. However, the standard method of determining eigenvector centrality does not work very well for directed networks, as is the case here. Using the Katz centrality, which works well for directed networks, also has limitations which can be overcome by using a variation, viz., the

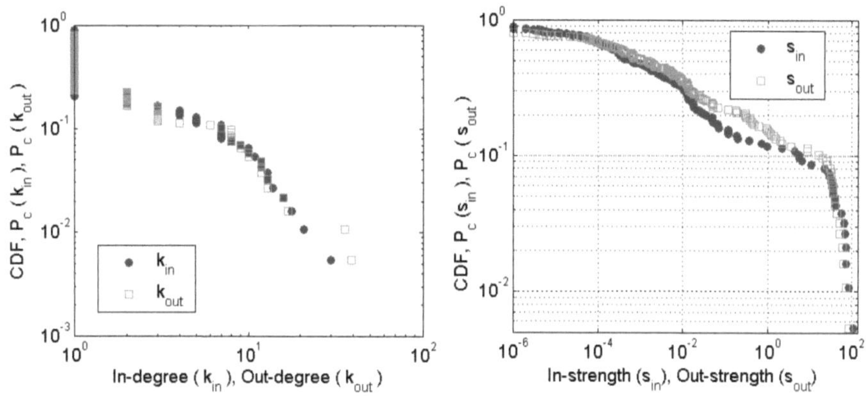


**Fig. 5.2** The cumulative distribution function for (*left*) the in-degree $k_{in}$ (*circles*) and out-degree $k_{out}$ (*squares*) and (*right*) the in-strength $s_{in}$ (*circles*) and out-strength $s_{out}$ (*squares*) of the 202 FIs

Page Rank centrality measure used for example to assign importance to web pages. In particular, we use the Arnoldi iteration algorithm for Page Rank [40] and implemented in a package by David Gleich [41]. The free parameter $\alpha$ is set equal to 0.85 (as used for heuristic reasons by the Google search engine [39]). We find that the bank having highest Page Rank is JP Morgan Chase (Node 4), while the next nine banks in decreasing order of Page Rank are Societe Generale (Node 17), Bank of America (1), Morgan Stanley (10), Deutsche Bank (9), Royal Bank of Scotland (2), Lloyds (7), Goldman Sachs (22), HSBC (16) and BNP Paribas (5). Thus, banks with high centrality (as measured by Page Rank) not only have large core capital but are also the ones belonging to the innermost core for both in-degree and out-degree. There is thus a large degree of agreement among the topological measures used to identify the most crucial nodes of the network.

A recent paper that has looked at systemic risk from the perspective of ecosystem stability has stated that two topological features of inter-bank networks that are crucial are "First, diversity across the financial system... homogeneity bred fragility. ...Second, modularity within the financial system... Modular configurations prevent contagion infecting the whole network in the event of nodal failure" [20]. We now proceed to verify whether the FI network indeed shows evidence of (a) heterogeneity in node properties, e.g., in terms of degree, strength, Tier-I capital, etc. and (b) the existence of modularity (i.e., multiple communities of nodes, with members of each community being more densely and/or strongly connected amongst themselves than with members of other communities).

We have already shown in Fig. 5.1 above the distributions for the Tier-I capital and the netted bilateral liabilities (scaled by the core capital), both of which span several orders of magnitude but exhibit sharply decaying tails. Figure 5.2(left) shows the distributions for both the in-degree and the out-degree of each node in the FI network, while the distributions for the in-strength and out-strength (corresponding to the aggregate of the net amounts lent and borrowed by a bank, respectively) are shown in Fig. 5.2(right). All the distributions have long tails; however, preliminary statistical tests do not appear to suggest a scale-free nature for them. JP Morgan Chase bank (Node 4) has the highest in-degree (30) and out-degree (39), as well as the largest out-strength (85.42), while Deutsche Bank (Node 9) has the highest in-strength (108.76). We note that there is a strong linear correlation between the in- and out-degrees of the nodes ($r = 0.88$ with $p$-value of 0) as well as between their in- and out-strengths ($r = 0.77$, $p$-value = 0). The degree and strength of nodes also show strong linear correlation of $r = 0.75$ and $r = 0.73$ respectively for the incoming and outgoing connections. Not surprisingly, the nodes having large Tier-I capital have high in-degree and out-degree (their linear correlation coefficients with $C_{T1}$ being 0.78 and 0.83 respectively with zero $p$-values), as well as, high in-strength and out-strength (the linear correlation coefficients with $C_{T1}$ being 0.80 and 0.85 respectively with zero $p$-values). In the LCC of 186 nodes, 21 have no in-degree, i.e., they are net borrowers in all of their bilateral interactions, while 36 have no out-degree, i.e., they are net lenders in all their bilateral interactions. 129 nodes (i.e., about 70 % of the LCC) has both incoming and outgoing connections so that they are net borrowers in some bilateral interactions while being net lenders in others.

In order to look for modularity in the FI network, we have used community detection techniques on both the unweighted and weighted LCC of the network. The spectral method for module determination [42] has yielded 13 communities in the unweighted network, the largest having 54 nodes (comprising all of the top 10 % of FIs according to their core capital except JP Morgan Chase) and the smallest containing 3 nodes. The smaller modules are seen to have a star-like topology with all other nodes having connections only to a central hub node of the module which links the community to the rest of the network. A generalized version of the spectral method has been used in the case of the weighted network, which results in the network being split into two modules: one containing 8 nodes and another having the remaining 178 nodes. The FIs in the smaller module (Royal Bank of Scotland, Lloyds, UniCredit, Deutsche Bank, Credit Suisse, Societe Generale, DEXIA and Standard Chartered) are all based in Europe, and this points to large credit flows between FIs whose base of operations are geographically close.

## 5.4 Dynamics of Failure Propagation

The topological properties of the FI network investigated above can alert us to the prominent role played by a small set of banks in the system, but do not by themselves explain how a series of failures can propagate through the network in a cascading process. In order to relate the static information contained in the weighted adjacency matrix $J$ to a dynamic picture of how perturbing certain "keystone" nodes can trigger a significant fraction of the network to break down, we need to assume a specific mechanism for the propagation of the effects of the default of a particular FI to other FIs connected to it via credit relations. We have used a simple and intuitive model where the failure of a node results in the loss of the net sums lent to it by all its creditors (assuming the existence of bilateral netting agreements between all pairs of FIs). This can cause another node to fail if the total loss it faces as a result of all other failures in the network so far, exceeds a critical fraction of its core capital. In this process, as more and more nodes fail, the total loss faced by the remaining creditor nodes increases substantially thereby making it more likely for them to fail in subsequent time steps. Understandably, all nodes will not have similar impact upon the network; we are particularly interested in identifying "super-spreader" nodes, whose collapse will result in a system-wide breakdown in the network (or at least that of a large fraction of nodes belonging to it).

To describe the model of failure propagation, we first define the dynamical state of each node in terms of a binary variable $s_i$. At any time step $t$, if $s_i(t) = 1$ the node is solvent, whereas if $s_i(t) = 0$ it is understood to have failed (once a node has failed, it will remain so for all subsequent time steps). The netted bilateral exposures $J_{ij}$ (i.e., how much is owed by bank $i$ to bank $j$) describes the interactions between the nodes. In the event of a node $i$ defaulting, all its creditors $j$ lose the net sum $J_{ij} (> 0)$ lent to it. If the total loss of any node $j$ as a result of such failures exceeds a critical fraction ($q$, a parameter in our simulations) of its Tier-I capital, $C_{T1}(j)$, it also fails. Thus, the dynamical evolution of each node $i$ in the FI network is described by the discrete-time equation:

$$s_i^{t+1} = 1 - \Theta\left[\sum_j J_{ji}\left(1 - s_j^t\right) - q \cdot C_{T1}(i)\right], \tag{5.1}$$

where, $\Theta$ is the Heaviside step function (i.e., $\Theta(z) = 1$ if $z > 0$ and $= 0$, otherwise). The parameter $q$ depends on the ease of credit availability in the system, with a liquidity crisis corresponding to a sharp decline in the value of $q$.

Initially, all nodes in our model are in the solvent state ($s = 1$). To simulate the propagation of failures, we then change the state of any one node to failed ($s = 0$) and observe whether this causes any of its neighbors to fail, and if so, whether the effect can propagate further along the network. We carry out the process repeatedly, choosing each node in the network in turn to be the initial failed node. While most nodes do not trigger any failure events among their neighbors, in a few cases the initial event can cause a series of failures to cascade along the network. We wait until the system reaches an equilibrium (i.e., the state of every node remains unchanged with time) and count the total fraction $f_{failed}$ of nodes which have failed as a result of the initial single node failure. Figure 5.3(left) shows the largest of such cascade events for $q = 0.01$, when the initial default of Node 23 results in a total of 67 nodes to fail by the end of the cascade. We observe that there are several such nodes whose collapse affects the entire core of strongly connected FIs and tentatively identify them as "super-spreader" nodes (i.e., FIs whose failure results in a network-wide disturbance, in contrast to most other nodes which have no effect). Looking at the distribution of $f_{failed}$ in Fig. 5.3(right) we note its strongly bimodal character. The large peak at extremely low values are due to the majority of nodes which have no effect on the rest of the network, while the smaller peak at higher values of $f_{failed}$ correspond to the "super-spreader" nodes. We also note that decreasing $q$ (corresponding to a constriction in the credit supply or capital buffers)

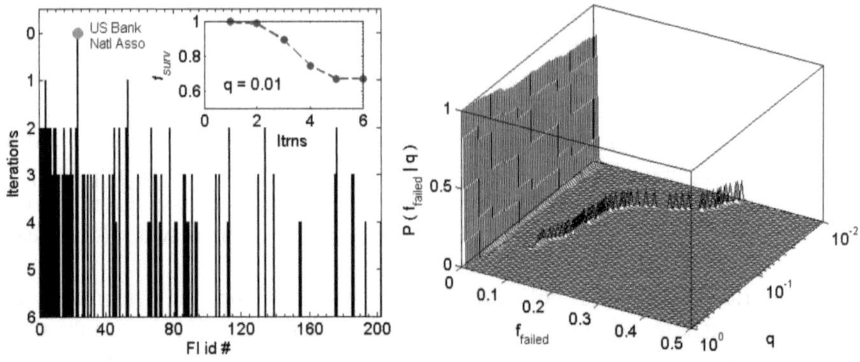


**Fig. 5.3** (*Left*) The failure of a single node in the FI network (node 23) can initiate a series of cascading failures propagating through the network that results in a total of 67 nodes failing by the end of the cascade ($q = 0.01$). The inset shows the time-evolution of the cascade process with the fraction of surviving nodes, $f_{surv}$, declining from 1 at the initial time to a final value of 0.67. (*Right)* The distribution of failure cascade sizes (measured in terms of the total fraction of nodes in the system that fail over the duration of the cascade, $f_{failed}$) shown as a function of the parameter $q$. The distribution has a strongly bimodal character with a node failure either resulting in no effect on the rest of the network, or, bringing down a significant number of other nodes ("many-or-nothing" behavior). The size of cascades increase significantly with decreasing $q$ (corresponding to tightening of credit availability)

increases the size of the cascades. However, the total number of nodes affected by an initial single node failure does not approach the size of the entire FI network, unless $q$ has extremely low (and possibly unrealistic) values. Thus the propagation of disturbances along the network of bilateral liabilities is unlikely to be the sole cause of a system-wide collapse of financial institutions. This agrees with an earlier study [31] which found that perturbations transferred via explicit financial linkages are not enough for triggering large scale breakdown of financial systems. However, our results also identify the vulnerability of the innermost core of broker-dealers, which, though few in number, can through their failure result in over 75 % loss of Tier-I capital in the system. Finally, the cascade of failures can initiate an accompanying liquidity crisis, as the simultaneous default of multiple FIs may restrict credit availability with lender institutions reluctant to give out large loans and adopting a wait-and-watch policy. The resulting decrease in the parameter $q$ will result in even more FIs failing, which in turn further decreases credit availability making the liquidity crisis more severe. Thus, a feedback process ensues with the failure propagation and liquidity crisis driving each other, eventually resulting in a global or system-wide collapse of the financial system. Thus, our results suggest that when evaluating the robustness of complex financial systems we need to focus not only on the network of explicit linkages between the institutions, but also on the overall environment in which they operate.

**Acknowledgements** We would like to thank S Raghavendra for earlier discussions on this topic and F Abergel, N Ganguly and S S Manna for useful comments on the work during presentations at meetings in Kolkata and Bangalore. We are grateful to E Schöll for stimulating discussions and hospitality at TU-Berlin where part of the work was done.

## References

1. Francis J (1850) Mammon and the money market. In: The church of England quarterly review, vol XXVII. W. E. Painter, Strand, London, p 142
2. Mackay C (1848) Extraordinary popular delusions and the madness of crowds. National Illustrated Library, London
3. Cassidy J (1998) Annals of finance: pricking the bubble. New Yorker. Aug 17, 1998. http://www.newyorker.com/archive/1998/08/17/
4. Sinha S, Chatterjee A, Chakraborti A, Chakrabarti BK (2011) Econophysics: an introduction. Wiley-VCH, Weinheim
5. Chancellor E (2000) Devil take the hindmost: a history of financial speculation. Papermac, London
6. Ormerod P (2005) Why most things fail: evolution, extinction and economics. Faber and Faber, London
7. Allen F, Gale D (2000) Financial contagion. J Polit Econ 108:1–33
8. Schweitzer F, Fagiolo G, Sornette D, Vega-Redondo F, Vespignani A, White DR (2009) Economic networks: the new challenges. Science 325:422–425
9. Tainter JA (1988) The collapse of complex societies. Cambridge University Press, Cambridge
10. McCann KS (2000) The diversity-stability debate. Nature 405:228–233
11. May RM (1973) Stability and complexity in model ecosystems. Princeton University Press, Princeton
12. Sinha S, Sinha S (2005) Evidence of universality for the May–Wigner stability theorem for random networks with local dynamics. Phys Rev E 71:020902
13. Watts DJ, Strogatz SH (1998) Collective dynamics of 'small-world' networks. Nature 393:440–442
14. de Solla Price DJ (1976) A general theory of bibliometric and other cumulative advantage processes. J Am Soc Inf Sci 27:292–306
15. Barabasi A-L, Albert R (1999) Emergence of scaling in random networks. Science 286:509–512
16. Sinha S (2005) Complexity vs stability in small-world networks. Physica A 346:147–153
17. Brede M, Sinha S (2005) Assortative mixing by degree makes a network more unstable. http://arxiv.org/abs/cond-mat/0507710
18. Sinha S (2010) Are large complex economic systems unstable? Sci Cult 76:454–458
19. May RM, Levin A, Sugihara G (2008) Ecology for bankers. Nature 451:893–895
20. Haldane AG, May RM (2011) Systemic risk in banking ecosystems. Nature 469:351–355
21. Soramäki K, Bech Arnold J ML, Glass RJ, Beyeler WE (2007) The topology of interbank payment flows. Physica A 379:317–333
22. Chatterjee N, Sinha S (2008) Understanding the mind of a worm: hierarchical network structure underlying nervous system function in *C. elegans*. Prog Brain Res 168:145–153
23. Ashraf I, Sinha S (2012) Core-periphery organization of graphemes in written sequences: decreasing positional rigidity with increasing core order. In: Gelbukh A (ed) CICLing 2012 part I. Springer lecture notes in computer science, vol 7181. pp 142–153
24. Boss M, Elsinger H, Summer M, Thurner S (2004) Network topology of the interbank market. Quant Finance 4:677–684
25. Sui P (2009) Financial contagion in the coreperiphery network. http://www2.warwick.ac.uk/fac/soc/economics/
26. Inaoka H, Takayasu H, Shimizu T, Ninomiya T, Taniguchi K (2004) Self-similarity of banking network. Physica A 339:621–634

27. Iori G, De Masi G, Precup OV, Gabbi G, Caldarelli G (2008) A network analysis of the Italian overnight money market. J Econ Dyn Control 32:259–278
28. Fujiwara Y, Aoyama H, Ikeda Y, Iyetomi H, Souma W (2009) Structure and temporal change of credit network between banks and large firms in Japan. Econ E-J 3:2009-7
29. De Masi G, Fujiwara Y, Gallegati M, Greenwald B, Stiglitz JE (2010) An analysis of the Japanese credit network. Evol Inst Econ Rev 7:209–232
30. http://en.wikipedia.org/wiki/Financial_intermediary
31. Furfine CH (2003) Interbank exposure: quantifying the risk of contagion. J Money Credit Bank 35:111–128
32. Gai P, Kapadia S (2010) Contagion in financial networks. Proc R Soc Lond A 466:2401–2423
33. http://en.wikipedia.org/wiki/Bank
34. Jiang L (2005) Mathematical modeling and methods of option pricing. World Scientific, Singapore
35. Quarterly OCC Report on bank trading and derivatives fourth quarter 2009. Comptroller of the currency administrator of national banks. Washington DC. http://www.occ.treas.gov/topics/capital-markets/financial-markets/trading/derivatives/dq409.pdf
36. http://riskinstitute.ch/00010538.htm
37. Markose S, Giansante S, Gatkowski M, Shaghaghi AR (2010) Too interconnected to fail: financial contagion and systemic risk in network model of CDS and other credit enhancement obligations of US banks. University of Essex Discussion Paper 683. http://www.essex.ac.uk/economics/discussion-papers/papers-text/dp683.pdf
38. http://en.wikipedia.org/wiki/Tier_1_capital
39. Newman MEJ (2010) Networks: an introduction. Oxford University Press, Oxford
40. Golub GH, Greif C (2006) An Arnoldi-type algorithm for computing page rank. BIT Numer Math 46:759–771
41. Gleich D (2006) Pagerank package for MATLAB. http://www.mathworks.com/matlabcentral/fileexchange/11613-pagerank
42. Leicht EA, Newman MEJ (2008) Community structure in directed networks. Phys Rev Lett 100:118703
43. Markose S, Giansante S, Shaghaghi AR (2010) 'Too interconnected to fail' financial network of US CDS market: Topological fragility and systemic risk. J Econ Behav Organ, in press.

# Chapter 6
# Study of Statistical Correlations in Intraday and Daily Financial Return Time Series

**Gayatri Tilak, Tamás Széll, Rémy Chicheportiche, and Anirban Chakraborti**

**Abstract** The aim of this article is to briefly review and make new studies of correlations and co-movements of stocks, so as to understand the "seasonalities" and market evolution. Using the intraday data of the CAC40, we begin by reasserting the findings of Allez and Bouchaud 2011: the average correlation between stocks increases throughout the day. We then use multidimensional scaling (MDS) in generating maps and visualizing the dynamic evolution of the stock market during the day. We do not find any marked difference in the structure of the market during a day. Another aim is to use daily data for MDS studies, and visualize or detect specific sectors in a market and periods of crisis. We suggest that this type of visualization may be used in identifying potential pairs of stocks for "pairs trade".

## 6.1 Introduction

Many complex features, including multi-fractal behavior, of financial markets have been studied for a long time, and constitute today a collection of empirical "laws", the so-called "stylized facts" [2]. The questions: "How efficient is the market? To what extent?" have been long debated on by economists, econometricians and practitioners of finance [3]. It is now accepted that the market is weakly efficient (at

G. Tilak
Université Paris Dauphine, Place du Maréchal de Lattre de Tassigny, 75016 Paris, France

G. Tilak (✉) · R. Chicheportiche · A. Chakraborti
Laboratoire MAS, École Centrale Paris, 92290 Châtenay-Malabry, France
e-mail: gayatri.tilak@ecp.fr

R. Chicheportiche
e-mail: remy.chicheportiche@ecp.fr

A. Chakraborti
e-mail: anirban.chakraborti@ecp.fr

T. Széll
Budapest University of Technology and Economics, Muegyetem rkp. 3-9, H-1111 Budapest, Hungary
e-mail: tomi@digiflex.hu

F. Abergel et al. (eds.), *Econophysics of Systemic Risk and Network Dynamics*, New Economic Windows,
DOI 10.1007/978-88-470-2553-0_6, 

least to some extent and in certain time scales), and that several quantities like the price returns, volatility, traded volume, etc. do exhibit "seasonal patterns";[1] why these "market anomalies" appear is, of course, not well-understood. One reason for their appearance could be that the markets operate in synchronization with human activities and so the financial time series of returns of many assets reveal the related statistical "seasonalities". Identifying such anomalies in order to make statistical arbitrage is a usual practice. Another related practice is estimating market co-movements, which is certainly relevant in several areas of finance, including investment diversification [5] and risk management [6].

In this paper, we first present some notations, definitions and methods. We then review existing results on intraday patterns concerning both individual and collective stock dynamics. We compare the cross-sectional "dispersion" of returns and its typical evolution during the day, with the intraday pattern of the leading modes of the cross-correlation matrix between stock returns, following the studies of Allez and Bouchaud [1]. Then, we make additional plots of the pair-wise cross-correlation matrix elements and study their typical evolution during the day. Finally, we use multidimensional scaling (MDS) in generating maps and visualizing the dynamic evolution of the stock market during the day. When the MDS studies are repeated with daily data, we find that it is easier to visualize or detect specific sectors and market events. We suggest that this type of plots may be used in identifying potential pairs of stocks for "pairs trade".

## 6.2 Some Data Specifications, Notations, and Definitions

In order to measure co-movements in the time series of stock prices, the popular Pearson correlation coefficient is commonly used. However, it is now known that several factors viz., the statistical uncertainty associated with the finite-size time series, heterogeneity of stocks, heterogeneity of the average inter-transaction times, and asynchronicity of the transactions may affect the reliability of this estimator. The investigation of high-frequency "tick-by-tick" data does enable one to monitor market co-movements and price formation in real time. However, high-frequency data have the drawback of aggravating the above mentioned factors even further, raising the need to adequately evaluate their impact through proper correlation measures, such as the Hayashi-Yoshida estimator [7]. In this section, we introduce such concepts, along with notations and definitions, and also specify the details of the datasets used.

[1] "The existence of seasonal asset returns may be an indicator of market inefficiencies... The presence of seasonal returns, however, does not necessitate market inefficiency" [4].

We have considered three data sets.

- Daily returns: we have used the freely downloadable daily closure prices from Yahoo for $N = 54$ companies in the New York Stock Exchange, over a period spanning from January 1, 2008 to May 31, 2011.
- Intraday tick-by-tick: $N = 40$ companies of the CAC40 stock exchange for March 2011, between 10:00–16:00 CET. We have purposefully avoided the opening and closing hours of the market, so as to avoid certain anomalies.
- Intraday sampled returns: Same universe as the tick-by-tick but sampled in bins of 5 minutes or 30 minutes. Thus, the total number of 5 minute bins is 72 per day and total number of 30 minute bins is 12 per day. The total number of trading days in one month is around $T = 21$.

### 6.2.1 *Cross-sectional "Dispersion" of the Binned Data*

In this section we introduce the notations and definitions used by the authors of Ref. [1] for their study of sampled intraday data; we will use the same notations when reproducing their results for our own dataset.

Stocks are labeled by $i = 1, \dots, N$, days by $t = 1, \dots, T$ and bins by $k = 1, \dots, K$. The return of stock $i$ in bin $k$ of day $t$ will be denoted as $r_i(k;t)$. The temporal distribution of stock $i$ in bin $k$ is characterized by its moments: mean $\mu_i(k)$ and standard deviation (volatility) $\sigma_i(k)$, which are defined as:

$$\mu_i(k) = \langle r_i(k;t) \rangle, \tag{6.1a}$$

$$\sigma_i^2(k) = \langle r_i(k;t)^2 \rangle - \mu_i^2(k), \tag{6.1b}$$

where averages over days for a given stock and a given bin are expressed with angled brackets: $\langle \dots \rangle$.

The cross-sectional "dispersion" of the returns of the $N$ stocks for a given bin $k$ in a given day $t$ is as well characterized by its moments:

$$\mu_d(k;t) = [r_i(k;t)], \tag{6.2a}$$

$$\sigma_d^2(k;t) = [r_i(k;t)^2] - \mu_d^2(k;t), \tag{6.2b}$$

where the averages over the "ensemble" of stocks for a given bin in a given day are expressed with square brackets: $[\dots]$. We note that $\mu_d(k;t)$ may be interpreted as the "return of an index", equiweighted on all stocks. We will be more interested in the average of $\sigma_d^2(k;t)$ over all days, as a way to characterize the typical intraday evolution of the "dispersion" between stock returns. Detailed studies of this dispersion and other such measures, concerning both stock prices and returns, will be presented elsewhere [8].

Although the dispersion, described above, indicates the "co-movements" of stocks, a more common and direct characterization is through the standard "correlation" of returns. In order to measure the correlation matrix of the returns, each return is *normalized* by the dispersion of the corresponding bin, to reduce the intraday seasonality and also take into account the fluctuation of the volatility in the considered time period $T$. Therefore, following the same prescription as in Ref. [1], we define: $\widehat{r}_i(k;t) = r_i(k;t)/\sigma_d(k;t)$ and study the correlation matrix defined for a given bin $k$:

$$\rho_{ij}(k) := \frac{\langle \widehat{r}_i(k;t)\widehat{r}_j(k;t)\rangle - \langle \widehat{r}_i(k;t)\rangle\langle \widehat{r}_j(k;t)\rangle}{\widehat{\sigma}_i(k)\widehat{\sigma}_j(k)}. \tag{6.3}$$

The largest eigenvalue of the $N \times N$ correlation matrix $\mathbf{C}(k)$ composed of the elements $\rho_{ij}(k)$, is denoted by $\lambda_1(k)$ and is equal to the risk of the corresponding eigenmode, the "market mode" with all entries positive and close to $1/\sqrt{N}$. In fact, $\lambda_1(k)/N$ can be seen as a measure of the average correlation between stocks. We will be interested in the intraday evolution or the bin-dependence of the largest eigenvalue.[2]

### 6.2.2 Correlation Matrix with Tick-by-Tick Data

Computing correlations using these intraday data, raises lots of issues concerning usual estimators, as already indicated above. Let us assume that we observe $T$ time series of prices or log-prices $p_i$ $(i = 1, \dots, T)$, observed at times $t_m$ $(m = 0, \dots, M)$. The usual estimator of the covariance of prices $i$ and $j$ is the *realized covariance estimator*, which is computed as:

$$\hat{\Sigma}_{ij}^{RV}(t) = \sum_{m=1}^{M} \big(p_i(t_m) - p_i(t_{m-1})\big)\big(p_j(t_m) - p_j(t_{m-1})\big).$$

The problem is that high-frequency tick-by-tick data record changes of prices when they happen, i.e. at times not predefined and not equidistant. Multivariate tick-by-tick data are thus asynchronous, contrary to daily close prices for example, which are by construction synchronous for all the assets on a given exchange. Using standard estimators without caution, could be one cause for the "Epps effect", first observed in [9], which stated that "correlations among price changes in common stocks of companies in one industry are found to decrease with the length of the interval for which the price changes are measured." Hence, here we use the Hayashi-Yoshida estimator [7] also, which takes (part of) the Epps effect into account. There are many other estimators that may be used in general, and a comparison of such estimators has been performed in Ref. [10].

---

[2] A similar study about the intraday evolution of the first eigenvector is of great interest and has been performed as well in [1].

**Hayashi-Yoshida (HY) Estimator** In [7], the authors introduced a new estimator for the linear correlation coefficient between two asynchronous diffusive processes. Given two Itô processes $X, Y$ such that

$$dX_t = \mu_t^X \, dt + \sigma_t^X \, dW_t^X, \tag{6.4}$$

$$dY_t = \mu_t^Y \, dt + \sigma_t^Y \, dW_t^Y, \tag{6.5}$$

$$d\langle W^X, W^Y \rangle_t = \rho_t \, dt, \tag{6.6}$$

and observation times $0 = t_0 \le t_1 \le \cdots \le t_{n-1} \le t_n = T$ for $X$, and $0 = s_0 \le s_1 \le \cdots \le s_{m-1} \le s_m = T$ for $Y$, which must be independent for $X$ and $Y$, they showed that the following quantity:

$$\sum_{i,j} r_i^X r_j^Y \mathbb{1}_{\{O_{ij} \neq \emptyset\}}, \tag{6.7}$$

$$O_{ij} = ]t_{i-1}, t_i] \cap ]s_{j-1}, s_j],$$

$$r_i^X = X_{t_i} - X_{t_{i-1}},$$

$$r_j^Y = Y_{s_j} - Y_{s_{j-1}},$$

is an unbiased and consistent estimator of $\int_0^T \sigma_t^X \sigma_t^Y \rho_t \, dt$, as the largest mesh size goes to zero. In practice, it amounts to *summing every product of increments as soon as they share any overlap of time*. In the case of constant volatilities and correlation, it provides a consistent estimator for the correlation

$$\rho_{ij}^t = \frac{\sum_{i,j} r_i^X r_j^Y \mathbb{1}_{\{O_{ij} \neq \emptyset\}}}{\sqrt{\sum_i (r_i^X)^2 \sum_j (r_j^Y)^2}}. \tag{6.8}$$

## 6.2.3 *Pearson Correlation Coefficient and Correlation Matrix with Daily Returns*

In order to study the *equal time* cross-correlations between $N$ stocks, we first denote the closure price of stock $i$ in day $\tau$ by $P_i(\tau)$, and determine the logarithmic return of stock $i$ as $r_i(\tau) = \ln P_i(\tau) - \ln P_i(\tau - 1)$. For the sequence of $T$ consecutive trading days, encompassing a given window $t$ with width $T$, these returns form the **return vector** $\boldsymbol{r}_i^t$. In order to characterize the synchronous time evolution of assets, we use the equal time Pearson correlation coefficients between assets $i$ and $j$ defined as

$$\rho_{ij}^t = \frac{\langle \boldsymbol{r}_i^t \boldsymbol{r}_j^t \rangle - \langle \boldsymbol{r}_i^t \rangle \langle \boldsymbol{r}_j^t \rangle}{\sqrt{[\langle \boldsymbol{r}_i^{t\,2} \rangle - \langle \boldsymbol{r}_i^t \rangle^2][\langle \boldsymbol{r}_j^{t\,2} \rangle - \langle \boldsymbol{r}_j^t \rangle^2]}}, \tag{6.9}$$

where $\langle \ldots \rangle$ indicates a time average over the $T$ consecutive trading days included in the return vectors. These correlation coefficients fulfill the usual condition of $-1 \leq \rho_{ij} \leq 1$ and form an $N \times N$ correlation matrix $\mathbf{C}^t$, which serves as the basis of further analyses [11–13].

For analysis, the data is divided time-wise into $M$ *windows* ($t = 1, 2, \ldots, M$) of width $T$, corresponding to the number of daily returns included in the window. The consecutive windows may be overlapping/non-overlapping with each other, the extent of which is dictated by the window step length parameter $\delta t$, describing the displacement of the window, measured also in trading days. The sizes of window width $T$, and window step width $\delta t$, are to be chosen cleverly: for example, $T$ must be long enough to grasp any signal with a certain statistical power, but not cover too long a period over which the signal could have varied.

### 6.2.4 Distance Matrix

To obtain "distances", a non-linear transformation

$$d_{ij} = \sqrt{2(1 - \rho_{ij})}, \tag{6.10}$$

is used, with the property $2 \geq d_{ij} \geq 0$, forming an $N \times N$ distance matrix $\mathbf{D}^t$, such that all distances are "ultrametric". The concept of ultrametricity is discussed in detail by Mantegna [14]. Out of the several possible ultrametric spaces, the subdominant ultrametric is opted for due to its simplicity and remarkable properties. The choice of the non-linear function is again arbitrary, as long as all the conditions of ultrametricity are met.

### 6.2.5 Multidimensional Scaling (MDS)

Multidimensional scaling is a set of data analysis techniques that display the structure of "distance"-like data as a "geometrical picture", where each object is represented by a point in a multidimensional space. The points are arranged in this space, such that the distances between pairs of points have the strongest possible relation to the "similarities" among the pairs of objects—two similar objects are represented by two points that are close together, and two dissimilar objects are represented by two points that are far apart. The space is usually a two- or three-dimensional Euclidean space, but may be non-Euclidean and may have more dimensions.

MDS is a generic term that includes many different types—classified according to whether the similarities data are "qualitative" (called non-metric MDS) or "quantitative" (metric MDS). The number of similarity matrices and the nature of the

MDS model can also classify MDS types. This classification yields classical MDS (one matrix, unweighted model), replicated MDS (several matrices, unweighted model), and weighted MDS (several matrices, weighted model). For a general introduction and overview, please see Ref. [15].

The collection of objects to be analyzed in our case, is $N$ stocks, on which a distance function is defined using (6.10). These distances are the entries of the similarity matrix

$$\mathbf{D}^t := \begin{pmatrix} d_{11} & d_{12} & \cdots & d_{1N} \\ d_{21} & d_{22} & \cdots & d_{2N} \\ \vdots & \vdots & \ddots & \vdots \\ d_{N1} & d_{N2} & \cdots & d_{NN} \end{pmatrix}. \tag{6.11}$$

Given $\mathbf{D}^t$, the aim of MDS is to find $N$ vectors $x_1, \ldots, x_N \in \mathbb{R}^D$, such that

$$\|x_i - x_j\| \approx d_{ij} \quad \forall i, j \in N, \tag{6.12}$$

where $\|\cdot\|$ is a vector norm. In classical MDS, this norm is typically the Euclidean distance metric.

In other words, MDS tries to find a mathematical embedding of the $N$ objects into $\mathbb{R}^D$ such that distances are preserved. If the dimension $D$ is chosen to be 2 or 3, we are able to plot the vectors $x_i$ to obtain a visualization of the similarities between the $N$ objects. It may be noted that the vectors $x_i$ are *not unique*—with the Euclidean metric, they may be arbitrarily *translated* and *rotated*, since these transformations do not change the pairwise distances $\|x_i - x_j\|$.

There are various approaches to determining the vectors $x_i$. Generally, MDS is formulated as an *optimization* problem, where $(x_1, \ldots, x_N)$ is found as a minimization of some cost function, such as

$$\min_{x_1,\ldots,x_N} \sum_{i<j} \left(\|x_i - x_j\| - d_{ij}\right)^2. \tag{6.13}$$

A solution may then be found by numerical optimization techniques. In our case, we used simulated annealing as the optimization procedure.

## 6.3 Results

### *6.3.1 U-Effect in Volatility*

In financial studies, among the periodicities or "seasonalities" is the "U-effect" [16, 17], which describes the intraday pattern of average volatility $\sigma(k) = [\sigma_i(k)]$ of individual stocks: the average volatility is high during the market opening hours, then decreases so as to reach a minimum around lunch time, and increases again

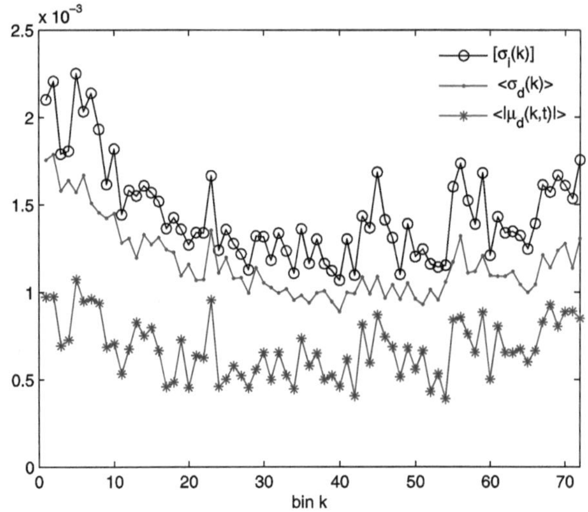


**Fig. 6.1** Plots of the average volatility of stocks $\sigma(k)$, the average cross sectional dispersion $\sigma_d(k)$ and the average absolute value of the index return $\langle|\mu_d(k,t)|\rangle$ as a function of the 5-minute bins denoted by $k$, from 10h00–16h00 CET, for the period March, 2011. Courtesy: E. Guevara H. et al. [8]

steadily until the market closes. We show a similar result in Fig. 6.1, computed with the CAC40 intraday data for the period March, 2011. The average of $|\mu_d(k;t)|$ is a proxy for the "index volatility", and is displayed in Fig. 6.1: it also shows a U-shaped pattern similar to that of $\sigma(k)$.

### 6.3.2 *The Eigenvalues of the Correlation Matrix and Average Correlations*

The largest eigenvalue $\lambda_1$ of the correlation matrix of stock returns, is well known to be associated with the "market mode", i.e. all stocks moving more or less in a synchronized manner. We show in the top panel of Fig. 6.2 the magnitude of $\lambda_1/N$ computed from (6.3) on 5-min data, as a function of the bin $k$. Interestingly, the average correlation clearly *increases* as time elapses. As mentioned earlier, the quantity $\lambda_1/N$ captures the behavior of the average correlation between stocks, which can be seen in the bottom panel of Fig. 6.2.

The evolution of the next six eigenvalues $\lambda_i(k)$, $i = 2, \ldots, 7$ is also shown in Fig. 6.2. We see that the amplitudes of these *decrease* with time. It may be appropriate to quote the authors of Ref. [1]: "Although by construction the trace of the correlation matrix, and therefore the sum of all $N$ eigenvalues is constant (and equal to $N$), this decrease is not a trivial consequence of the increase of $\lambda_1$... What we see here is that as the day proceeds, more and more risk is carried by the market factor, while the amplitude of sectorial moves shrivels in relative terms (but remember that the correlation matrix is defined after normalizing the returns by the local volatility, which increases in the last hours of the day)."

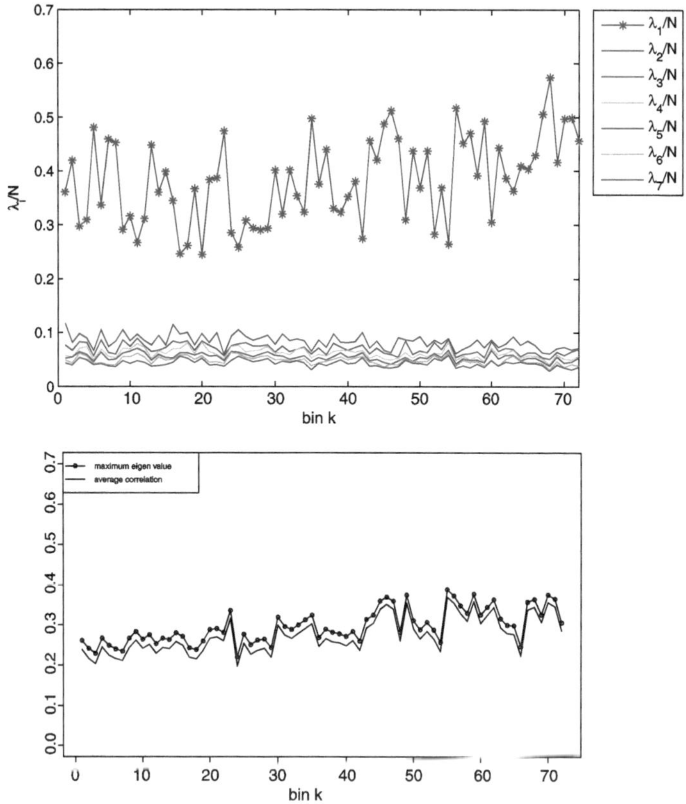


**Fig. 6.2** *Top*: Top eigenvalues of the correlation matrix, $\lambda_i(k)/N$, $i = 1, \ldots, 7$, as a function of the 5-minute bins denoted by $k$, from 10h00–16h00 CET, in March, 2011. [5-min sampled prices, courtesy E. Guevara H. et al. [8]]. *Bottom*: The largest eigenvalue $\lambda_1/N$ (circles) is a proxy for the average correlation (*plain*) [HY correlations for every pair and every bin of every day, then averaged over days for visual comfort and comparison with previous figure]

We also compute using (6.8) the cross-correlation matrices with tick-by tick data, for all 72 bins per day and 20 days in a month. The temporal evolution of the pairwise average correlation coefficients as a function of bins, for different days, and further averaged over all the days, are plotted below in Fig. 6.3.

## *6.3.3 MDS Using Intraday Data*

In order to *visually* capture the co-movement of stocks, we used the MDS plots of the 40 stocks of the CAC40 index (see list of CAC40 stocks in Table 6.1), for the period of March 2011. We used 30 minute bins to compute the correlations, using the Hayashi-Yoshida estimator. We used the period 10h00–16h00 CET, so as

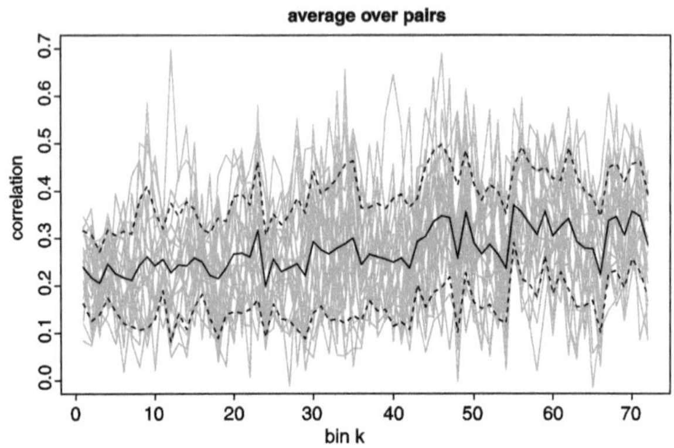


**Fig. 6.3** Plot of the (pairwise) average correlations as functions of bins $k$, for different days. *Thick solid line*: Plot of the average correlation coefficients, further averaged over all the days, which shows that the average correlation between stocks increases throughout the day. *Thick dashed lines*: Plots of the standard deviations on either side of the average correlation

to get 12 bins per day for the 22 days. Using the correlation matrices as input, we made the distance transformations (using (6.10)) to produce the distance matrices. These distance matrices were then used as inputs to the standard MDS function in MATLAB. We used the method of simulated annealing to optimize the cost function of a particular bin. The first bin starts with an initial set of coordinates chosen at random; for the following bins, we used the final results of the previous bins as the initial states.[3] The output of the MDS were the coordinates, which were plotted as the MDS maps. The coordinates were plotted in a manner such that the centroid of the map coincided with the origin $(0,0)$. We then computed the mean distance of all the coordinates from the center, and plotted this measure as a function of time.

During the course of any day, since for every bin the correlation matrix changes, the MDS map also changes. Just as it is interesting to study how the average correlation between the stocks varies during the day, we thought it would be also interesting to study how the MDS map evolves "on an average" during the day. We had two choices: (i) Run the MDS algorithm for every bin for 22 days, and take the average of the coordinates over all the 22 maps, and plot this map for every bin. (ii) Take the average of the correlations over the 22 days for each bin, and plot a single MDS map for every bin. We executed both, to see the variations. In choice (i), for every bin $k$ we take an average of the coordinates generated by the 22 MDS runs (for different days) and plot them stock by stock. Some stocks fluctuate a lot on a day to day basis, in the same time bin; others fluctuate less. On the whole we expected to see the average structure (clustering) of the market. In choice (ii), we expected to see less structure, since when we take the average of correlations over all 22 days, and then run the MDS once for every bin, the variances in the correlations disappear and so the MDS plots look more uniform.

---

[3]This is to avoid too drastic a change in the MDS plots from one bin to another, keeping in mind that the vectors $x_i$ are *not unique*—with the Euclidean metric, they may be arbitrarily *translated* and *rotated*.

**Table 6.1** RICS list of the stocks in the CAC 40

| Names | RICS |
|---|---|
| ACCOR FICTIVE | ACCP.PA |
| AIR LIQUIDE | AIRP.PA |
| ALCATEL LUCENT | ALUA.PA |
| ALSTOM | ALSO.PA |
| ARCELOR MITTAL FICTIVE | ISPA.AS |
| AXA | AXAF.PA |
| BNP PARIBAS | BNPP.PA |
| BOUYGUES | BOUY.PA |
| CAP GEMINI | CAPP.PA |
| PERNOD RICARD | PERP.PA |
| VALLOUREC | VLLP.PA |
| CARREFOUR | CARR.PA |
| PEUGEOT SA | PEUP.PA |
| VEOLIA ENVIRONNEMENT | VIE.PA |
| CREDIT AGRICOLE SA | CAGR.PA |
| PPR | PRTP.PA |
| VINCI | SGEF.PA |
| DANONE | DANO.PA |
| PUBLICIS | PUBP.PA |
| VIVENDI | VIV.PA |
| EADS PEA FICTIVE | EAD.PA |
| RENAULT | RENA.PA |
| EDF | EDF.PA |
| SAINT GOBAIN | SGOB.PA |
| ESSILOR INTERNATIONAL | ESSI.PA |
| SANOFI | SASY.PA |
| FRANCE TELECOM | FTE.PA |
| SCHNEIDER ELECTRIC SA | SCHN.PA |
| GDF SUEZ | GSZ.PA |
| SOCIETE GENERALE | SOGN.PA |
| LOREAL | OREP.PA |
| STMICROELECTRONICS PEA FICTIVE | STM.PA |
| LVMH | LVMH.PA |
| SUEZ ENVIRONNEMENT SA | SEVI.PA |
| LAFARGE | LAFP.PA |
| TECHNIP | TECF.PA |
| MICHELIN | MICP.PA |
| TOTAL | TOTF.PA |
| NATIXIS | CNAT.PA |
| UNIBAIL-RODAMCO SE | UNBP.PA |

**Fig. 6.4** MDS plots for bins 1–6. Each point on a plot represents a stock (see list of CAC40 stocks in Table 6.1), designated by two coordinates $(x_i, y_i)$, $i = 1, \ldots, N$. We took the average of the coordinates (output of the MDS) of each company over all 22 days, for a particular bin. We then plotted the MDS maps using these averaged coordinates for the different bins to see the evolution during the day

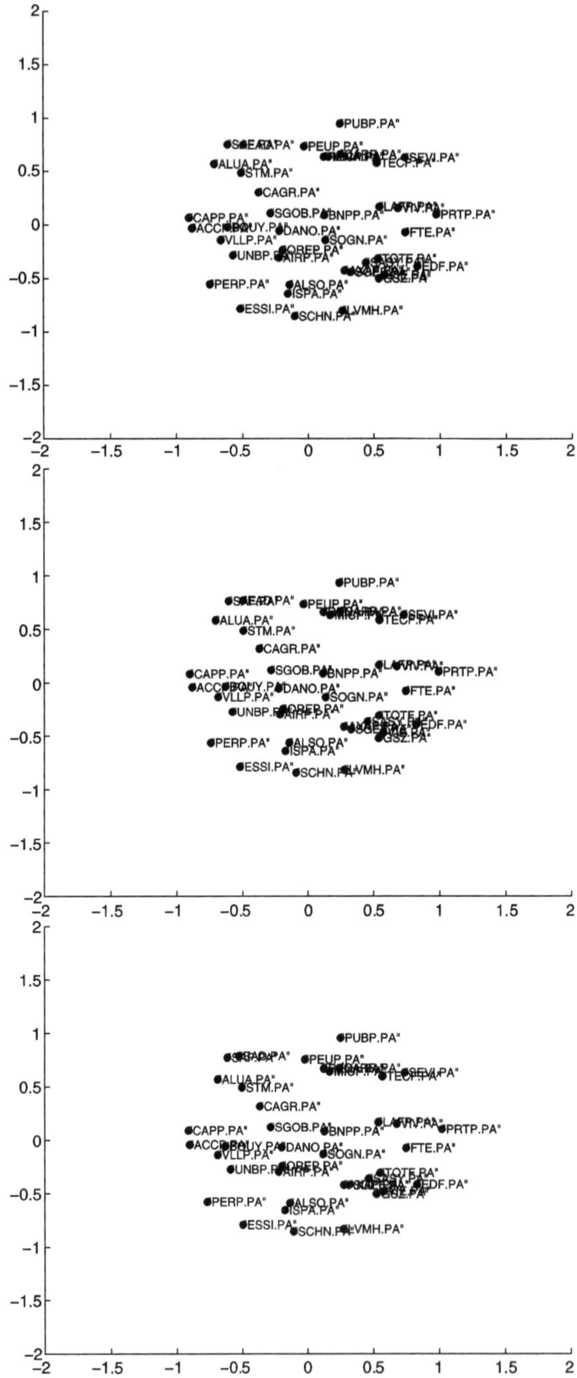


(a)

**Fig. 6.4** (Continued)

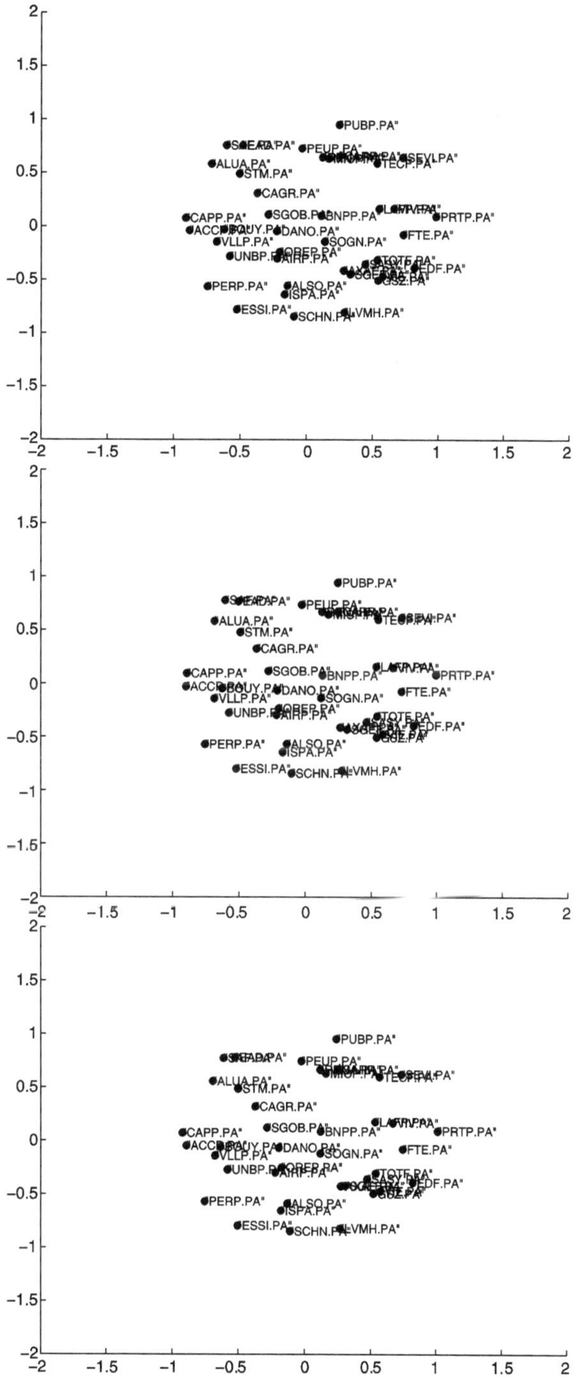


(b)

#### 6.3.3.1 Averaged (over Days) Coordinates in Different Bins

We took the average of the coordinates (output of the MDS) of each company over all 22 days, for a particular bin. We then plotted the MDS maps using these averaged coordinates for the different bins to see the evolution during the day, as shown in Fig. 6.4 (for first six bins) and Fig. 6.5 (for last six bins). We find that there is some structure, and particular companies always stay together in a cluster or a group.

#### 6.3.3.2 Averaged (over Days) Correlations in Different Bins

We also took the average of the correlation coefficients for each pair over all 22 days, and then used them to generate the MDS plot for a particular bin. We then plotted the MDS maps for the different bins to see the evolution during the day, as shown in Fig. 6.6 (for first six bins) and Fig. 6.7 (for last six bins). We find that there is less structure than the previous plots (as average of correlations "smoothen out" the dissimilarities). The structures of the maps and positions of the companies do not change drastically during the course of the day.

We further plotted the variation of the mean distance of all the coordinates from the center of the map, over the different bins to see the temporal evolution during the day, in Fig. 6.8. This follows exactly the opposite trend of the average correlations as shown in Fig. 6.2 or Fig. 6.3—the mean distance *decreases* during the day. This result is as expected, and not very surprising.

### *6.3.4 MDS Using Daily Data*

In order to capture the co-movement of stocks visually, we again used the MDS plots of 54 stocks from Yahoo daily data, for the period of January 2008–May 2011. We computed the correlations using non-overlapping windows of $T$ consecutive trading days, using (6.9). The choice of $T$ is important because if $T/N$ is small, then according to the Random Matrix Theory we cannot distinguish between noise and the true signal. Since MDS needs a full rank correlation matrix, the noise needs to be cleaned with appropriate statistical measures before applying MDS.

As before, using the correlation matrices as input, we made the distance transformations (using (6.10)) to produce the distance matrices. These distance matrices were then used as inputs to the MDS code in `MATLAB`. We used the method of simulated annealing to optimize the cost function of a particular day. The first day (time-step) starts with an initial set of coordinates chosen at random; for the following days (time-steps), we used the final results of the previous day (time-step) as

**Fig. 6.5** MDS plots for bins 7–12. Each point on a plot represents a stock (see list of CAC40 stocks in Table 6.1), designated by two coordinates $(x_i, y_i)$, $i = 1, \ldots, N$. We took the average of the coordinates (output of the MDS) of each company over all 22 days, for a particular bin. We then plotted the MDS maps using these averaged coordinates for the different bins to see the evolution during the day

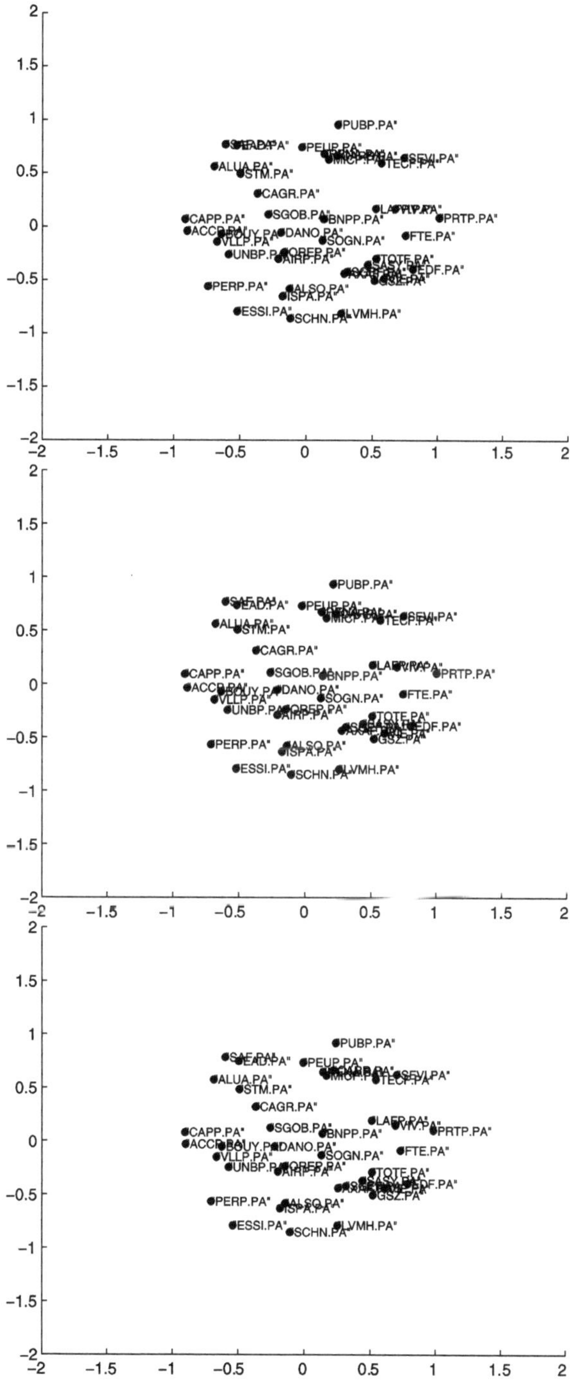


(a)

**Fig. 6.5** (Continued)

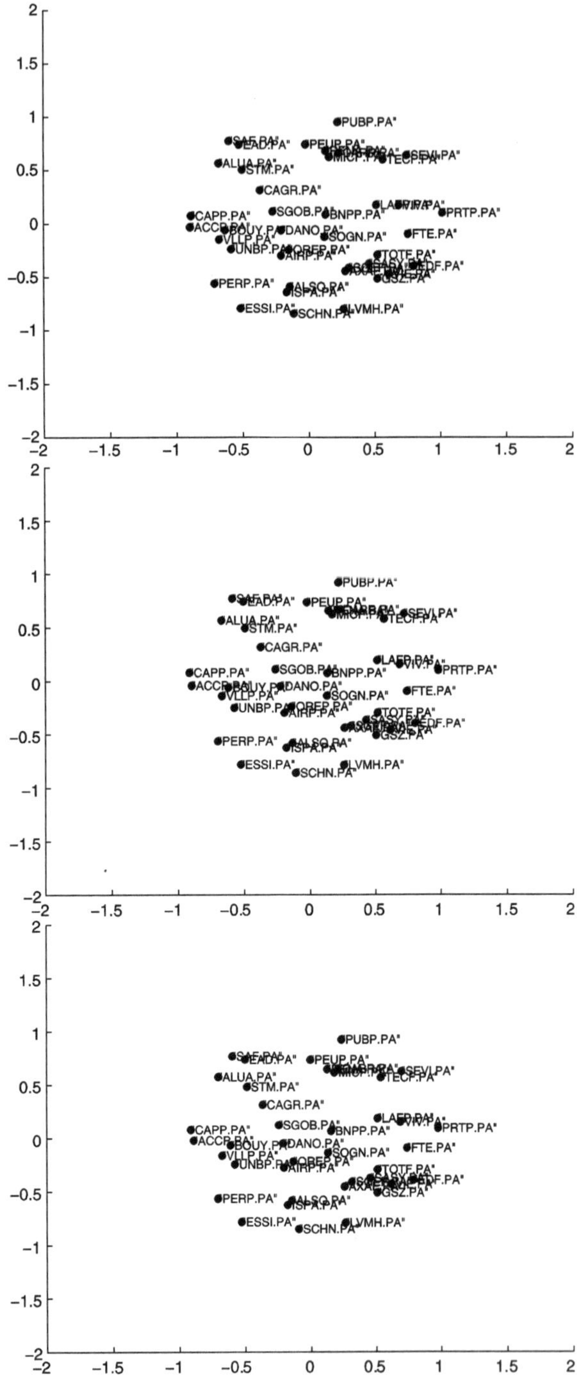


(b)

Fig. 6.6 MDS plots for bins 1–6. Each point on a plot represents a stock (see list of CAC40 stocks in Table 6.1), designated by two coordinates $(x_i, y_i)$, $i = 1, \ldots, N$. We took the average of the correlation coefficients for each pair over all 22 days, and then used them to generate the MDS plot for a particular bin. We then plotted the MDS maps for the different bins to see the evolution during the day

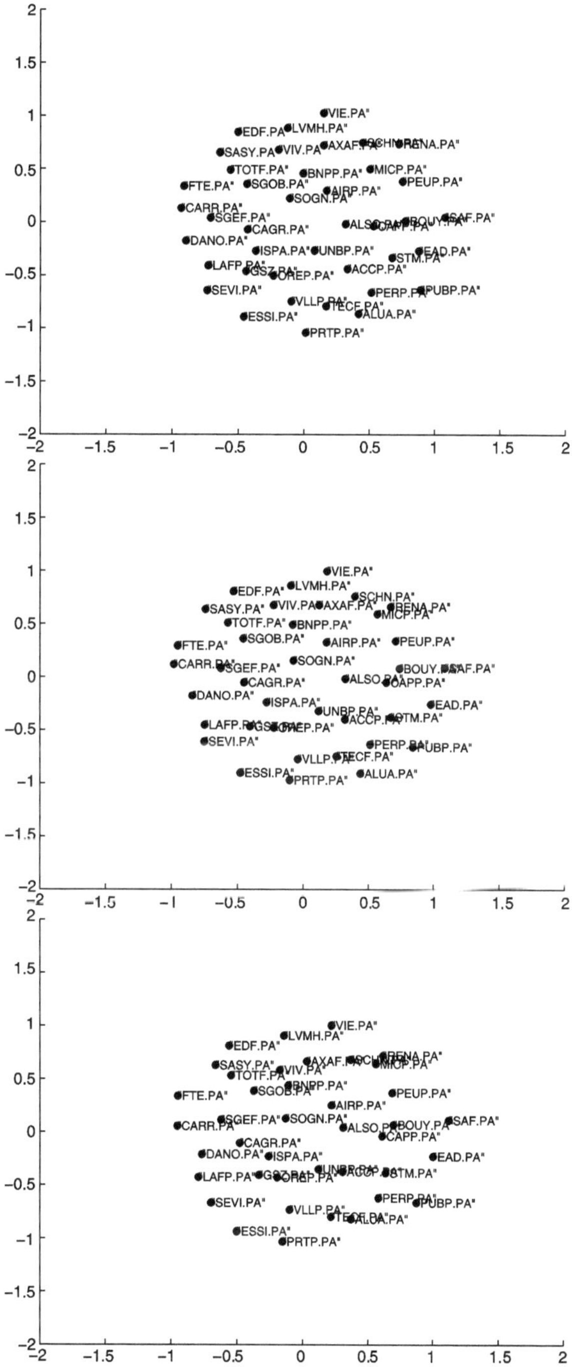


(a)

**Fig. 6.6** (Continued)

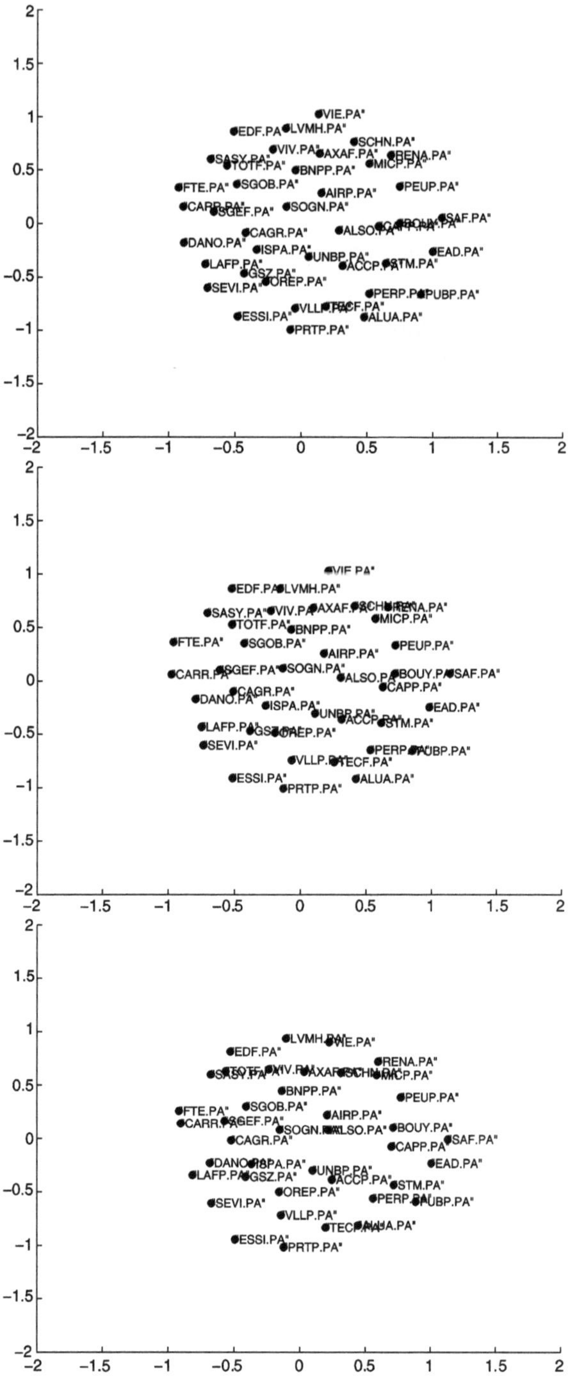


(b)

**Fig. 6.7** MDS plots for bins 7–12. Each point on a plot represents a stock (see list of CAC40 stocks in Table 6.1), designated by two coordinates $(x_i, y_i)$, $i = 1, \ldots, N$. We took the average of the correlation coefficients for each pair over all 22 days, and then used them to generate the MDS plot for a particular bin. We then plotted the MDS maps for the different bins to see the evolution during the day

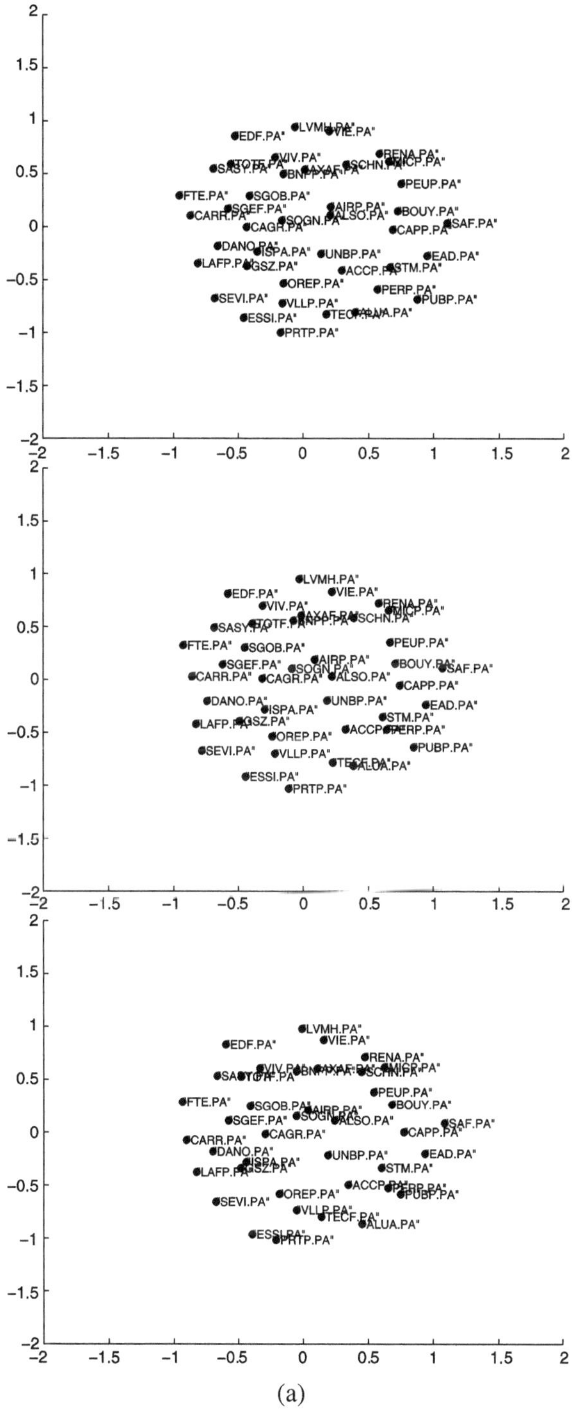

**Fig. 6.7** (Continued)

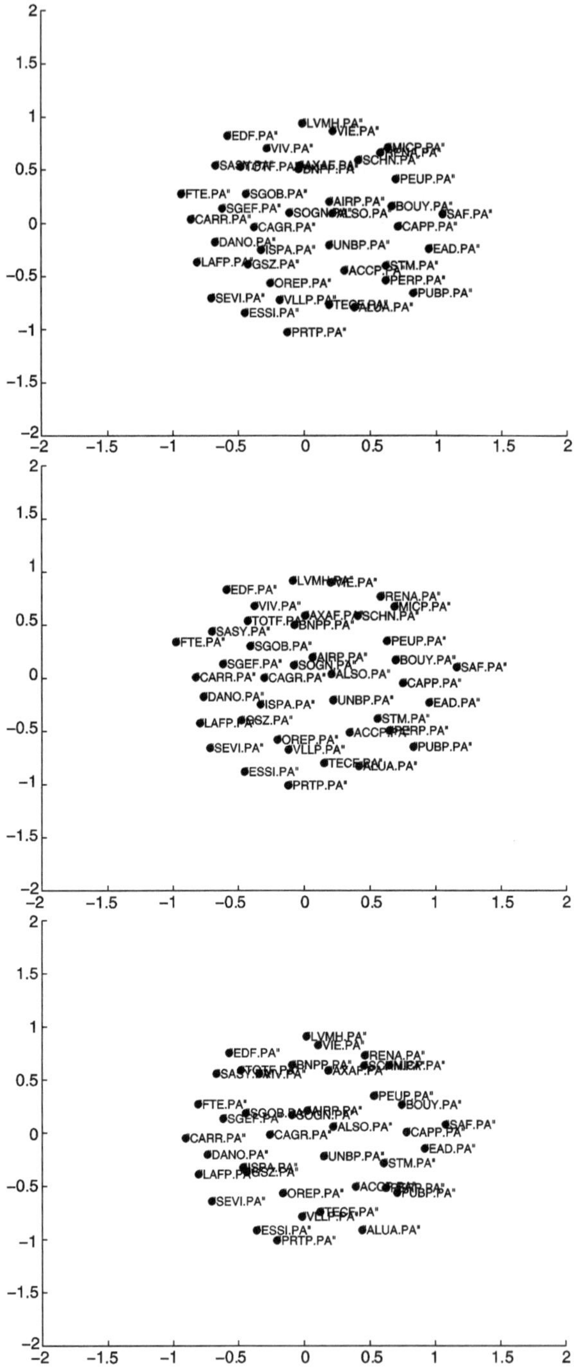


(b)

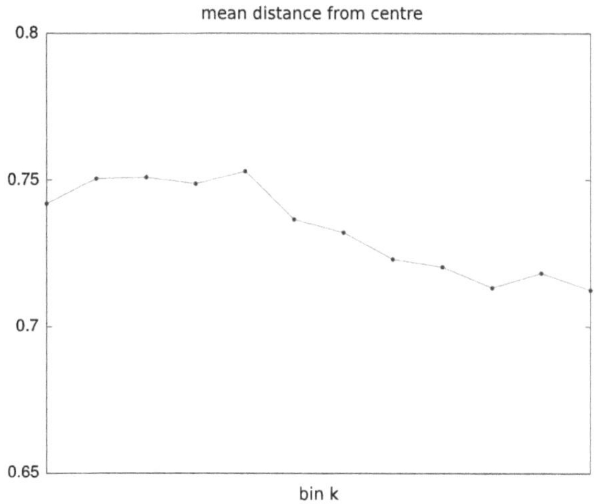


**Fig. 6.8** Mean distance of coordinates of all the points (40 stocks) from center of the map, as a function of the bin $k$. There are 12 bins of 30 minutes between 10:00 and 16:00 CET

the initial state.[4] The output of the MDS were the coordinates, which were plotted as the MDS maps. The coordinates were plotted in a manner such that the centroid of the map coincided with the origin $(0, 0)$. We then computed the mean distance of all the coordinates from the center, and plotted this measure as a function of time.

In Fig. 6.9 we plot MDS maps for sample dates: 28/05/2008 (pre-Subprime crisis), 27/10/2008 (onset of Subprime crisis) and 28/06/2010 (post-Subprime crisis). In these plots we do see the difference in the positions of the companies. The position of Lehman brothers in the plot of the MDS during the post-Subprime crisis is noteworthy.

We also plot in Fig. 6.9, the mean distance of coordinates from center for the period 01/01/2008 to 31/12/2009. There is certainly a noticeable variation in this entire period, and the period of the Subprime crisis can be identified with the low value of mean distance.

In order to examine carefully whether any clusters can be identified, we worked with a subset of 18 companies. In Fig. 6.10 and Fig. 6.11, we plot MDS maps for different sample dates: 03/06/2008, 25/07/2008, and 05/09/2008 (pre-Subprime crisis); 17/10/2008, 28/11/2008 and 13/01/2009 (during Subprime crisis); 24/02/2009, 07/04/2009, 12/09/2009 and 04/11/2009 (post-Subprime crisis).

In these plots we do see the considerable differences in the positions of the companies. However, it is interesting to follow the positions of certain pairs:

(i) JP Morgan and Bank of America
(ii) Nissan and Toyota

[4]This is to avoid too drastic a change in the MDS plots from one time step to another, keeping in mind that the vectors $x_i$ are *not unique*—with the Euclidean metric, they may be arbitrarily *translated* and *rotated*. We imposed a small penalty in the cost function for deviation from the initial state.

**Fig. 6.9** The correlation matrices are computed from Yahoo daily closure price data using (6.9) and 54 trading day window, for the set of 54 companies. The points on each MDS plot represent stocks, each designated by two coordinates $(x_i, y_i)$, $i = 1, \ldots, 54$. (**a**) *Top*: MDS plot for date 28/05/2008. (**a**) *Middle*: MDS plot for date 27/10/2008. (**a**) *Bottom*: MDS plot for date 28/06/2010. (**b**) Mean distance of coordinates from center for the two year period 01/01/2008 to 31/12/2009

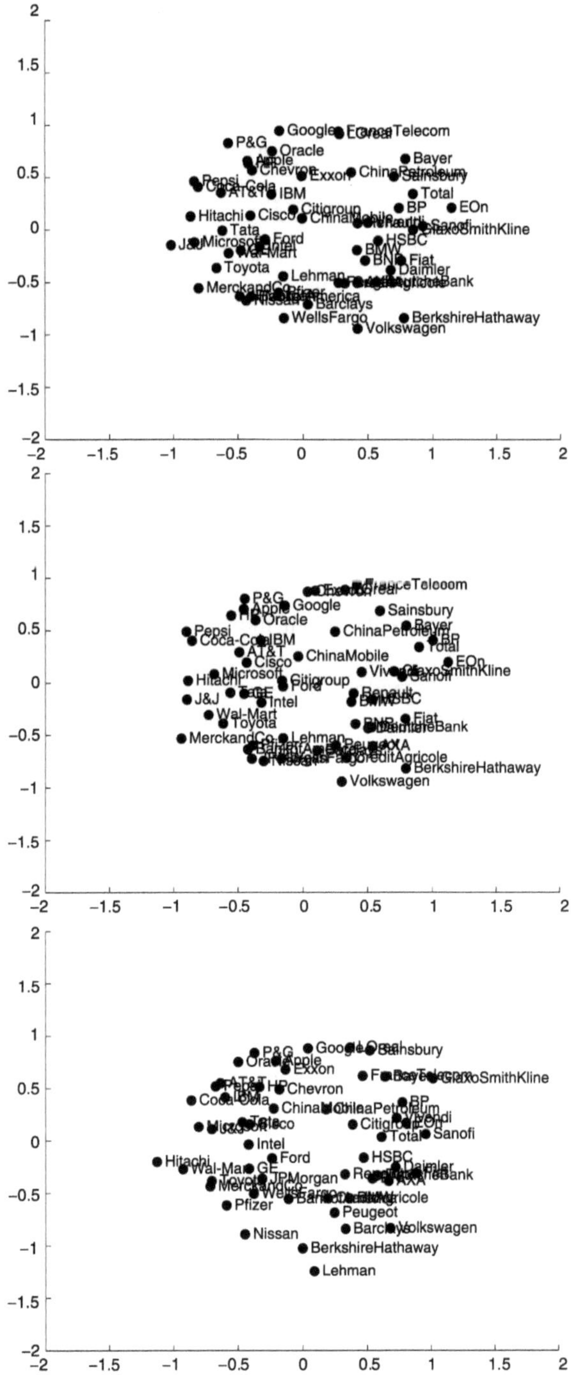


(a)

**Fig. 6.9** (Continued)

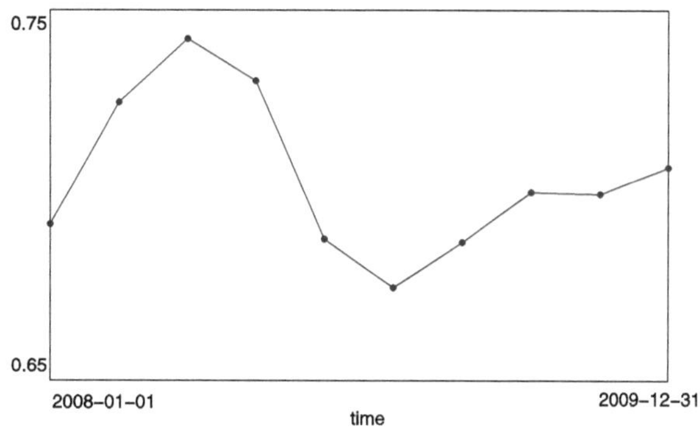


(b)

(iii) Chevron and Exxon
(iv) Pepsi and Coca Cola.

This type of visual plot may therefore be used in identifying potential pairs of stocks for "pairs trade". Such a strategy monitors the performances of two historically correlated stocks: when the correlation between the two securities temporarily weakens, i.e. one stock moves up while the other moves down, the pairs trade strategy would be to *short* the outperforming stock and to *long* the underperforming one, betting that the "spread" between the two would eventually converge. Further analysis is of course necessary to devise such a strategy.

We also find that there is some noticeable clustering effect, e.g. as all the European banks are in one cluster and all the European automobiles are in another cluster.

## 6.4 Concluding Remarks

In this paper, we first reviewed existing results on intraday patterns concerning both individual and collective stock dynamics. We studied the cross-sectional "dispersion" of returns and its typical evolution during the day, and found that the average volatility is high during the market opening hours, then decreases so as to reach a minimum around lunch time, and increases again steadily until the market closes. The average of $|\mu_d(k;t)|$, which is a proxy for the "index volatility", also displayed a U-shaped pattern similar to that of $\sigma(k)$. Studying the intraday pattern of the leading modes (eigenvalues) evaluated using the cross-correlation matrix between stock returns, we found that the maximum eigenvalue $\lambda_1(k)$ (corresponding to the market mode or *average correlation*) clearly *increases* as time elapses. However, the evolution of the next six eigenvalues $\lambda_i(k)$, $i = 2, \dots, 7$ showed that the amplitudes of these *decrease* with time. Then, we made additional plots of the pair-wise cross-correlation matrix elements and studied their typical evolution during the day. Finally, we used multidimensional scaling (MDS) in generating maps

**Fig. 6.10** MDS plots for different dates.
(**a**) *Top*: 03/06/2008
(**b**) *Top*: 25/07/2008
(**a**) *Middle*: 05/09/2008
(**b**) *Middle*: 17/10/2008
(**a**) *Bottom*: 28/11/2008
(**b**) *Bottom*: 13/01/2009.
The correlation matrices are computed from Yahoo daily closure price data using (6.9) and 30 trading day window, for the subset of 18 companies. The points on each plot represent stocks, each designated by two coordinates $(x_i, y_i)$, $i = 1, \ldots, 18$

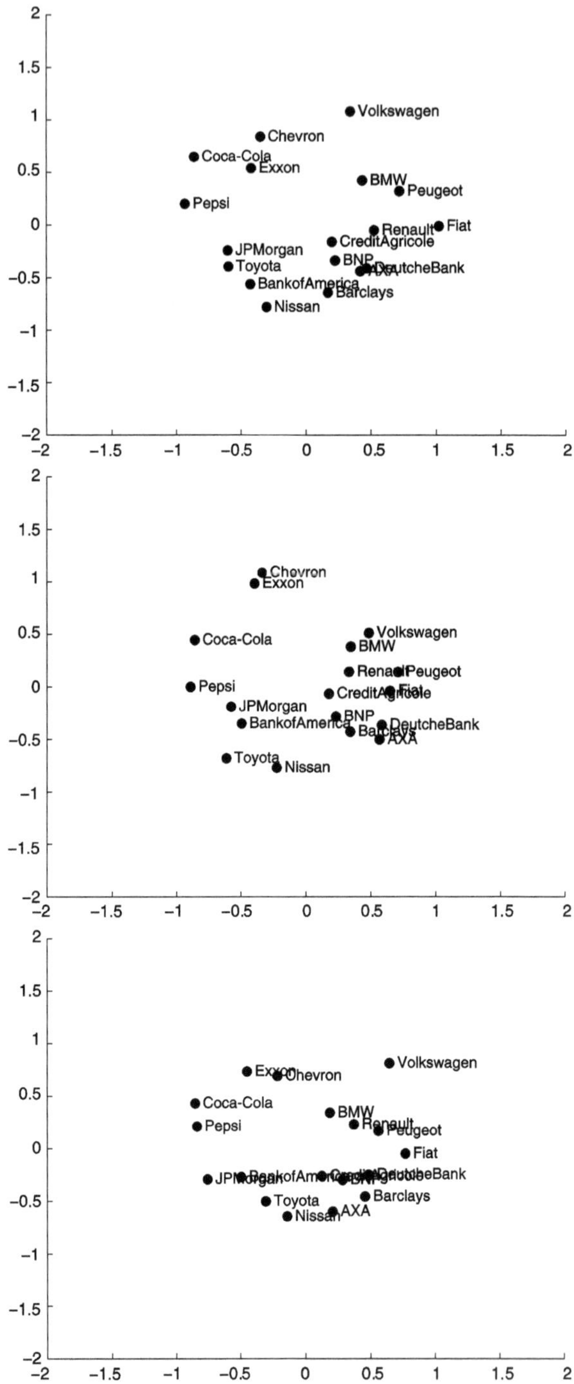


(a)

**Fig. 6.10** (Continued)

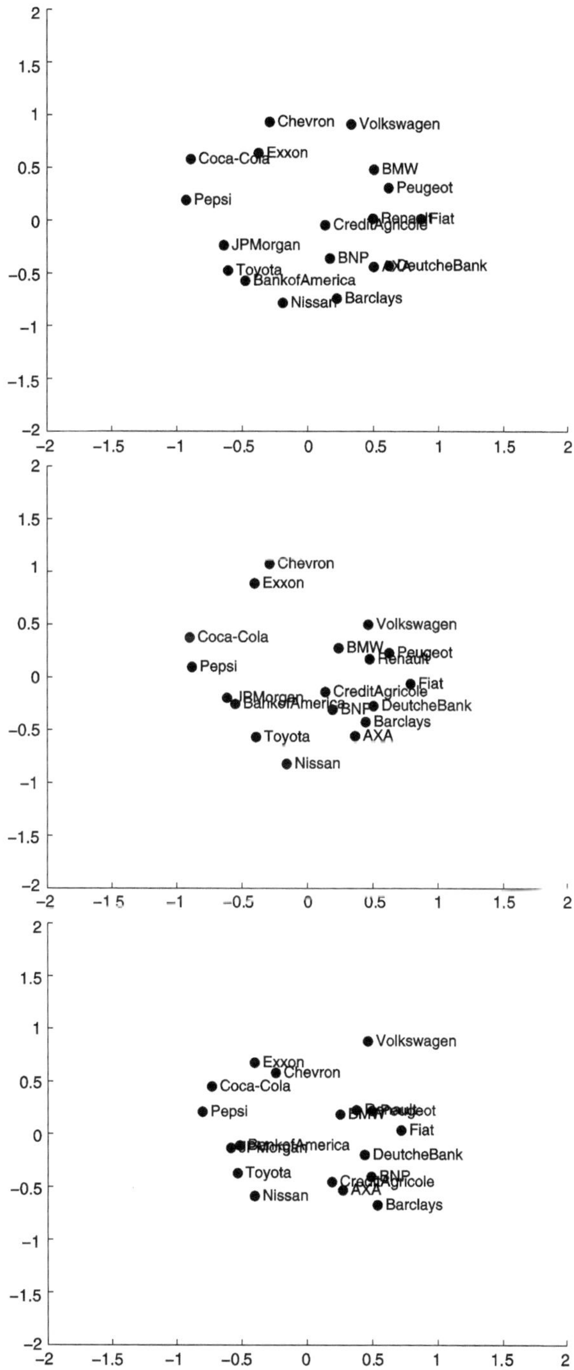


(b)

**Fig. 6.11** MDS plots for different dates. (**a**) *Top*: 24/02/2009 (**b**) *Top*: 07/04/2009 (**a**) *Bottom*: 12/09/2009 (**b**) *Bottom*: 04/11/2009. The correlation matrices are computed from Yahoo daily closure price data using (6.9) and 30 trading day window, for the subset of 18 companies. The points on each plot represent stocks, each designated by two coordinates $(x_i, y_i)$, $i = 1, \ldots, 18$

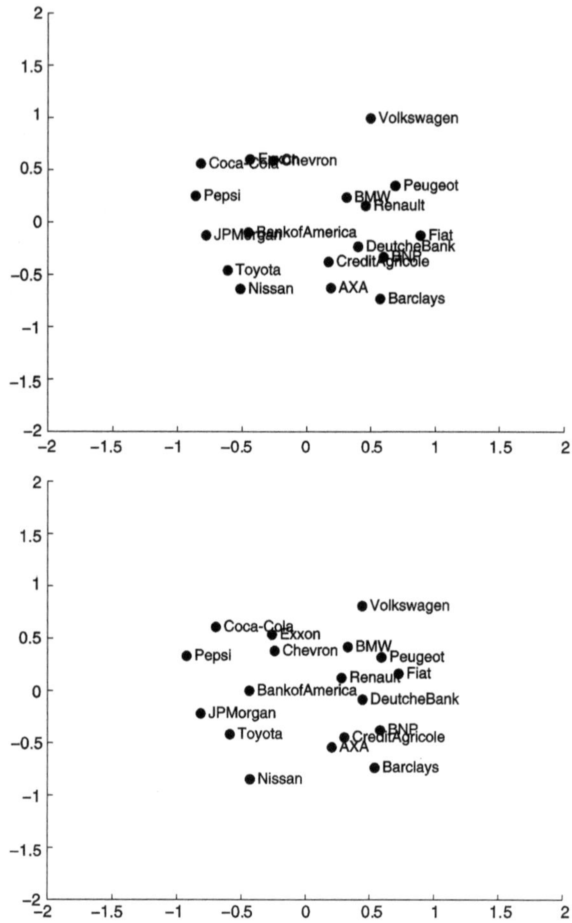


(a)

and visualizing the dynamic evolution of the stock market during the day. When the MDS studies were repeated with daily data, we found that it was easier to visualize or detect specific sectors, strongly correlated pairs and market events. We suggest that this type of plots using daily data may be used in designing strategies of “pairs trade” as explained earlier, or identifying clusters or detecting market trends.

**Acknowledgements** The authors acknowledge F. Abergel, N. Huth, A. Jedidi and F. Pomponio for critical discussions, and thank E. Guevara H. for Fig. 6.1 and Fig. 6.2 (Top). A.C. and T.S. are also grateful to R. Malla and P. Nguemo for some preliminary exercises and computations.

**Fig. 6.11** (Continued)

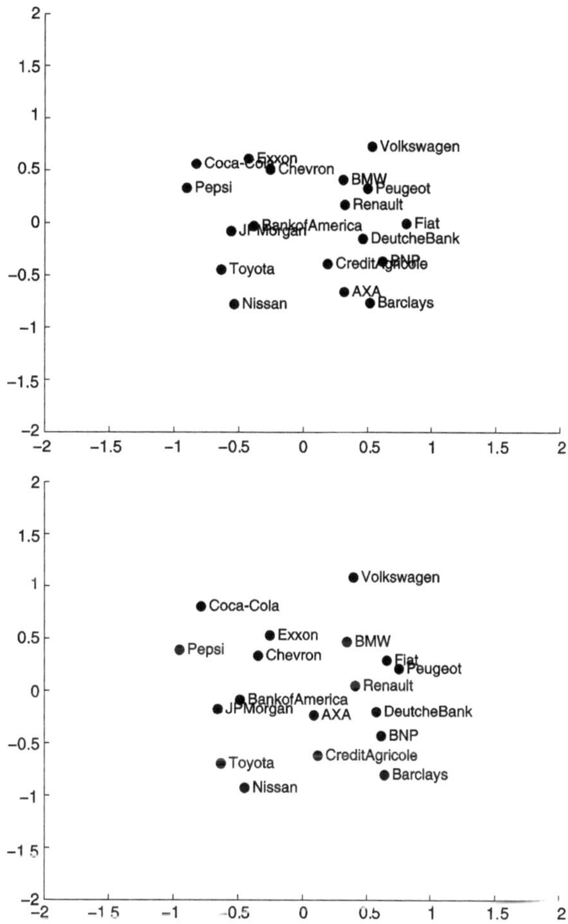


(b)

## References


1. Allez R, Bouchaud J-P (2011) New J Phys 13:025010
2. Chakraborti A, Muni Toke I, Patriarca M, Abergel F (2011) Quant Finance 11(7):991
3. Chakraborti A, Patriarca M, Santhanam MS (2007) Financial time-series analysis: a brief overview. In: Chatterjee A, Chakrabarti BK (eds) Econophysics of markets and business networks. Springer-Verlag, Italia, Milan. Available at arXiv:0704.1738v1 [physics.soc-ph]
4. Schneeweis T, Woolridge JR (1979) J Financ Quant Anal 14:939
5. Markowitz HM (1959) Portfolio selection: efficient diversification of investments. Cowles Foundation Monograph 16. Yale University Press, New Haven
6. Embrechts P, McNeil A, Straumann D (2001) Correlation and dependency in risk management: properties and pitfalls. In: Dempster M, Moffatt H (eds) Risk management: value at risk and beyond. Cambridge University Press, Cambridge, pp 176–223
7. Hayashi T, Yoshida N (2005) Bernoulli 11:359
8. Guevara HE, Chakraborti A, Abergel F (2012). In preparation
9. Epps TW (1979) J Am Stat Assoc 74:291–298

10. Mattiussi V, Tumminello M, Iori G, Mantegna RN (2011) Comparing correlation matrix estimators via Kullback–Leibler divergence. November 30, 2011. Available at SSRN: http://ssrn.com/abstract=1966714 or http://dx.doi.org/10.2139/ssrn.1966714
11. Onnela J-P, Chakraborti A, Kaski K, Kertesz J, Kanto A (2003) Phys Rev E 68:056110
12. Onnela J-P, Chakraborti A, Kaski K, Kertesz J, Kanto A (2003) Phys Scr T 106:48
13. Chakraborti A (2006) An outlook on correlations in stock prices. In: Chatterjee A, Chakrabarti BK (eds) Econophysics of stock and other markets. Springer-Verlag, Milan, Italia. Available at arXiv:physics/0605246
14. Mantegna RN (1999) Eur Phys J B 11:193
15. Borg I, Groenen P (2005) Modern multidimensional scaling: theory and applications. Springer, New York
16. Admati A, Pfleiderer P (1988) A theory of intraday patterns: volume and price variability. Rev Financ Stud 1:3–40
17. Andersen T, Bollerslev T (1997) Intraday periodicity and volatility persistence in financial markets. J Empir Finance 4:115–158

# Chapter 7
# A Robust Measure of Investor Contrarian Behaviour

**Damien Challet and David Morton de Lachapelle**

**Abstract** Using the transaction history of all the clients of an on-line broker, we analyse the daily aggregated investment fluxes of individual investors, companies, and asset managers. Computing the probability that price returns and daily investment fluxes have the same sign provides a robust characterisation of contrarian behaviour. The three categories are found to be contrarian, but with widely different intensities. Individual investors are by far the most contrarian of the three, followed by companies. Asset managers are only mildly contrarian with respect positive price returns.

## 7.1 Introduction

Despite the availability of exhaustive data about the dynamics of some markets and large-scale data analysis, relatively little is known about the statistical behaviour of the traders themselves. And yet, it is exactly what is missing to make giant leaps in agent-based modelling. While it is relatively easy to obtain market-like be-

D. Challet (✉)
Chaire de finance quantitative, Laboratoire de mathématiques appliquées aux systèmes, École Centrale Paris, Grande Voie des Vignes, 92295 Châtenay-Malabry Cedex, France
e-mail: damien.challet@ecp.fr

D. Challet
Encelade Capital SA, EPFL / PSE-C, 1015 Lausanne, Switzerland

D. Morton de Lachapelle
LBS-Institute of Theoretical Physics, École Polytechnique Fédérale de Lausanne, 1015 Lausanne, Switzerland

D. Morton de Lachapelle
Department of Quantitative Asset Management, Swissquote Bank SA, Geneva Office, 1196 Gland, Switzerland
e-mail: david.mortondelachapelle@swissquote.ch

F. Abergel et al. (eds.), *Econophysics of Systemic Risk and Network Dynamics*, New Economic Windows,
DOI 10.1007/978-88-470-2553-0_7, 

haviour from interacting heterogeneous adaptive agents (see e.g. [1–4]), validating such models with fully aggregated data (e.g. price, volume) is mostly hopeless.[1]

The current state of knowledge is of course due to the necessary secrecy that protects to a point strategic behaviour and to which brokers are bound. Thus, data about the actions of single traders are hard to obtain. Some less detailed data sets have used, for instance daily individual net investment [6], daily aggregate investment fluxes of the individual traders from NYSE [7], the same type of data with a weekly resolution from the Australian stock exchange [8], or intraday data of market participants (i.e. banks, brokers) in the Spanish stock market [9]. On the other hand, some authors had access to the trader identity of every single transaction in Taiwan [10], India [11], and China [12, 13]. Quite tellingly, when having access to individual time series, researchers have mainly focused on trader gains/losses. The best established stylised fact is the (on-average) contrarian nature of individual investors who buy when the asset price has decreased and vice-versa [7, 8]; intraday trader behaviour was analysed in [9, 13]. Traders types do share however some common behaviours: in a previous work [14], we showed that the relationship between the average transaction values and the average portfolio values was determined, on average, by transaction fee structure and was the same for the three kinds of traders. Here we investigate contrarian behaviour of the latter at a daily time scale and find out how independent it can be from price dynamics and from the underlying company earnings.

## 7.2 Dataset

Our dataset consists of all orders sent by the clients of Swissquote Bank SA (thereafter referred to as Swissquote). Client accounts belong to three categories: individual investors, companies investing their liquidities, and asset managers supervising the destiny of other people's accounts. We kept only the five most traded stocks in order to have reliable statistics. Table 7.1 reports the size of this database; UBS data starts in 2008; the rank of the most traded stocks is as follows: UBS, ABB, Crédit Suisse Group, Novartis, Nestlé; the tickers of these stocks end with an N; we will hence refer to them as ABBN, CSGN, NESN, NOVN, UBSN, respectively. We have double checked our dataset by matching each Swissquote transaction with Thomson-Reuters tick-by-tick data.

As Swissquote market share went from 0 % to about 3 % in 10 years, raw measures are necessarily non-stationary. As an illustration, we plot the 5 cumulated number of new traders, that is, the number of traders who have made at least one transaction at time $t$ in Fig. 7.1. The rate of trader addition is not constant (it is linked for example to marketing campaigns) and increased for all stocks around 2009, except for UBSN.

[1]It is however possible to attempt market reverse-engineering with agent-based models, as pioneered in [5].

**Table 7.1** Size of the studied dataset

| | ABB | Crédit Suisse Group | NEStlé | NOVartis | UBS |
|---|---|---|---|---|---|
| # clients | 30203 | 20327 | 13442 | 15580 | 40111 |
| # events | 698992 | 348096 | 140420 | 168127 | 873446 |
| # trades | 166916 | 95596 | 36784 | 39632 | 250230 |

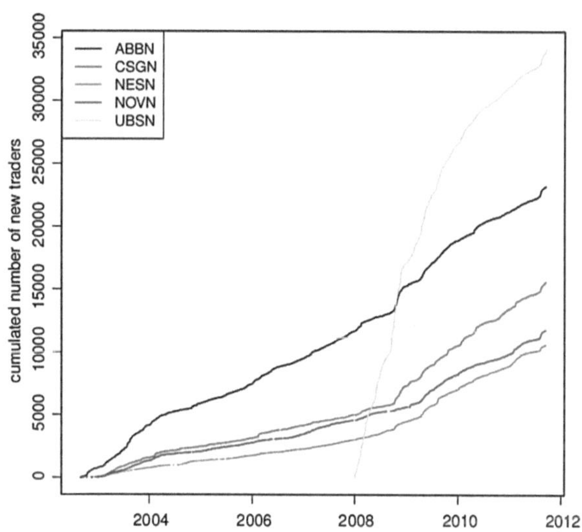


**Fig. 7.1** Cumulated number of known traders as a function of time for all five assets

## 7.3 Investment Fluxes

Every financial transaction has a buyer and a seller, thus the net flux of money is zero, neglecting transaction costs. However, grouping transactions according to some criterion, e.g. investor type, broker, etc., opens the possibility of disentangling their respective contributions. Finance literature has studied, mostly at a weekly time horizon the fluxes of individual clients, mutual funds, on-line and full-service brokers in various markets; the contrarian behaviour of individual client at this horizon is well documented, as well as their trend-following tendency at a yearly horizon, in several markets, provided that their respective weight is small enough; whether individual client actions predict or not future asset returns is still discussed (see Ref. [7] for an overview).

Let us now define the way we have aggregated our dataset. The buy and sell flux in the time interval $[t, t+1[$, where the unit can be fixed arbitrarily, is defined as

$$B(t) = \sum_{h \in [t,t+1[} \Theta\big[v(h)\big] v(h) p(h) > 0,$$

$$S(t) = \sum_{h \in [t,t+1[} \Theta\big[-v(h)\big] v(h) p(h) < 0,$$

where $h$ is a timestamp of a given transaction, $v(h)$ the volume exchange at price $p(h)$, and $\Theta(x)$ the Heaviside function that selects buy or sell transactions. These quantities, when measured for a given client category $g$ is denoted as $B_g(t)$, similarly for $S_g(t)$. One can then define the net flux as $f_g(t) = B_g(t) + S_g(t)$, and the integrated net flux as $F_g(t) = \sum_{t'=0}^{t} f_g(t')$.

The interpretation of an integrated flux is simple: it is the sum of realised losses and net transaction values of open positions. The flux of money of a successful round-trip is indeed negative, since money has been extracted from the market.

### 7.3.1 Global Behaviour

Figure 7.2 reports $F$ for the five assets under study. Individual traders generally produce an increasing function of time for two reasons: first the total number of Swissquote individual clients has been steadily increasing and second, they cannot easily take short positions on stocks (in practice, they do not). The increase is very marked when the price of an asset keeps falling. This behaviour was also seen at the end of the first Internet bubble [15]; it is a signature of contrarian behaviour as discussed below.

The same plots performed for companies and asset managers (Figs. 7.3 and 7.4) show a markedly different dynamics.

Yet this does not remedy the fact that the absolute value of fluxes also depends on the total number of active traders during a given time window, denoted as $N(t)$. The simplest way to overcome this problem is to introduce normalised quantities $b_g(t) = B_g(t)/N(t)$ and $s_g(t) = S_g(t)/N(t)$. In order to remove the dependence on $N(t)$, we will study normalised net fluxes defined as

$$\phi_g(t) = \frac{b_g(t) + s_g(t)}{b_g(t) + |s_g(t)|} = \frac{B_g(t) + S_g(t)}{B_g(t) + |S_g(t)|} \in [-1, 1];$$

these quantities can be interpreted as the net buy/sell tendency in a given time interval. In addition, since these quantities are not fat-tailed, they lead to less noisy measurements. Figures 7.5, 7.6 and 7.7 confirm the behaviour found for the unnormalised fluxes; since normalised fluxes are not fat-tailed, the resulting plots are much less noisy and clearly indicate that institutions and individual investors have a markedly different behaviour.

### 7.3.2 Fluxes vs Price Returns

Let us now examine how trader behaviour depends on price returns. A comparison between cumulated fluxes and price history of individual traders in the case of

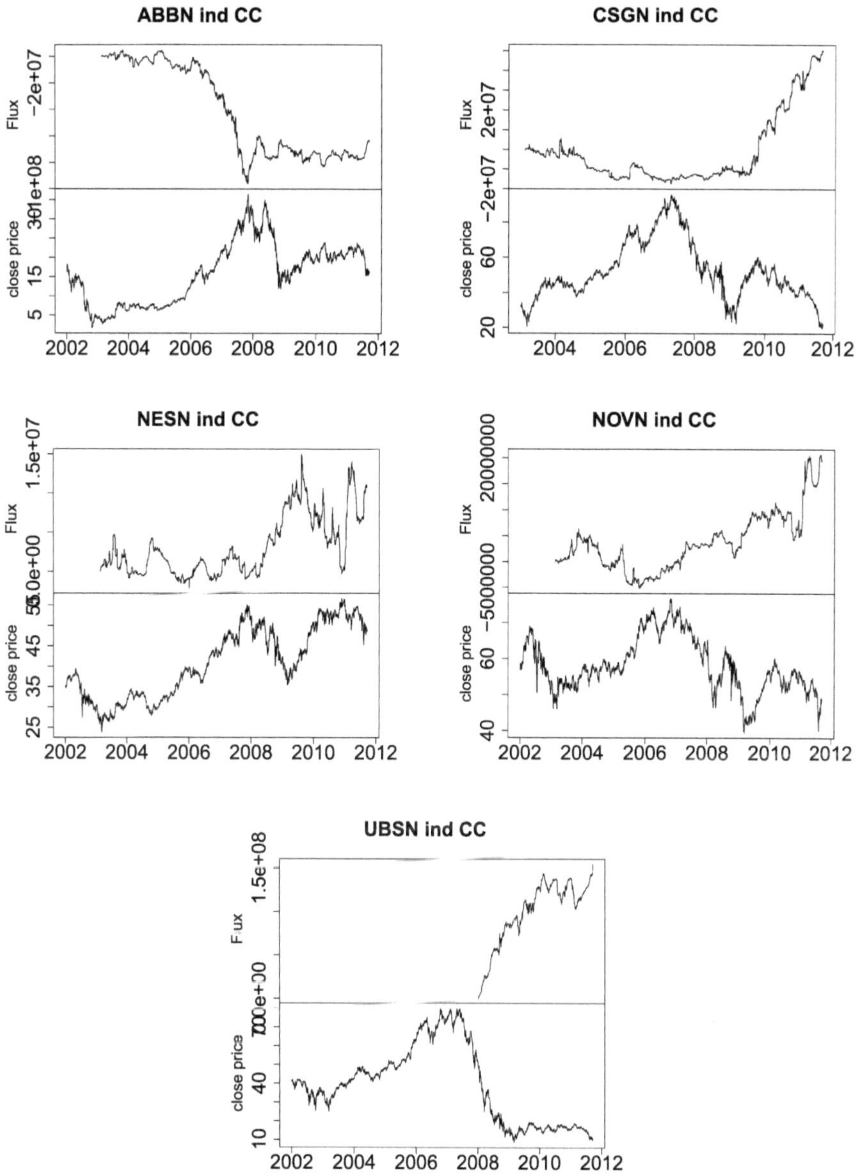


**Fig. 7.2** Cumulated net daily investment fluxes $F$ of individual clients as a function of time

UBS strongly suggest that some kind of contrarian behaviour: as long as UBS stock price plunges, the individual traders have been keen on buying more shares. This behaviour is not mainly due to the total draw-down or draw-up, but to an effective behaviour that depends only the price returns of a few past traded days, including the then-present trading day.

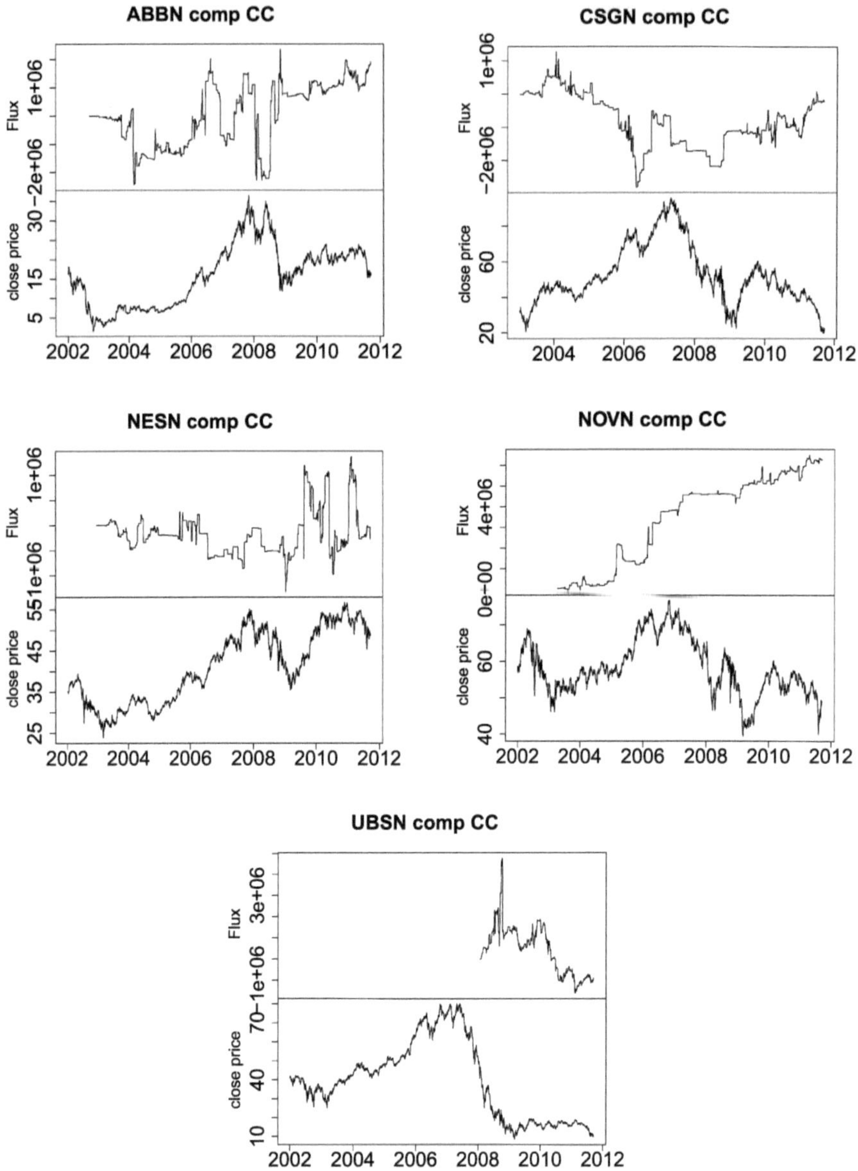


**Fig. 7.3** Cumulated net daily investment fluxes $F$ of companies as a function of time

Relating raw investment fluxes and daily returns $r(t)$ is hard because of their fat-tailed nature. This is why this contribution focuses on a more robust measure: the cumulated product of the sign of these quantities, defined as

$$R_g(t) = \sum_{t'=1}^{t} \operatorname{sign}\big[\phi_g(t')\big] \operatorname{sign}\big[r(t')\big],$$

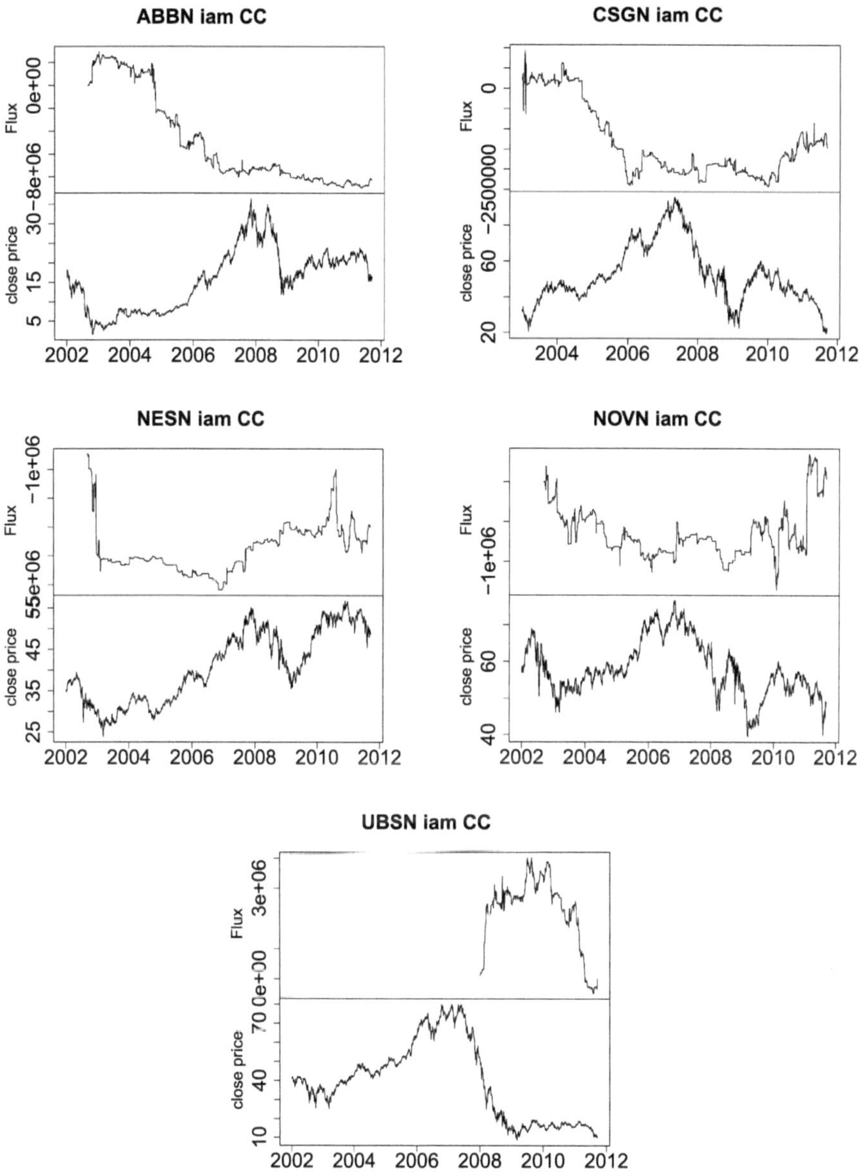


**Fig. 7.4** Cumulated net daily investment fluxes $F$ of asset managers as a function of time

provides a good visual check of the evolution of the correlation between the signs of these quantities as a the time goes on.

This quantity is reported in Fig. 7.8. Remarkably the probability that individual traders as a group is contrarian remains constant to a good approximation for all the stocks over all the years spanned by our data, despite the tremendous, lasting,

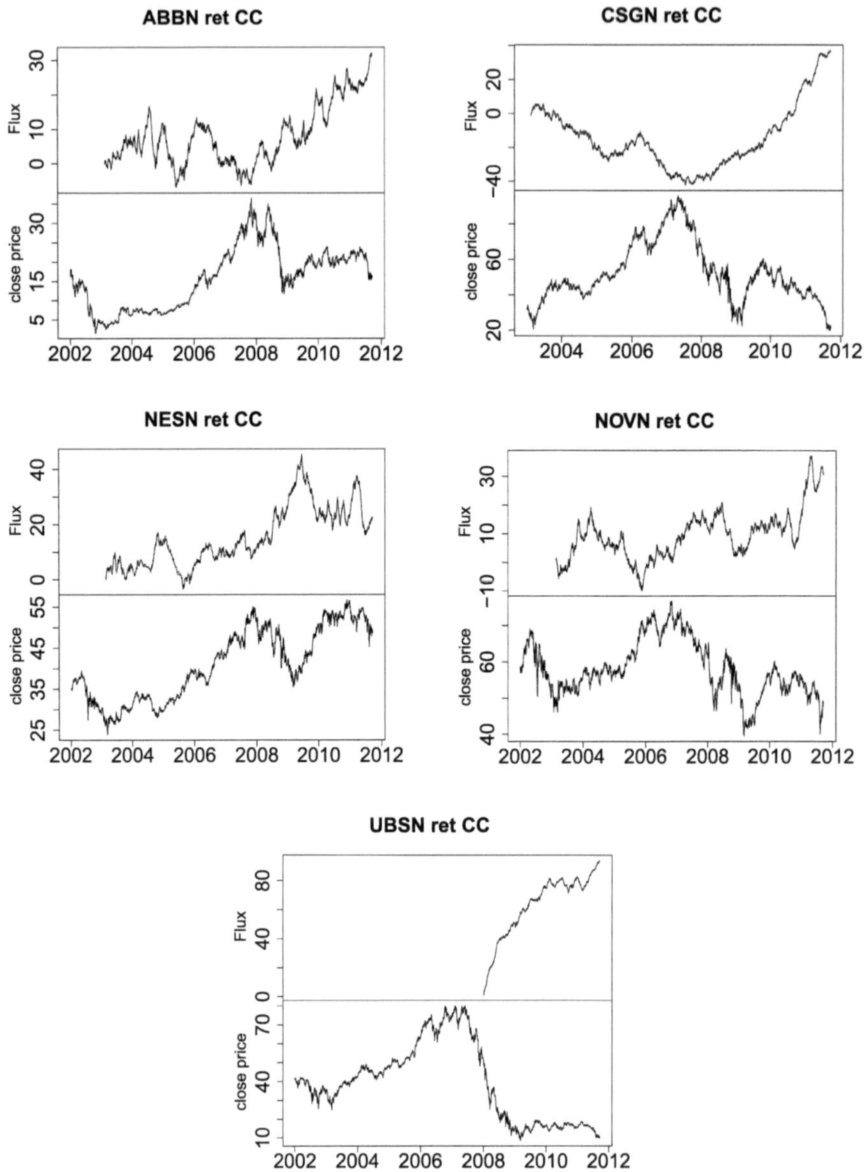


**Fig. 7.5** Cumulated daily normalised investment fluxes $\phi$ of individual clients as a function of time

and sometimes fast price variations. ABBN and UBSN, the two stocks for which one has the most data, are also the ones for which this behaviour is visually most clearly constant. The case of UBSN illustrates particularly well how impervious to economics fundamentals individual traders can be. 2008 was a terrible year for UBS:

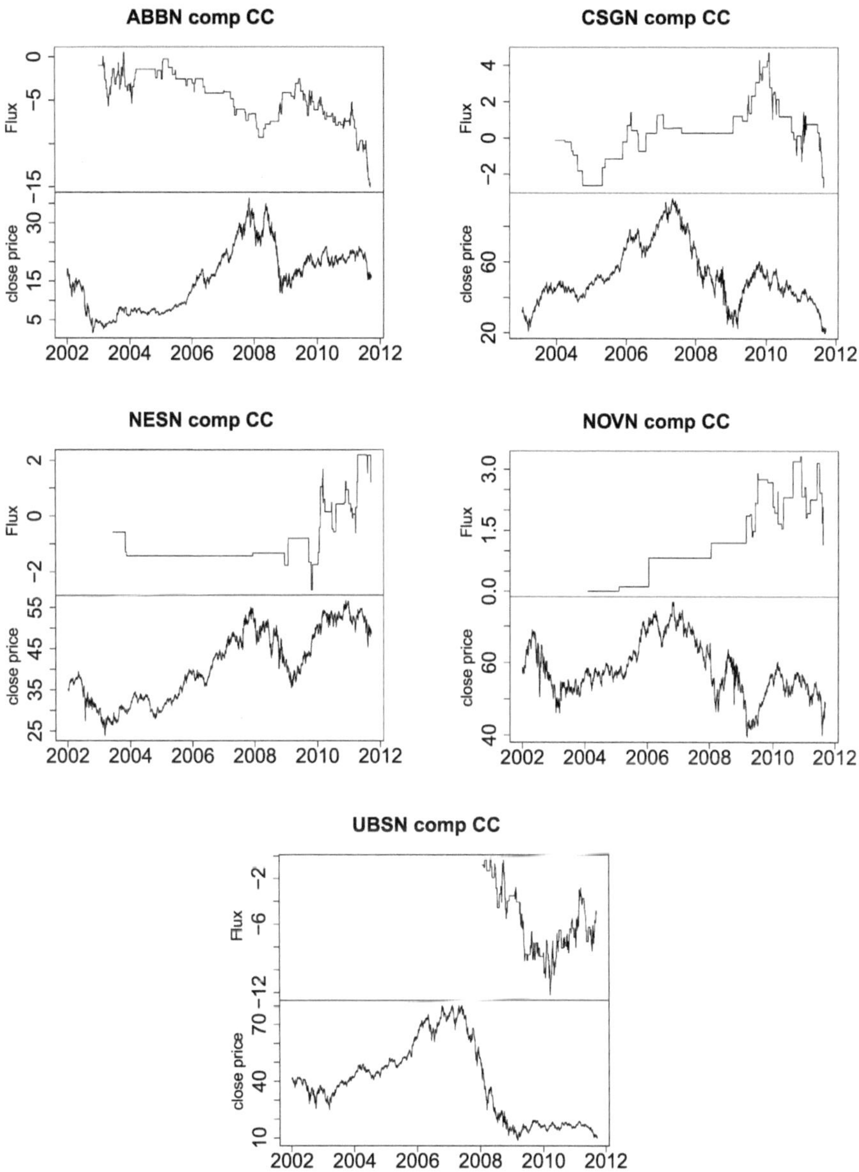


**Fig. 7.6** Cumulated daily normalised investment fluxes $\phi$ of companies as a function of time

operating losses were very high; it had to be bailed out by the Swiss government; in addition it was accused by a USA Senate Panel of facilitating tax evasion and its licence to operate there was menaced; when the situation cleared up from 2009 onwards, the behaviour of the sign of individual investments did not change (see Fig. 7.9).

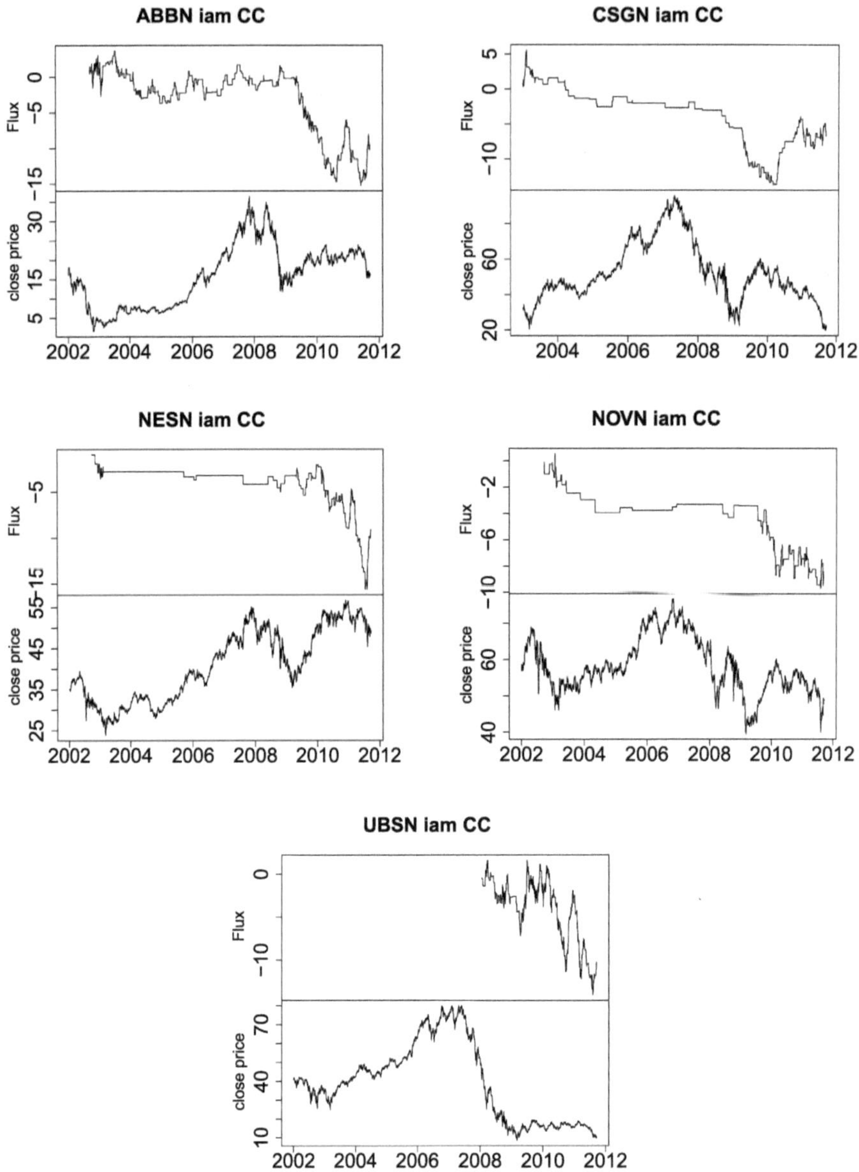


**Fig. 7.7** Cumulated daily normalised investment fluxes $\phi$ of asset managers as a function of time

Table 7.2 reports the measured probability that the signs of net fluxes and price returns are opposite. Every single figure reported in this table is statistically very significant. It is quite revealing to split the probability to act as a contrarian into two parts: one when the price return is negative and one when it is positive. It turns out that the probability of being contrarian is higher when the price return is negative,

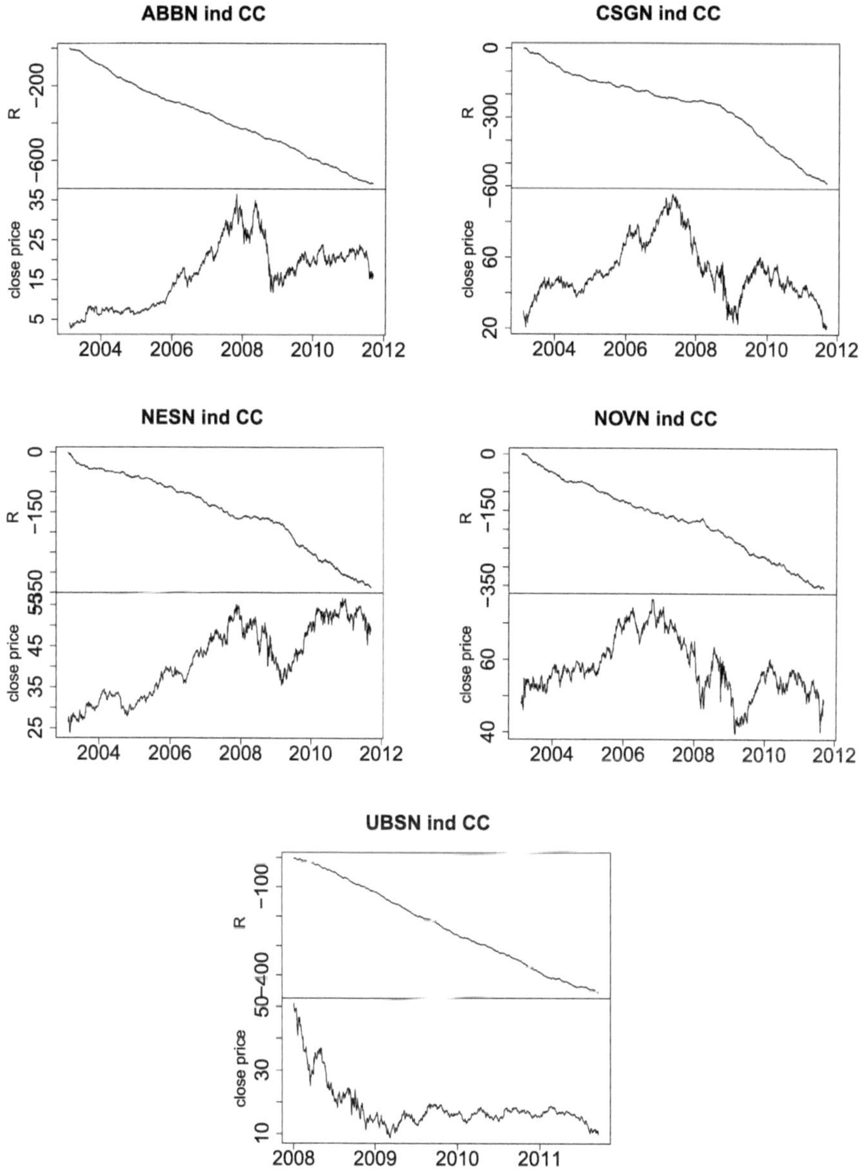


**Fig. 7.8** Cumulated sign product $R_{\text{individual}}(t)$ as a function of time; close-to-close price returns

except for NESN and NOVN for which they do not differ significantly. One possible explanation is that these are the only two assets that did not experience tumultuous times: their annualised share price volatility was about 20 %, whereas the ones of other tickers are larger than 40 %. It may hence be that above a given volatility

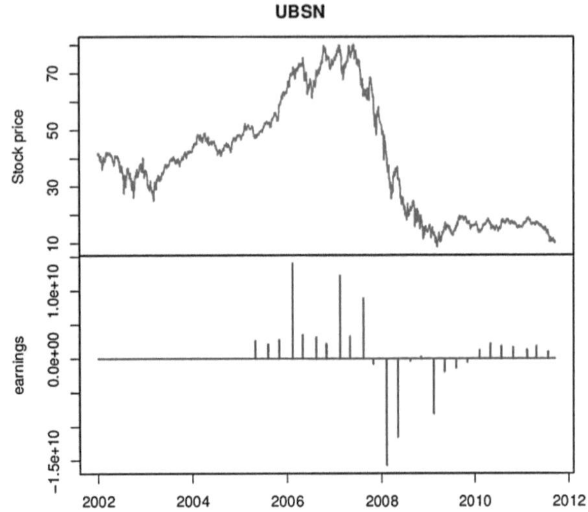


**Fig. 7.9** Share price and earnings in CHF of UBS. Source: quarterly reports of UBS, available on ubs.com

threshold, individual investors react in a different way with respect to negative and positive price returns.

We have much less data about companies and asset managers. We overcame this difficulty by aggregating the data for all the stocks in the last columns of Tables 7.3 and 7.4. Companies turn out to be also contrarian, particularly with respect to negative price returns. Asset managers on the other hand are also contrarian, but with respect to positive returns.

## 7.4 Discussion

The above results show that the conditional signs of the investments of the three categories of traders have different average behaviours. Individual investor behaviour is markedly contrarian and stays roughly constant as a function of time, even if the asset prices changes wildly, as exemplified by the extreme case of UBSN. Companies follow the same pattern, albeit in a much less pronounced way. Asset managers are contrarian with respect to positive price returns. We also found that individual traders are more likely contrarian with respect to negative price returns if the asset prices have a high volatility. The difference between the categories of traders is probably due to the use of investment tools and/or strategies by more sophisticated traders.

This emphasises the need of individual investors for financial advice and decision tools, since their behaviour is so much determined by the current price returns. Some brokers offer help building portfolios and actively managing them automatically.

Future work will include more assets that were plagued by a long series of bad news in order to understand on what the contrarian nature depends on, in particular with respect to the price volatility.

**Table 7.2** Probabilities that individual investment fluxes $\phi(t)$ and close-close returns $r(t)$ have opposite signs; uncertainties are equal to two standard deviations

| Individual clients | ABBN | CSGN | NESN | NOVN | UBSN |
|---|---|---|---|---|---|
| $P(\text{sign}[\phi(t)] = -\text{sign}[r(t)])$ | $0.69 \pm 0.01$ | $0.69 \pm 0.01$ | $0.62 \pm 0.01$ | $0.61 \pm 0.01$ | $0.76 \pm 0.01$ |
| $P(\phi(t) < 0 \mid r(t) > 0)$ | $0.63 \pm 0.01$ | $0.65 \pm 0.01$ | $0.62 \pm 0.01$ | $0.59 \pm 0.01$ | $0.65 \pm 0.02$ |
| $P(\phi(t) > 0 \mid r(t) < 0)$ | $0.76 \pm 0.01$ | $0.73 \pm 0.01$ | $0.63 \pm 0.01$ | $0.63 \pm 0.01$ | $0.85 \pm 0.01$ |

**Table 7.3** Probabilities that companies investment fluxes $\phi(t)$ and close-close returns $r(t)$ have opposite signs; uncertainties are equal to two standard deviations

| Companies | ABBN | CSGN | NESN | NOVN | UBSN | All tickers |
|---|---|---|---|---|---|---|
| $P(\text{sign}[\phi(t)] = -\text{sign}[r(t)])$ | $0.56 = 0.03$ | $0.54 \pm 0.05$ | $0.57 \pm 0.06$ | $0.64 \pm 0.07$ | $0.55 \pm 0.03$ | $0.56 \pm 0.01$ |
| $P(\phi(t) < 0 \mid r(t) > 0)$ | $0.55 \pm 0.04$ | $0.60 \pm 0.06$ | $0.53 \pm 0.08$ | $0.68 \pm 0.09$ | $0.61 \pm 0.04$ | $0.54 \pm 0.01$ |
| $P(\phi(t) > 0 \mid r(t) < 0)$ | $0.58 \pm 0.04$ | $0.46 \pm 0.07$ | $0.61 \pm 0.08$ | $0.60 \pm 0.10$ | $0.50 \pm 0.04$ | $0.58 \pm 0.01$ |

**Table 7.4** Probabilities that asset manager investment fluxes $\phi(t)$ and close-close returns $r(t)$ have opposite signs; uncertainties are equal to two standard deviations

| Asset managers | ABBN | CSGN | NESN | NOVN | UBSN | All tickers |
|---|---|---|---|---|---|---|
| $P(\text{sign}[\phi(t)] = -\text{sign}[r(t)])$ | $0.51 \pm 0.03$ | $0.57 \pm 0.04$ | $0.53 \pm 0.04$ | $0.51 \pm 0.05$ | $0.54 \pm 0.04$ | $0.54 \pm 0.01$ |
| $P(\phi(t) < 0 \mid r(t) > 0)$ | $0.55 \pm 0.04$ | $0.62 \pm 0.05$ | $0.56 \pm 0.06$ | $0.53 \pm 0.08$ | $0.57 \pm 0.06$ | $0.56 \pm 0.01$ |
| $P(\phi(t) > 0 \mid r(t) < 0)$ | $0.48 \pm 0.04$ | $0.52 \pm 0.06$ | $0.50 \pm 0.06$ | $0.48 \pm 0.07$ | $0.52 \pm 0.06$ | $0.51 \pm 0.01$ |

**Acknowledgements** DC warmly thanks Fabrizio Pomponio and Riadh Zaatour for their help with Thomson-Reuters data; DMdL is grateful to Swissquote Bank SA for financial support. DC acknowledges useful discussions with the participants of Kolkata Econophys VI conference.

## References

1. Caldarelli G, Marsili M, Zhang Y-C (1997) A prototype model of stock exchange. Europhys Lett 50:479–484
2. Lux T, Marchesi M (1999) Scaling and criticality in a stochastic multi-agent model of a financial market. Nature 397:498–500
3. Jefferies P, Hart ML, Hui PM, Johnson NF (2001) From market games to real-world markets. Eur Phys J B 20:493–502. cond-mat/0008387
4. Challet D, Chessa A, Marsili M, Zhang Y-C (2000) From minority games to real markets. Quant Finance 1:168. cond-mat/0011042
5. Lamper D, Howison S, Johnson N (2002) Predictability of large future changes in a competitive evolving population. Phys Rev Lett 88:017902–017905. cond-mat/0105258
6. Tumminello M, Lillo F, Piilo J, Mantegna RN (2011) Identification of clusters of investors from their real trading activity in a financial market. Arxiv preprint. arXiv:1107.3942
7. Kaniel R, Saar G, Titman S (2008) Individual investor trading and stock returns. J Finance 63(1):273–310
8. Jackson A (2004) The aggregate behaviour of individual investors
9. Lillo F, Moro E, Vaglica G, Mantegna RN (2008) Specialization and herding behavior of trading firms in a financial market. New J Phys 10:043019
10. Barber BM, Lee YT, Liu YJ, Odean T (2009) Just how much do individual investors lose by trading? Rev Financ Stud 22(2):609–632
11. De S, Pochiraju B, Gondhi N, Chhabra R (2011) Does sign matter more than size? An investigation into the source of investor overconfidence
12. Zhou WX, Mu GH, Chen W, Sornette D (2011) Strategies used as spectroscopy of financial markets reveal new stylized facts. Arxiv preprint. arXiv:1104.3616
13. Zhou WX, Mu GH, Kertész J (2012) Random matrix approach to the dynamics of stock inventory variations. Arxiv preprint. arXiv:1201.0433
14. Morton de Lachapelle D, Challet D (2010) Turnover, account value and diversification of real traders: evidence of collective portfolio optimizing behavior. New J Phys 12:075039
15. Griffin JM, Harris JH, Shu T, Topaloglu S (2011) Who drove and burst the tech bubble? J Finance 66(4):1251–1290

# Chapter 8
# Evolution of Zipf's Law for Indian Urban Agglomerations Vis-à-Vis Chinese Urban Agglomerations

**Kausik Gangopadhyay and Banasri Basu**

**Abstract** We investigate into the rank-size distributions of urban agglomerations for India between 1981 to 2011. The incidence of a power law tail is prominent. A relevant question persists regarding the evolution of the power tail coefficient. We have developed a methodology to meaningfully track the power law coefficient over time, when a country experience population growth. A relevant dynamic law, Gibrat's law, is empirically tested in this connection. We argue that these empirical findings for India are in contrast with the findings in case of China, another country with population growth but monolithic political system.

## 8.1 Introduction

It is the job of a scientist to find a mathematical rigor in a natural system, which is apparently anomalous to a casual observer. When a scientist does this in the physical world, the laws discovered are called physical laws. On the other hand, the human society and the institutions created by human beings seem somewhat vulnerable for such laws being maintained. Since human beings have their own desires and wishes, human institutions are often kept outside the purview of physical laws. Nevertheless, if we find some physical laws being observed in the context of human society, not only it will widen the scope of physical laws, but also usher a novel dimension in the study of social sciences.

We discuss the case of Indian city size distribution and Zipf's law [1] (alternatively known as Pareto distribution or simply power law) in this backdrop. Zipf's law, named after linguist George Kingsley Zipf, is a simple empirical law which is often successful in describing the distribution of populations for various cities in a nation. Zipf noted that the second most common word in the English language ('of')

K. Gangopadhyay (✉)
Indian Institute of Management Kozhikode, IIMK P.O., Kozhikode DT, Kerala-673570, India
e-mail: kausik@iimk.ac.in

B. Basu
Physics and Applied Mathematics Unit, Indian Statistical Institute, 203, B. T. Road, Kolkata-700108, India
e-mail: banasri@isical.ac.in

F. Abergel et al. (eds.), *Econophysics of Systemic Risk and Network Dynamics*, New Economic Windows,
DOI 10.1007/978-88-470-2553-0_8, 

appears at approximately half the rate of the most common word ('the'). The third most common word ('to') appears at approximately one third the rate of the most common word [1]. The fact that the law has been observed in many other spheres makes it even more mysterious. We can set many examples in this context: word usage in human language [2], size distribution of islands [3], websurfing [4], the distribution of wealth and income in many countries [5, 6], and the size distribution of lunar craters [7]. The examples also include forest fires [8], solar flares [9], and football goal distribution [10]. Recently in a quantitative analysis [11] of extensive chess databases it is shown that the pooled distribution of all opening weights follows Zipf's law with universal exponent.

In the context of urban economics and regional science, "Zipf's law" is synonymous to a remarkable regularity in the distribution of city sizes all over the world. It is also known as the "Rank-Size Distribution". This says that the population of a city is inversely proportional to the city's rank among all cities. This could be interpreted in multiple ways. Let us take a cut-off, say a population of fourteen million. Indeed according to the 2011 Census, there are three Indian metropolises over the population of fourteen million, Greater Mumbai (18,414,288), Delhi (16,314,838) and Kolkata (14,112,536). If we consider another cut-off, which is just half of the previous cut-off, there ought to be double the number of cities over the new cut-off compared to the previous cut-off. We verify that there are exactly six cities with a population of more than seven million—the other three cities being Chennai (8,696,010), Bangalore (8,499,399) and Hyderabad (7,749,334). If one calculates the natural logarithm of the rank and of the city size (measured in terms of the number of people) and plot the resulting data in a diagram, a remarkable log-linear pattern is obtained, this is the Rank-Size Distribution. If the slope of the line equals minus 1, (as is for example approximately the case for the USA, India, and France) the relationship is known as Zipf's Law. Zipf's law has repeatedly been shown to hold in the top tails of city size distribution across different countries and periods [12–17].

Of course, Zipf's law is really not a law at all. It's merely a simple mathematical model that appears to describe some human behavior. Even more amazingly, Zipf's law has apparently held for at least 100 years. Given the different social conditions from country to country, the different patterns of migration a century ago and many other variables that you'd think would make a difference; the generality of Zipf's law is astonishing. Keep in mind that this pattern emerged on its own, that is, it is "self-organized". No city planner imposes it, and no citizen conspires to make it happen. Something is enforcing this invisible law, but we are still in the dark about what that something might be. Many inventive theorists working in disciplines ranging from economics to physics have taken a whack at explaining Zipf's law, but no one has completely solved it. Paul Krugman, who has tackled the problem himself, wryly noted [15] that "the usual complaint about economic theory is that our models are oversimplified—that they offer excessively neat views of complex, messy reality. In the case of Zipf's law the reverse is true: we have complex, messy models, yet reality is startlingly neat and simple." Zipf's law is popular and thrilling because of its mysterious nature despite being simple. In the complex human decision of choice of a dwelling place, the existence of such a simple relationship being held is a mystery in itself.

The evolution of Zipf's law coefficient have been studied for countries such as Japan [18], USA [19]. In this article, we investigate the evolution of Zipf's law coefficient in case of India during 1981 to 2011 using Indian Census data. Even though, Zipf's law is a static phenomenon, it is important to investigate into its dynamic evolution over time partly because that enables us with a clearer understanding of the process of growth of urban agglomerations. Also it is pertinent for the reason that we want to relate the evolution of Zipf's law coefficient with another law related to the dynamic process—Gibrat's law. More specifically, Gibrat's law postulates that the mean and variance of the growth rate of an urban agglomeration are independent of its size. It has been demonstrated [8] that Zipf's law is an outcome of Gibrat's law. We will expound the case of India—a country with remarkable population growth and contrast the findings to the Chinese experience.

The organization of this paper is as follows. Section 8.2 elaborates the empirical analysis. Section 8.3 summarizes our results comparing and contrasting them with the empirical findings in case of China. This section uses findings from an earlier work [20].

## 8.2 Empirical Analysis

### *8.2.1 Zipf's Law for Indian Urban Agglomerations*

We gather our data from the Indian Census [21], which is conducted once in a decade. We have used data from four different waves, namely censuses conducted in the years of 1981, 1991, 2001, and 2011. According to the census conducted on the first day of March, 2011, the population of India stood at 1,210,193,422 persons. The details of Indian census is tabulated in Table 8.1. Figure 8.1 shows the rank-size distribution for urban agglomerations of India in 1981, 1991, 2001, and 2011, respectively.

The slope of the rank-size line in log scale determines the power law coefficient for any set of urban agglomerations. The estimation for Zipf's law requires a statement on the threshold of the power law region. We can visualize the power law tails for each census year data. However, the starting point of such a tail varies across years. For a country with population growth, it is intuitive that a shift in the threshold of power law region is bound to happen over years. The pertinent issue is: how to handle this question of finding an "appropriate" threshold level? We need a measure, which is irrespective of the absolute number of the threshold. This could be materialized, in one way, by considering the population-proportions of urban agglomerations, namely, proportion of population in each urban agglomeration of the total population of India. This will stabilize the threshold value of power law obeying tail over time in a world with considerable population growth. We expound the mathematical form of Zipf's law:

$$\log R(x) = \alpha - \beta \log x, \quad \text{for all } x > x_0. \tag{8.1}$$

**Table 8.1** Indian Census during 1981–2011. Figures in parenthesis represent the corresponding figure as percentage of the total population in that particular year

| Census year | Total population | Urban population | Rural population |
|---|---|---|---|
| 2011 | 1,210,193,422<br>(100) | 377,105,760<br>(31.16) | 833,087,662<br>(68.84) |
| 2001 | 1,027,015,247<br>(100) | 285,354,954<br>(27.78) | 741,660,293<br>(72.22) |
| 1991 | 844,324,222<br>(100) | 217,177,625<br>(25.72) | 627,146,597<br>(74.28) |
| 1981 | 683,329,097<br>(100) | 159,462,547<br>(23.34) | 523,866,550<br>(76.66) |

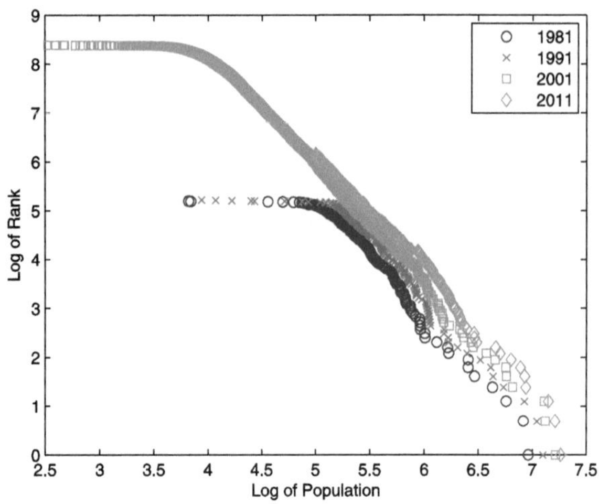


**Fig. 8.1** Rank-size distribution of urban agglomerations in India during 1981–2011

The parameter $\beta$ will be close to unity, under Zipf's law and $x_0$ is the threshold size. For a theoretical abstraction, let us suppose that population of all urban agglomerations grow in equal proportions. Therefore, an urban agglomeration with size $x$ has become one with a population of $A \cdot x$. The above equation, in that case, boils down to:

$$\log R(x) = (\alpha + \beta a) - \beta \log Ax, \quad \text{for all } Ax > Ax_0, \tag{8.2}$$

where $a = \log A$. This (8.2) represents the structure of Zipf's law in case a population growth happens. This also implies that the minimum cut-off has to be set upward and the estimated power-law line will have a higher intercept, in such a scenario. Indeed, once the threshold is adjusted, the slope of this line will not change even slightly. Alternatively, if we look into population proportions, we note no change both in the intercept and slope in case of a population growth. Let the

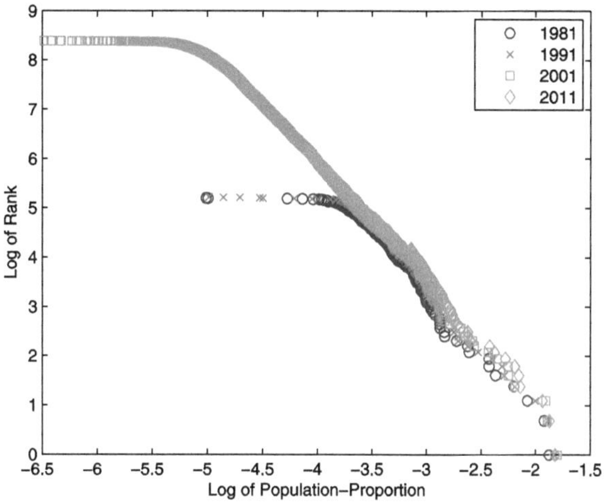


**Fig. 8.2** Rank-population proportion distribution of urban agglomerations in India during 1981–2011

population of a country be N. We can calculate the population proportions of urban agglomerations of this country, $\frac{x}{N}$. From a plot of rank-population proportions in the log scale, we can derive a form of (8.1). This equation remains unchanged in case the populations of all urban agglomerations and consequently that of the entire country grow by a factor of $A$:

$$\begin{aligned} \log R(x) &= (\alpha - \beta N) - \beta \log \frac{x}{N}, \quad \text{for all } x > x_0, \\ \log R(x) &= (\alpha - \beta N) - \beta \log \frac{Ax}{AN}, \quad \text{for all } Ax > Ax_0. \end{aligned} \tag{8.3}$$

We illustrate our case in Fig. 8.2 with plots of ranks of urban agglomerations against population-proportions. The curves coincide on one another, mostly. This shows the stability of Zipf's law with respect to relative population over time. We undertake this strategy of dealing with population growth. The issue boils down to the choice of threshold for urban agglomerations to consider under power law in 1981. Once that decision has been made, we will adjust that threshold for subsequent years by multiplying the initial value by the factor of population growth. Our focus is to study evolution of Zipf's law coefficient over time. Since we are interested in the evolution of power law tail, the initial value of the power law coefficient hardly matters. We set the threshold for 1981 in such a manner so that the slope of the rank-size plot (in log scale) is as close as possible to unity. This is the value postulated by Zipf's law.

The basic way to estimate the parameters of Pareto distribution is called the "linear fit method". Under this method, we regress the log of rank of an urban agglomeration on the log of its population. The coefficient of the regression line yields the estimate for the exponent of power law. Though widely used, this method produces a biased estimate of the power law exponent [22].The alternative approach lies in estimation of the Maximum Likelihood Estimator (MLE). This is also known as the "hill method" in the econophysics literature. For a sample consisting of a finite

**Table 8.2** Zipf's law verified for Indian urban agglomeration. The standard errors of estimates are noted in the parentheses

| Census year | Minimum size | Linear fit estimate | Maximum likelihood estimate |
|---|---|---|---|
| 2011 | 212,523 | 0.935 | 1.018 |
| | | (0.007) | (0.069) |
| 2001 | 180,355 | 0.921 | 1.044 |
| | | (0.007) | (0.075) |
| 1991 | 148,272 | 0.899 | 0.976 |
| | | (0.008) | (0.075) |
| 1981 | 120,000 | 0.889 | 0.991 |
| | | (0.009) | (0.080) |

number of data points, we can calculate the probability of observing this sample, in entirety, by employing the probability density function and cumulative density function. Given a particular process, this probability is a function of parameters inherent in the particular probability distribution used in the calculation of the sample probability, commonly known as the likelihood of that sample. We maximize this likelihood with respect to the distribution parameters. The set of parameter-values, for which the likelihood is maximized, is collectively known as the maximum likelihood estimate of the parameters. We have rendered the analytical expressions of the maximum likelihood estimate for the Pareto distribution in our earlier work [23].

We have computed both estimates using linear fit method and maximum likelihood method. Our estimates are tabulated in Table 8.2 for 1981 to 2011 for all four census rounds. As discussed, we have chosen a cut-off of 120,000 as the minimum size of urban agglomeration to be included in 1981 sample so that the Zipf's law coefficient—estimated value 0.991—is close to the theoretical prediction of 1.000. Afterwards, we multiply this minimum value by the population growth rate to arrive at the subsequent figures for the years 1991–2011. For example, the growth rate of population was 23.56 % during 1981–1991. Therefore, we add 23.56 % to the threshold value of 120,000 to arrive at the figure of 148,272 (rounded off to whole number). From Fig. 8.3 it is apparent that the coefficient of Zipf's law is remarkably close to the theoretical value of two for all these years. Also, the movement of Zipf's law coefficient is negligible which indicates little change in the process of formation of urban agglomerations over the course of three decades.

### *8.2.2 Empirical Validation of Gibrat's Law for India*

Next pertinent empirical issue is the validity of Gibrat's law in this context. We examine whether the relation between mean and variance of growth rate of urban

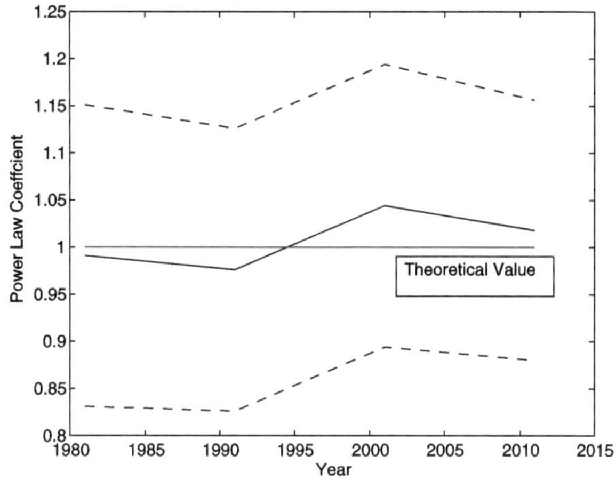


**Fig. 8.3** Evolution of Zipf's law coefficient for India during 1981–2011

agglomerations depend on size. We consider the population growth rate of all available urban agglomerations for the period of 1991–2001 and 2001–2011. The size for these urban agglomerations are their populations in the initial year of the considered period, say 1991 and 2001. A non-parametric way to summarize the population growth rates is through the Kernel estimates of local mean, which we elaborate hereby. Suppose, the growth rate of a city, $g_i$, bears some relation with the size of the city, $S_i$, modeled as:

$$g_i = m(S_i) + \epsilon_i,$$

for all $i = 1, 2, \dots, n$, $n$ being the total number of urban agglomerations with available data. $g_i$ is the growth rate of the $i$th urban agglomeration in a time period and $S_i$ is its size in the initial year of the period considered.

The objective is to find a smooth estimate of local mean of growth rate over size, say $m(S)$. The Kernel estimate often give rise to inaccuracies in the boundary region. We choose a particular interval, say $[1.2 \cdot \min_i S_i, 0.8 \cdot \max_i S_i]$ to exclude the effect of the boundaries. We perform a kernel density regression in the support of $S_i$. The local average smooths around a point $s$, and the smoothing is done using a kernel, which is a continuous weight function symmetric around $s$. We use a popular kernel, namely Epanechnikov, for which: $K(x) = \frac{3}{4}(1 - \psi^2) \cdot \mathbf{1}_{|\psi| \le 1}$. The function, $m(\cdot)$, does depend on the size the bandwidth $h$ of a kernel determines the scale of smoothing. The Nadaraya-Watson estimate [24] of $m(\cdot)$ is given by the following expression,

$$\hat{m}(s) = \frac{n^{-1} \sum_{i=1}^{n} K_h(s - S_i) g_i}{n^{-1} \sum_{i=1}^{n} K_h(s - S_i)}.$$

The means and variances of the growth rate of the urban agglomerations are illustrated in Fig. 8.4. A visual inspection verifies the veracity of Gibrat's law. There is no significant trend—either in the mean or in the variance—over size for both time periods. Thereby, it verifies that since Gibrat's law is satisfied in this case, Zipf's law should hold true, dynamically.

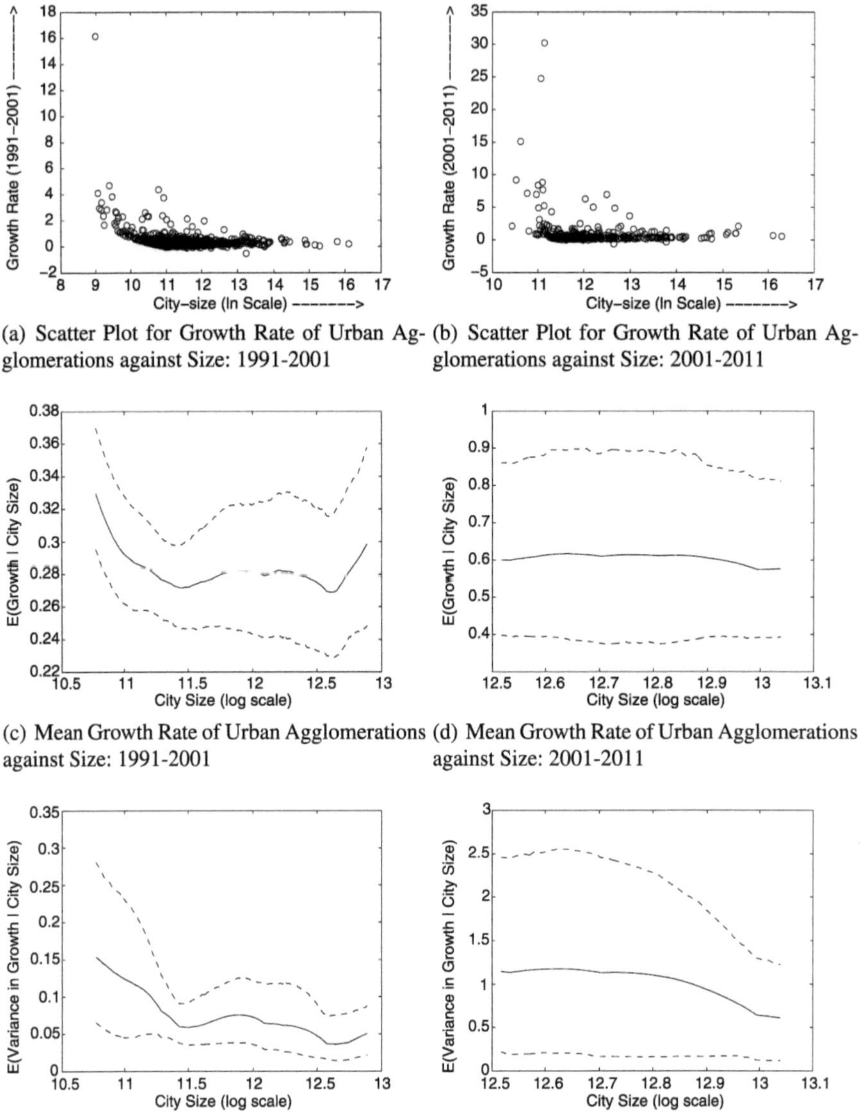


(a) Scatter Plot for Growth Rate of Urban Agglomerations against Size: 1991-2001

(b) Scatter Plot for Growth Rate of Urban Agglomerations against Size: 2001-2011

(c) Mean Growth Rate of Urban Agglomerations against Size: 1991-2001

(d) Mean Growth Rate of Urban Agglomerations against Size: 2001-2011

(e) Growth Rate Variance of Urban Agglomerations against Size: 1991-2001

(f) Growth Rate Variance of Urban Agglomerations against Size: 2001-2011

**Fig. 8.4** The growth rates of Indian urban agglomerations have been plotted against their sizes. The scatter plot, mean growth rate (calculated through Nadaraya-Watson estimate) and the variance of growth rates have been plotted against size of urban agglomerations. Panels (**a**), (**c**) and (**e**) illustrates plots for 1991–2001; whereas panels (**b**), (**d**), and (**f**) are for 2001–2011

## 8.3 Discussion

India and China are two countries in Asia with a lot of similarities. These are traditionally less urbanized nations and the impetus for urbanization is fairly recent for both India and China. For India, the urban population augmented from 23.34 % in 1981 to 31.16 % in 2011. Urbanization has taken place in China since the 1980s in an even more rapid scale. The urbanization rate increases [25] from 23.01 % in 1984 to 43.90 % by the end of 2006. Definitely, this immense growth opens up a plethora of questions about the morphology in general. For example, cities in special economic zones may have prospered due to favorable government policies unlike their counterparts in non-industrialized zones. First, in this context we demonstrated [23] that Zipf's law can be equally valid for these countries like urbanized western nations. The subsequent question is about dynamic authenticity of this empirical phenomenon. The Indian experience has already been narrated in Sect. 8.2. We sum up our empirical findings: Power law coefficient stays near the theoretically prediction over time, 1981–2011, and Gibrat's law is also satisfied. We draw heavily from our past study [20] regarding the empirical facts on China. The power law coefficient grew between 1990 and 2000. The underlying reason has been detected to a violation in Gibrat's law in case of China. We reproduce the graphs in Fig. 8.5, in which mean and variance of growth rates of the Chinese urban agglomerations during 1990–2000 have been plotted against their sizes. The plots indicate that the large cities grew rather less in China compared to medium and small sized urban agglomerations.

Why this anomaly between India and China? Is the cause rooted in the policies for formation of Special Economic Zones (SEZs) as surmised before [20]? Obviously, the growth rate of urban agglomerations is pegged for China at the upper end. The following model of job creation and migration to the newly created SEZs could match the empirical reality in a simulation experiment. The government introduce the feature of Special Economic Zones by giving special privileges to some urban agglomerations. The privileged urban agglomerations are chosen in such a way that they are not from the most populous cities. The other elements of that mathematical model are as follows. The probability of an additional job being created at a location is proportional to the number of already existing jobs at that location. There is a scale parameter, which for certain ranges favors growth of big cities and for other ranges, favors the growth of small cities. A number of new jobs are created in the locations of the SEZs. These new jobs require higher skill levels compared to the previously existing jobs. A worker matched with these jobs leave their old locations of work and move to the new location. Also higher skilled workers are primarily from the top ranking cities. Under these circumstances, a simulation study [20] demonstrated the augment of power law exponent on account of creation of SEZs. A verification of the role of SEZ could be experimented with India adopting favorable policies in creating SEZs in India. The Special Economic Zone Act was passed by the Government of India in 2005. Subsequent formation of SEZs [26, 27] could have induced some bias against the growth of urban agglomerations in the upper end. Nevertheless, we fail to notice any significant difference between mean growth rate of urban

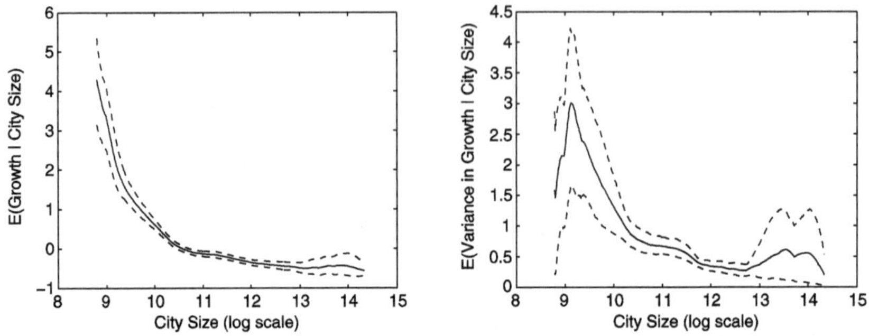


(a) Mean Growth Rate of Urban Agglomerations against Size: 1990-2000

(b) Growth Rate Variance of Urban Agglomerations against Size: 1990-2000

**Fig. 8.5** The mean and variance of growth rates of Chinese urban agglomerations have been plotted against their sizes for 1990–2000

agglomerations across different sizes in panels (a) and (b) of Fig. 8.4. Therefore, the hypothesis stating the role of SEZs is ineffectual in reconciling this issue, when Indian evidence is considered alongwith.

We propose another avenue to resolve this anomaly. It might depend on the government policies on the growth of cities as it is in the case of China. It has been observed [25] that the Chinese governmental policies favored restriction of populous cities to a specific size and expansion of non-populous cities since 1980s. On a more elaborate note, the Chinese government started the urban planning policy in 1980. This policy and its successor the Urban Planning Law enacted in 1990 meticulously controls the size of large cities and appropriately develops medium-sized and small cities. This policy introduces a bias against the growth of large cities and could very well be responsible for non-observance of Gibrat's law. However, that is hardly the case for India. The expansion of rather small cities and restriction of populous cities are not at all promoted in a democratic society like India compared to the policies pursued in a monolithic society like China. Therefore, the distinction may lie in the political system rooted in these two countries.

## References

1. Zipf GK (1949) Human behaviour and the principle of least effort. Addison-Wesley, Cambridge
2. Zipf GK (1935) The psychobiology of language. Houghton-Mifflin, Boston
3. Sasaki Y et al (2006) J Phys Soc Jpn 75:074804
4. Huberman BA et al (1998) Science 280:95
5. Yakovenko VM, Rosser JB Jr (2009) Rev Mod Phys 81:1703
6. Chatterjee A, Chakrabarti BK (2007) Eur Phys J B 60:135
7. Baldwin RB (1964) Astron J 69:377
8. Malamud BD et al (1998) Science 281:1840
9. Boffetra G et al (1999) Phys Rev Lett 83:4662
10. Malacarne LC, Mendes RS (2000) Physica A 286:391

11. Blasuis B, Tonjes R (2009) Phys Rev Lett 103:218701
12. Zanette D, Manrubia SC (1997) Phys Rev Lett 79:523
13. Moura NJ Jr, Ribeiro MB (2006) Physica A 367:441
14. Gabaix X, Ioannides Y (2004) Handbook of regional and urban economics 4. Enderson V, Thisseb J-F (eds) North-Holland, Amsterdam, p 2341
15. Krugman P (1996) The self organising economy. Blackbell Publishers, Oxford, UK and Cambridge
16. Gabaix X (1999) Q J Econ 114:739
17. Soo KT (2005) Reg Sci Urban Econ 35(3):239
18. Kuninaka H, Matsushita M (2008) J Phys Soc Jpn 77:114801
19. Eeckhout J (2004) Am Econ Rev 94(5):1429
20. Gangopadhyay K, Basu B (2010) Econophysics and economics of games, social choices and quantitative techniques. ISBN: 978-88-470-1500-5, p 90
21. http://www.censusindia.gov.in
22. Clauset A, Shalizi CR, Newman MJ (2009) SIAM Rev 51:661
23. Gangopadhyay K, Basu B (2009) Physica A 388:2682
24. Pagan A, Ullah A (1999) Nonparametric econometrics. Cambridge University Press, Cambridge
25. Chen Z, Fu S, Zhang D (2010) Searching for the parallel growth of cities. SSRN Working Paper
26. Nema P, Pokhariyal P (October 5, 2008) SEZs as growth engines – India vs China. SSRN Working Paper
27. Sen A (Nov 2006) Will India recreate China's SEZ magic? The Economic Times

# Part II
# Model-Based Studies

# Chapter 9
# Reaction to Extreme Events in a Minimal Agent Based Model

**Andrea Zaccaria, Matthieu Cristelli, and Luciano Pietronero**

**Abstract** We consider the issue of the overreaction of financial markets to a sudden price change. In particular, we focus on the price and the population dynamics which follows a large fluctuation. In order to investigate these aspects from different perspectives we discuss the known results for empirical data, the Lux-Marchesi model and a minimal agent based model which we have recently proposed. We show that, in this framework, the presence of a overreaction is deeply linked to the population dynamics. In particular, the presence of a destabilizing strategy in the market is a necessary condition to have an overshoot with respect to the exogenously induced price fluctuation. Finally, we analyze how the memory of the agents can quantitatively affect this behavior.

## 9.1 Introduction

The interest of the physics community in the study of financial markets has been mainly focused on two aspects: the analysis of the large amount of the recently available data [4–6, 13, 14, 17] and a number of different proposals for the modelization of the dynamics. Concerning this last aspect, the most used way to describe financial markets is represented by Agent Based Models (ABM) [5, 7–10, 16]. This approach is probably the closest to physicists' tastes, being grounded on the non trivial connection between traders' strategies and price dynamics. However, a number of ABMs, even starting from very different hypothesis, are able to reproduce the main statistical evidences of real markets. This is mainly due to the lack of strict empirical results, which led to a severe difficulty to discriminate between the different models and, in turn, to perform the standard falsification procedure which is so important from a physicist's point of view. As a consequence, we believe that a big effort must be made in order to investigate i) the main ingredients an ABM must

A. Zaccaria · M. Cristelli · L. Pietronero
ISC-CNR, Via dei Taurini 19, 00185 Roma, Italy

A. Zaccaria (✉) · M. Cristelli · L. Pietronero
Dipartimento di Fisica, Sapienza, Università di Roma, P.le Aldo Moro 2, 00185 Roma, Italy
e-mail: andrea.zaccaria@roma1.infn.it

F. Abergel et al. (eds.), *Econophysics of Systemic Risk and Network Dynamics*, New Economic Windows,
DOI 10.1007/978-88-470-2553-0_9, 

contain to reproduce the empirical evidence (for an attempt in this direction, see [1–3]) ii) the existence of more statistical regularities in financial data and iii) the non trivial connections between ABMs' and price dynamics. The present paper is focused on this last issue and, in particular, on a peculiar behavior which has been detected after an extreme price fluctuation: the overreaction with respect to large price movements. At present we do not know if agents update their strategies after a shock and which direction takes their short-term reaction: this kind of study might be performed, for example, using the data analyzed in [18] and [19]. In the present paper, after a brief review of some known result about this issue, we study the reaction to large price movements in the framework of a minimal ABM which permits a detailed investigation of agents' reactions in terms of their strategy update. We find that a large fraction of the agents must use a trend-following, i.e. chartist strategy in order to reproduce a realistic price dynamics.

This article is organized as follows. Section 9.2 summarizes some of the results obtained by Zawadowski et al. with respect to the empirical price dynamics after a shock [20] and in the context of a specific ABM [21]. Section 9.3 studies with some details the price and the population dynamics in similar conditions. Section 9.4 concludes.

## 9.2 Known Results

### *9.2.1 Overshoots in Real Financial Markets*

Zawadowski et al. [20] have analyzed the behavior of real financial markets after a big intra-day price jump. In particular, they have studied two TAQ (Trades and Quotes) datasets from NYSE and NASDAQ. In order to select the events in a precise way, they performed a detailed filtering procedure. They found a significant reversion tendency after both large increases and decreases: within a time of about 60 minutes after the initial price change, the market tends, on average, to overreact after the jump, to rebound and finally to retrieve a new stable value of the price. Moreover, they have found an evidence for a linear relation between the size of the triggering fluctuation and the size of the reversal and a sharp increase in volatility, volume and bid-ask spread in correspondence with the event. This empirical evidence has been recently confirmed by Preis et al. [15].

### *9.2.2 Lux-Marchesi Model*

In [21] the authors have analyzed the reaction to an extreme price movement in the framework of the Lux-Marchesi [11, 12] model. Here we briefly summarize the main assumptions of Lux and Marchesi. By the way, some of them will turn to be useful when we will describe the minimal ABM which we are going to study in the next section. The main feature of this model is its high degree of realism: agents' strategies, impact on the price and population dynamics are represented in detail and give rise to a reasonable price statistics. Three classes of agents are present. The *fundamentalists* represent a stabilizing tendency, because their market operations are driven by their perception about the price they believe to be fair for the stock. This value is called *fundamental price* $p_f$. As a consequence, they buy if the price is lower than $p_f$ and they sell if it is higher. Instead, the *chartist* use past price movements as a reference. They are divided in two categories: *optimists*, who always buy, and *pessimists*, who always sell. The population dynamics takes into account many parameters and a detailed discussion about this issue would go beyond the aims of the present paper; however, we point out that, on average, a change of class occurs if the agent finds a strategy which is more profitable than the one she is actually adopting.

Zawadowski et al. [21] have compared the response to an endogenous (that is, triggered by the market itself) fluctuation with the response due to an exogenous event.

In the first case, they have studied the events as a function of the distance $\Delta p = p - p_f$ between the actual price and the fundamental one. They have found an exponential decay of $\Delta p$ as a function of time, whose characteristic time depends on the initial price variation and, being the attractive force due to the presence of fundamentalists, on the fraction of chartists in the market.

In order to reproduce the effect of the sudden arrival of an exogenous news the authors changed the fundamental price $p_f \rightarrow p_f + \Delta p$ and recorded the average response finding, as expected, that $p$ quickly follows the movement of the fundamental price and then, more interestingly, an overshoot occurs, after which the price reaches $p_f$.

The authors suggest that the cause of the overshoot is linked to the population dynamics: the movements inside the chartists community (that is, the flows between optimists and pessimists) are much faster than the ones between chartists and fundamentalists. Since the transition rates between chartists depend on the time derivative of the price, the upward trend makes most of the chartists optimist (and vice versa, pessimist in the case of price drops) who act following the trend without considering the fundamental price and ultimately causing the overshoot.

We point out, however, that this explanation is, out of necessity, qualitative, because of the presence of a lot of parameters and the high complexity which characterizes the dynamics which underlies the transition rates from one class to the other. These two factors prevent a detailed analysis about the mechanisms which ultimately are causing the overshoot and a study of the reaction as a function of the different parameters which are involved in the dynamics. As a consequence this explanation, even if reasonable and shareable, can be hardly tested.

## 9.3 Overshoots in a Minimal Agent Based Model

As we have seen in the previous sections, the empirically founded presence of overshoots following large price fluctuations calls for an interpretation in terms of agents' strategies. A first attempt has been made by Zawadowski et al. [21], who analyzed this interesting aspect in the framework of the Lux-Marchesi model. In this section we study this problem in the framework of the minimal ABM recently introduced in [1–3]. As we will see, the low number of parameters and the simplicity which characterizes the population dynamics permits a detailed investigation of the phenomenon.

Now we briefly summarize the main features of the model.

The $N$ agents are divided in two classes: $N_f$ fundamentalists and $N_c = N - N_f$ chartists. Differently from the Lux-Marchesi model, there is no further division among the chartist class; however, chartists still represent the destabilizing tendency of the market by having, as a reference for the price, the moving average $p_M(t) \equiv \frac{1}{M}\sum_{i=0}^{M-1} p(t-i)$ and being trend-followers. They will buy if the actual price is higher than $p_M(t)$ and sell otherwise. This results in the following equation for the price evolution:

$$p(t+1) = p(t) + \frac{N_c}{N}\frac{b}{M-1}\big(p(t) - p_M(t)\big) + \frac{N_f}{N}\gamma\big(p_f - p(t)\big) + \sigma\xi(t) \quad (9.1)$$

where $b$ and $\gamma$ quantify the strength of the chartists' and fundamentalists' impacts on the price, respectively, and the factor $1/(M-1)$ makes the volume of chartists' trades independent on the time window $M$. The white noise $\sigma\xi(t)$ has $\sigma = 1$. The transition rates between the two classes are composed by an herding factor and a term which accounts for the signal readable from the price dynamics. In formulas, the probability a chartist has to become a fundamentalist reads

$$P_{cf} \propto \left(K + \frac{N_f}{N}\right)\exp\big(\gamma|p_f - p|\big) \quad (9.2)$$

and the rate from fundamentalists to chartists is

$$P_{fc} \propto \left(K + \frac{N_c}{N}\right)\exp\left(b\frac{|p_M - p|}{M-1}\right). \quad (9.3)$$

Let us now consider the response to an extreme price movement in the framework of this minimal model.

As a first step we varied only the fundamental price $p_f$. In this case no overshoot is present, because the sudden variation of $p - p_f$ gives a strong signal to fundamentalists, who rapidly become the majority and bring the price back to its fundamental value.

Then we varied only the price. A typical example of the result of our simulations is depicted in Fig. 9.1. We have plotted the price (black line) and the number of chartists (red line) as a function of time. Here the shock is represented by a sudden crash (or rise) of the price, that is, $p \to p \pm \delta p$ (with $\delta p = 200$), keeping the fundamental price constant. In this case only in the first two shocks have an overshoot (that is, the realized price movement is larger than $\delta p$). Looking at the population dynamics one can notice that this is due to the fact that chartists were already active in the market. However the distance between the price and the fundamental

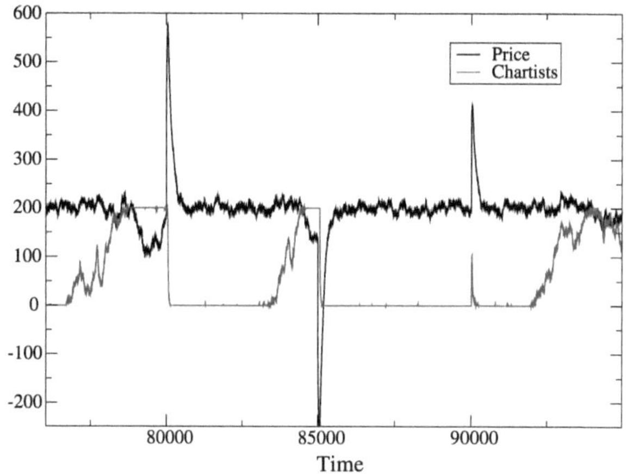


**Fig. 9.1** Model reaction to three sudden price changes $p \to p \pm 200$, keeping $p_f$ fixed. The overshoot is present in the first two cases, in which chartists were active before the extreme event. In the last case the fundamentalists quickly bring the price to its fundamental value, without any overshoot. (Color figure online)

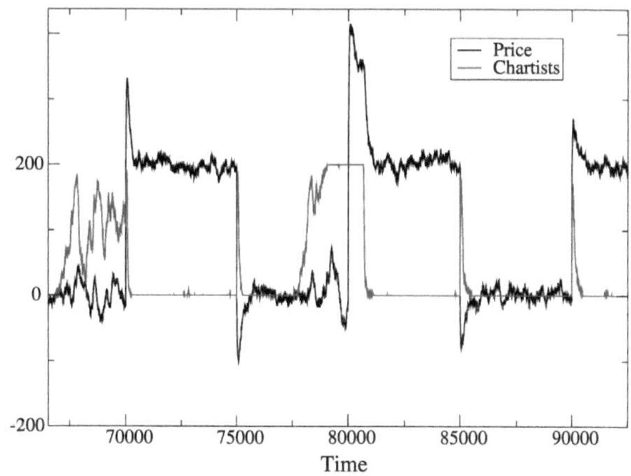


**Fig. 9.2** Model reaction to five sudden changes $p \to p \pm 200$ and $p_f \to p_f \pm 200$. The overshoot is always present, being the market signal due to the extreme price movement stronger for chartist. (Color figure online)

one quickly makes all the chartists fundamentalists and the price is quickly brought back to a value close to $p_f$. The third shock occurs when most of the agents were fundamentalists: in this case there is no overshoot.

Finally, we vary both $p$ and $p_f$ by adding or subtracting $\delta p = 200$, with results plotted in Fig. 9.2. The price (black line) quickly adjusts around the new value of $p_f$ always exhibiting an aftershock due to the trend followers (red line). In fact, the signal which fundamentalists perceive, $p - p_f$, is small compared to the chartists' one due to the trend and so, in this case, there are always bursts of chartism and, consequently, overshoots.

In conclusion, the overshoots are always caused by the trend following strategy of the chartists, whose presence in the market is a necessary condition to observe this phenomenon.

As a further step, we take into account how a variation of the parameters acts on the overshoot dynamics. Is it clear from (9.1) that only three variables are involved: $b$, $\gamma$ and $M$. While the first two are deeply linked and must be somehow balanced

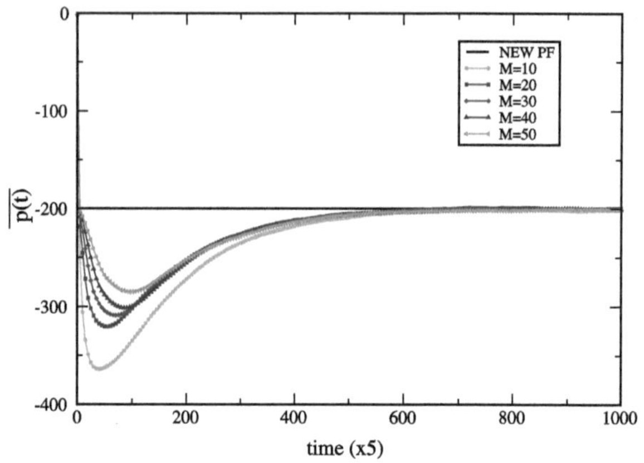


**Fig. 9.3** The average overshoot for different values of the agents' memory $M$

in order to give a realistic population dynamics, $M$ can be freely varied in order to study how the presence of different time scales affects the output [1–3]. To analyze the overshoot from a quantitative point of view we define $t_s$ as the time at which the market is perturbed by the signal and $t_m$ as the time at which the maximum price variation is reached with respect to the next equilibrium price. In the following we will consider $\Delta t = t_m - t_s$, the time interval between the signal and the start of the price reversal, and $\Delta p = |p(t_m) - p(t_s)|$, the size of the overshoot, defined as the maximum distance the price has reached. We find that these quantities may fluctuate from one realization to the other; thus we average them over a sufficient number of simulations, keeping the exogenous fluctuation fixed and equal to $\delta p = -200$ (the case $\delta p = +200$ shows identical results). The averaged price evolution is plotted in Fig. 9.3, with the time rescaled by a factor 5. One can see a strong dependence on the value of $M$: if the agents have a short memory, they produce a quicker and stronger overshoot. In Fig. 9.4 we show how $\Delta p$ and $\Delta t$ vary with $M$. In the second case one may notice a clear linear dependence.

## 9.4 Conclusions and Perspectives

We have analyzed the response to a large price fluctuation in the framework of a minimal agent based model. We believe this issue to be of great importance, being linked to the psychological dynamics of the traders. We have found a realistic behavior in terms of the price movements: agents act in such a way to cause an overshoot. In terms of agents' strategies, the presence of fundamentalists in the market cause a quick rearrangement of the price to the fair value; as a consequence, the overshoot can be reproduce only if chartists were already active when the exogenous signal arrived. Finally, we have analyzed the dependence of the size and the characteristic time of the overshoot as a function of the main strategic parameter of the model, the length of agents' memory $M$.

As a further analysis, we would like to compare events of different sizes and having different sources of fluctuations. In particular, we would like to compare the

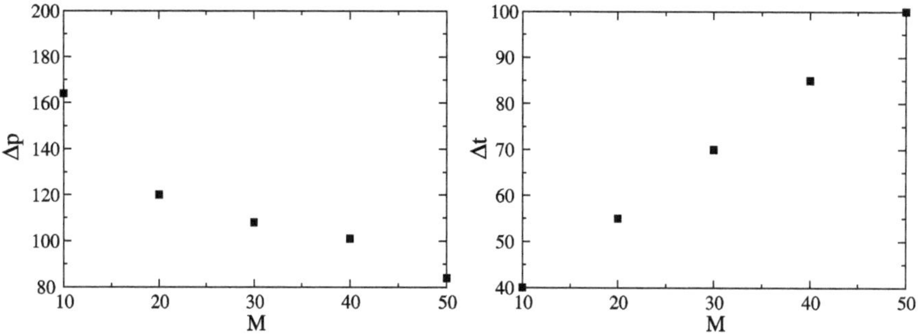


**Fig. 9.4** The size of the average overshoot (*left*) and reversal time (*right*) as functions of $M$. Agents with a longer memory produce smaller overshoots, but reach their maximum distance later

reactions to an endogenous versus an exogenous fluctuation. We believe that the empirically established differences between these reactions can be taken into account if one consider a simple mechanism of coherence in agents' strategies. These issue will be considered in future works.

**Acknowledgements** The authors thank János Kertész for the interesting discussions. We acknowledge support from the projects FET Open Project FOC nr. 255987 and the PNR national project CRISIS-Lab.

## References

1. Alfi V, Cristelli M, Pietronero L, Zaccaria A (2009) Minimal agent based model for financial markets I: origin and self-organization of stylized facts. Eur Phys J B 67:385–397
2. Alfi V, Cristelli M, Pietronero L, Zaccaria A (2009) Minimal agent based model for financial markets II: statistical properties of the linear and multiplicative dynamics. Eur Phys J B 67:399–417
3. Alfi V, Pietronero L, Zaccaria A (2009) Self-organization for the stylized facts and finite-size effects in a financial-market model. Europhys Lett 86(5):58003
4. Bouchaud JP, Potters M (2003) Theory of financial risk and derivative pricing: from statistical physics to risk management. Cambridge University Press, Cambridge
5. Chakraborti A, Toke IM, Patriarca M, Abergel F (2011) Econophysics review: II. Agent-based models. Quant Finance 11(7):1013–1041
6. Cont R (2001) Empirical properties of assets returns: stylized facts and statistical issues. Quant Finance 1:223–236
7. Cristelli M, Zaccaria A, Pietronero L (2011) Critical overview of agent-based models for economics. In: Mallamace F, Stanley EH (ed) Proceedings of the School of Physics "E. Fermi", course CLXXVI. Varenna (in press). arXiv:1101.1847
8. De Martino A, Marsili M (2006) Statistical mechanics of socio-economic systems with heterogeneous agents. J Phys A 39:465–540
9. Hommes CH (2006) Heterogeneous agent models in economics and finance. In: Handbook of computational economics. North-Holland, Elsevier, Amsterdam
10. LeBaron B (2006) Agent-based computational finance. In: Handbook of computational economics. North-Holland, Elsevier, Amsterdam
11. Lux T, Marchesi M (1999) Scaling and criticality in a stochastic multi-agent model of a financial market. Nature 397:498–500

12. Lux T, Marchesi M (2000) Volatility clustering in financial markets: a micro-simulation of interacting agents. Int J Theor Appl Finance 3:675–702
13. Mantegna RN, Stanley HE (2000) An introduction to econophysics: correlation and complexity in finance. Cambridge University Press, Cambridge
14. Mizuno T, Takayasu H, Takayasu M (2007) Analysis of price diffusion in financial markets using PUCK model. Phys A, Stat Mech Appl 382:187–192
15. Preis T, Schneider JJ, Stanley HE (2011) Switching processes in financial markets. Proc Natl Acad Sci 108(19):7674–7678
16. Samanidou E, Zschischang E, Stauffer D, Lux T (2007) Agent-based models of financial markets. Rep Prog Phys 70:409–450
17. Takayasu M, Mizuno T, Takayasu H (2006) Potential force observed in market dynamics. Phys A, Stat Mech Appl 370:91–97
18. Tóth B, Lempérière Y, Deremble C, de Lataillade J, Kockelkoren J, Bouchaud J-P (2011) Anomalous price impact and the critical nature of liquidity in financial markets. Phys Rev X 1:021006
19. Tumminello M, Lillo F, Piilo J, Mantegna RN (2011) Identification of clusters of investors from their real trading activity in a financial market. New J Phys 14:013041
20. Zawadowski AG, Andor G, Kertesz J (2006) Short-term market reaction after extreme price changes of liquid stocks. Quant Finance 6(4):283–295
21. Zawadowski AG, Karadi R, Kertesz J (2002) Price drops, fluctuations, and correlation in a multi-agent model of stock markets. Physica A 316(1–4):403–412

# Chapter 10
# Predatory Trading and Risk Minimisation: How to (B)Eat the Competition

**Anita Mehta**

**Abstract** We present a model of predatory traders interacting with each other in the presence of a central reserve (which dissipates their wealth through say, taxation), as well as inflation. This model is examined on a network for the purposes of correlating complexity of interactions with systemic risk. We suggest the use of selective networking to enhance the survival rates of arbitrarily chosen traders. Our conclusions show that networking with 'doomed' traders is the most risk-free scenario, and that if a trader is to network with peers, it is far better to do so with those who have less intrinsic wealth than himself to ensure individual, and perhaps systemic stability.

## 10.1 Introduction

The topic of predatory trading and its links with systemic risk is of great contemporary interest: at the time of writing this paper, these links have been mentioned repeatedly in the World Economic Forum at Davos, in addition to having formed the backbone of the 'Occupy' movements worldwide. Immense public anger has been expressed against corporate greed (with predatory trading forming a major way that this is manifested), and many intellectuals worldwide attribute this to the collapse of the world economic system. In this paper, we examine these ideas in a more technical way to see if rigorous mathematical links can be established between these two concepts.

In order to put our mathematical models in the context of current interdisciplinary literature, we quote the conclusions of two key papers. First we define predatory trading along the lines of a recent paper [1], as that which induces and/or exploits other investors' need to 'reduce' their positions. If one trader needs to sell, others also sell and subsequently buy back the asset, which leads to price overshooting and a reduced liquidation value for the distressed trader. In this way, a trader profits from triggering another trader's crisis; according to the authors of [1], the crisis can spill

A. Mehta (✉)
S. N. Bose National Centre for Basic Sciences, Block-JD Sector III, Salt Lake, Kolkata 700098, India
e-mail: anita@bose.res.in

F. Abergel et al. (eds.), *Econophysics of Systemic Risk and Network Dynamics*, New Economic Windows,
DOI 10.1007/978-88-470-2553-0_10, 

over 'across traders and across markets'. To model this scenario, we invoke a model of predatory traders in the presence of a central reserve [2], which is principally a source of wealth dissipation in the form of taxation. We assume that this dissipation acts uniformly across the traders' wealth, irrespective of their actual magnitudes. Among our findings [2] is the fact that when all traders are interconnected and interacting, the entire system collapses, with one or zero survivors. This finds resonance with the ideas of another key paper [3], where analogies with model ecosystems have led the authors to conclude that propagating complexity (via the increase of the *number* and *strength* of interactions between different units) can jeopardise systemic stability.

The original model [4] on which [2] is based, was introduced as a model of complexity. It embodies predator-prey interactions, but goes beyond the best-known predator-prey model due to Lotka and Volterra by embedding interacting traders in an active medium; this is a case where the Lotka-Volterra model cannot be simply applied. As mentioned above, a central reserve bank represents such an active medium in the case of interacting traders, whose global role is to reduce the value of held wealth as a function of time [2]. This forms a more realistic social backdrop to the phenomenon of predatory trading, and it is this model that we study in this paper. In order to relate it to the phenomenon of systemic risk, we embed the model on complex networks [5, 6]; these represent a compromise between the unrealistic extremes of mean field, where all traders interact with all others (too global) and lattice models, where interaction is confined to local neighbourhoods (too local). Many real world networks, in spite of their inherent differences, have been found to have the topology of complex networks [7–9]; and the embedding of our model on such networks [10] allows us to probe the relevance of predatory trading to systemic stability.

The plan of the paper is as follows. First, in Sect. 10.2, we introduce the model of interacting traders of varying wealth in the presence of a central reserve, and show that typically only the wealthiest survive. Next, in Sect. 10.3, we probe the effect of networks: starting with an existing lattice of interacting traders with nearest-neighbour interactions, we add non-local links between them with probability $p$ [5]. Survivor ratios are then measured as a function of this 'wiring probability' $p$, as $p$ is increased to reflect the topologies of small world and fully random networks ($p = 1$). In Sect. 10.4, we ask the following question: can the destiny of a selected trader be changed by suitable networking? We probe this systematically by networking a given trader non-locally with others of less, equal and greater wealth and find indeed that a trader who would die in his original neighbourhood, *is able to change his fate*, becoming a survivor via such selective non-local networking. In Sect. 10.5, we provide a useful statistical measure of survival, the pairwise probability for a trader to survive against wealthier neighbours, i.e. to win against the odds. Finally, we discuss the implications of these results to systemic stability in Sect. 10.6.

## 10.2 Model

The present model was first used in the context of cosmology to describe the accretion of black holes in the presence of a radiation field [11]. Its applications, however, are considerably more general; used in the context of economics [2], it manifests an interesting rich-get-richer behaviour. Here, we review some of its principal properties [4].

Consider an array of traders with time-dependent wealth $m_i(t)$ located at the sites of a regular lattice. The time evolution of wealth of the traders is given by the coupled deterministic first order equations,

$$\frac{dm_i}{dt} = \left(\frac{\alpha}{t} - \frac{1}{t^{1/2}} \sum_{j \neq i} g_{ij} \frac{dm_j}{dt}\right) m_i - \frac{1}{m_i}. \tag{10.1}$$

Here, the parameter $\alpha$ is called the wealth accretion parameter (modelling investments, savings etc) and $g_{ij}$ defines the strength of the interaction between the traders $m_i$ and $m_j$. The first parenthesis in the R.H.S of (10.1) represents the wealth gain of the $i$th trader, which has two components: his wealth gain due to investments/savings (proportional to $\alpha$) modulated by dissipation (at the rate of $1/t$) due to e.g. taxation, and his wealth gain due to predatory trading (the second term of (10.1)), also modulated by dissipation (at the rate of $1/t^{1/2}$) in the same way. Notice in the second term, that the loss of the other traders corresponds to the gain of the $i$th trader, so that each trader 'feeds off' the others, thus justifying the name 'predatory trading'. The last term, $-1/m_i$, represents the loss of the $i$th trader's wealth through inflation to the surroundings; we will see that this term ensures that those without a threshold level of wealth 'perish', as in the case of those individuals in society who live below the poverty line. Here and in the following we will use words such as 'dying' or 'perishing' to connote the bankruptcy/impoverishment, of a trader and conversely, 'life' will be associated with *financial* survival, i.e. solvency.

A logarithmic time is introduced in the study for convenience. We define a scaled time $s = \ln(t/t_0)$, where $t_0$ is some initial time. Similarly, for convenience, we rescale wealth to be $X_i = m_i/t^{1/2}$. Using the new variables, (10.1) can be rewritten as,

$$\frac{dX_i}{ds} \equiv X'_i = \left(\frac{2\alpha - 1}{2} - \sum_{j \neq i} g_{ij} \left(\frac{X_j}{2} + X'_j\right)\right) X_i - \frac{1}{X_i}, \tag{10.2}$$

where the primes denote differentiation performed with respect to $s$.

Continuing our recapitulation of the results of the model [4], we consider a scenario where there is a single isolated trader, whose initial ('inherited') wealth is $X_0$. Under the dynamics defined by (10.2), the trader will survive financially only if $X_0 > X_\star (= \sqrt{\frac{2}{2\alpha - 1}})$ (this imposes the condition $\alpha > 1/2$ [4]), else he will eventually go bankrupt (see Fig. 10.1). Next, consider a system of two traders with equal initial wealth; here, there exists a critical coupling $g_c$ such that for $g < g_c$ the two traders both survive, provided that their individual inherited wealth is greater than $X_\star$. For two unequally wealthy traders ($X_1 < X_2$, say), the poorer trader goes bankrupt first at time $s_1$; the richer one either survives (if his wealth at time $s_1$, $X_2(s_1)$, exceeds the threshold $X_\star$) or goes bankrupt (if $X_2(s_1) < X_\star$). The main inferences are twofold: the wealthier predatory trader 'consumes' the poorer one's

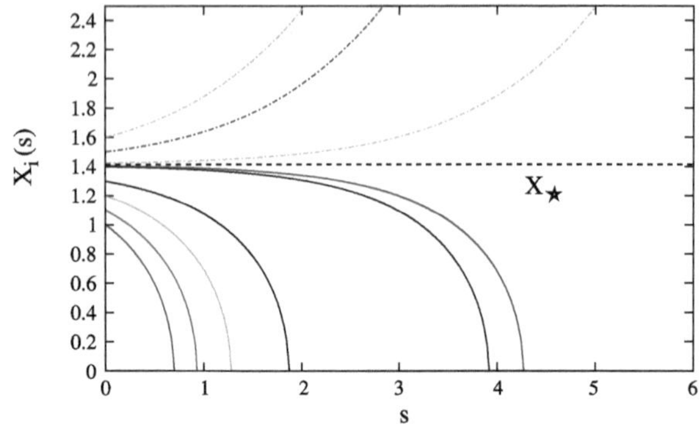


**Fig. 10.1** The plots show the evolution of individual non-interacting traders with a range of initial wealth obeying (10.2). Traders with initial wealth $X_0$ greater than $X_\star$ live forever, or else they die in time. Here $\alpha = 1.0$, therefore $X_\star = \sqrt{\frac{2}{2\alpha-1}} = \sqrt{2}$ (Color figure online)

wealth in due course, and then survives or not depending on whether his own wealth at that point is enough to tide him through its eventual dissipation through taxes and inflation. We thus see that this relatively simple model captures not just the mechanism of predatory trading, but also includes the flavour of more sophisticated concepts like inherited wealth, taxation and inflation.

Consider the limit of an infinitely large number of traders all connected to each other; this represents a limiting mean field regime, with fully collective behaviour involving long-range interactions. For $g > g_c$, all but the wealthiest will eventually go bankrupt. In the weak coupling regime ($g < g_c$) on the other hand, the dynamics consist of two successive stages [4]. In Stage I, the traders behave as if they were isolated from each other (but still in the presence of the reserve); they get richer (or go bankrupt ) quickly if their individual wealth is greater (or less) than the threshold $X_\star$. In Stage II, slow, collective and predatory dynamics leads to a scenario where again, only the single wealthiest trader survives. This weakly interacting mean field regime shows the presence of two well-separated time scales, a characteristic feature of glassy systems [12]. The separation into two stages embodies an interesting physical/sociological scenario: the first stage is fast, and each trader survives or 'dies' only on the basis of his inherited wealth, so that everyone without this threshold wealth is already eliminated before the second stage sets in. Competition and predatoriness enter only in the second stage, when the wealthier feast off their poorer competitors progressively, until there is only overlord left. This is a perfect embodiment of systemic risk, as the entire system collapses, with only one survivor remaining [2].

Similar glassy dynamics also arise when the model is solved on a periodic lattice with *only nearest-neighbour* interactions. The dynamical equations in (10.2) take the form:

$$X'_{\mathbf{n}} = \left(\frac{2\alpha-1}{2} + g\sum_{\mathbf{m}}\left(\frac{1}{X_{\mathbf{m}}} - \alpha X_{\mathbf{m}}\right)\right)X_{\mathbf{n}} - \frac{1}{X_{\mathbf{n}}}, \tag{10.3}$$

by keeping the terms upto first order in $g$ [4]. Here, $\mathbf{m}$ runs over the $z$ nearest neighbours of the site $\mathbf{n}$, where for a one-dimensional ring topology, $z = d$, while $z = 2d$

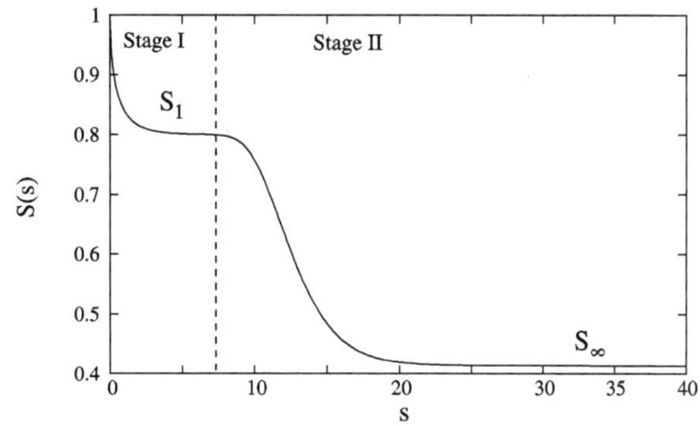


**Fig. 10.2** The survival ratio $S(s)$ plotted as a function of reduced time $s$, for traders distributed in a regular one-dimensional lattice of size 100, 000. In *Stage I*, traders grow independently, while in *Stage II* the growth is collective. Here, $S_1 = 0.8$ is the survival ratio at the end of *Stage I* and $S_\infty$ is the asymptotic survival ratio

for a two-dimensional lattice—these are the two cases we consider here. We summarise earlier results [4] on the dynamics: In Stage I, traders evolve independently and (as before) only those whose initial wealth $X_i(s=0)$ exceeds the threshold $X_\star$, survive. In Stage II, the dynamics are slow and collective, with competition and predatoriness setting in: however, an important difference with the earlier fully connected case is that there can be *several* survivors, provided that these are isolated from each other by defunct or bankrupt traders (i.e. no competitors remain *within their effective domain*). Their number asymptotes to a constant $S_\infty$ (Fig. 10.2), and these 'isolated overlords' survive forever. The moral of the story is therefore that in the presence of predatory dynamics, interaction-limiting 'firewalls' can help avoid systemic collapse: conversely, *full globalisation with predatory dynamics makes systemic collapse inevitable*. Apart from providing quantitative support for the conclusions of [3], this underscores the necessity of economic firewalls for the prevention of systemic risk [13].

Following the mean field scenario, where the wealthiest trader is the only survivor, it is natural to ask if this would also hold when the range of interactions is limited. Somewhat surprisingly, this turns out not always to be the case, with nontrivial and counter-intuitive survivor patterns being found often [10]. While one can certainly rule out the survival of a trader whose initial wealth is less than threshold ($X_\star$), many-body interactions can then give rise to extremely complex dynamics in Stage II for traders with $X > X_\star$. This points to the existence of ways of winning against the odds to (b)eat the competition, when interactions are limited in range; in the remainder of this paper, we use selective networking as a strategy to achieve this aim.

## 10.3 Traders in Complex Networks

In this section, we examine the mechanisms of selective networking: a particularly interesting example to consider is the class of small-world networks. These have the property that long- and short-range interactions can coexist; such networks also contain 'hubs', where certain sites are preferentially endowed with many connections.

Small-world networks can be constructed by starting with regular lattices, adding links randomly with probability $p$ to their sites [5] and then freezing them, so that the average degree of the sites is increased for all $p > 0$.

### *10.3.1 One-Dimensional Ring and Two-Dimensional Square Lattices*

Consider a regular one-dimensional ring lattice of size $N = 2000$. To start with, the wealth of the traders located on the lattice sites evolve according to (10.3), where the interactions are with nearest neighbours only. Next, the lattice is modified by adding new links between sites chosen randomly with an associated probability $p$. For $p = 0$, the network is ordered, while for $p = 1$, the network becomes completely random.

In the first scheme [10], we add links probabilistically starting with site $i = 1$ and end with $i = N$, only once: we call this the 1-cycle scheme. The survival ratios of traders as a function of reduced time $s$ for different values of wiring probability $p$ are presented in Fig. 10.3(a). Consider the $p = 0$ case, which corresponds to a regular lattice; here the survival ratio $S(s)$ shows two stages, Stage I and Stage II, in its evolution. For all values of $0 < p \leq 1$, the existence of these well-separated Stages I and II is also observed. There is a noticeable fall in the survivor ratio as $p$ is increased, though; this is clearly visible in the asymptotic values $S_\infty(p)$ plotted with respect to $p$ in the inset of Fig. 10.3(a). As $p$ increases, the number of links increases, leading to more interaction and competition between the traders, and hence a decrease in the number of survivors [3]. As expected, this behaviour interpolates between the two characteristic behaviours relevant to the regular lattice and mean field scenarios.

Next we implement the 5-cycle scheme [10], where the rewiring is done five times. Figure 10.3(b) shows the survivor ratio $S(s)$ as a function of $s$, with a clear decrease of $S(s)$ for increasing $p$. The asymptotic survival ratios $S_\infty$ for the 1- and 5-cycle schemes are shown in the insets of Fig. 10.3(a) and (b) respectively, with a clear decrease of $S_\infty$ as $p$ increases. In addition, the survivor ratio in the 5-cycle scheme is consistently smaller than in the 1-cycle scheme, for all $p$.

The above procedures are repeated for a two-dimensional square lattice of size $50 \times 50$ [10], and survival ratios obtained as a function of $s$ for the 1- and 5-cycle schemes (see Figs. 10.3(c)–(d)). The asymptotic survivor ratios for these two cases are shown in the insets of Figs. 10.3(c)–(d); they follow a decreasing trend with increasing $p$, similar to the $1d$ case. Finally, we plot the asymptotic survival ratios of the 1- and 5-cycle schemes in $1d$ and $2d$, in Fig. 10.4.

All the above simulations reinforce one of the central themes of this paper, that economic firewalls are good ways to avoid systemic collapse in a predatory scenario, since the more globalised the interactions, the greater the systemic risk [3].

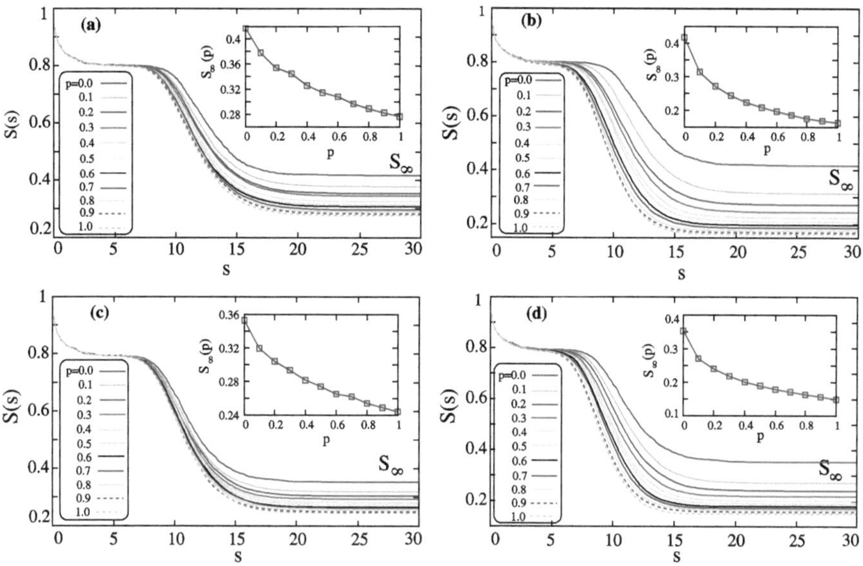


**Fig. 10.3** The plots of $S(s)$ as a function of reduced time $s$, for increasing values of wiring probabilities $p$. The 1-cycle scheme is shown in (**a**) for a 1-dimensional ring and (**c**) for a 2-dimensional square lattice. The 5-cycle scheme is shown in (**b**) for a 1-dimensional ring and (**d**) for a 2-dimensional square lattice. The insets in all the figures show the asymptotic survival ratios $S_\infty$ as a function of the probability $p$. Here, the system size for the 1-d ring is 2000 and for the 2-d square lattice it is $50 \times 50$—all our data is averaged over 10 random network configurations (Color figure online)

**Fig. 10.4** The plot of asymptotic survival ratios $S_\infty(s)$ as a function of the probabilities $p$ for 1- and 5-cycle schemes in 1 and 2 dimensions. The error bars for all the graphs do not exceed 0.003 and are smaller than the plot symbols

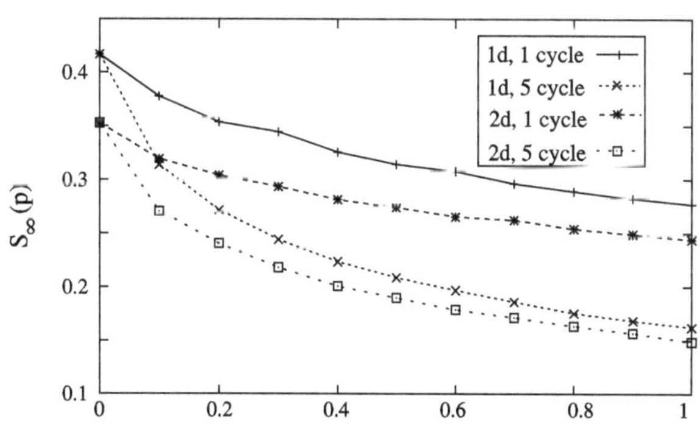


## 10.4 Networking Strategies: The Lazarus Effect

Since the central feature of this model is the survival of traders against the competition, it is of great interest to find a smart networking strategy which can change the fate of a trader, for example, by reviving a 'dying' trader to life—we call this the Lazarus [14] effect.

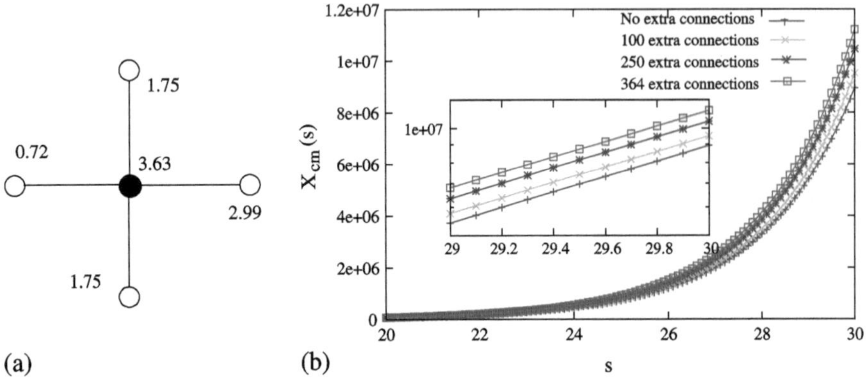


**Fig. 10.5** (**a**) The central trader here is a survivor in his original configuration. (**b**) The asymptotic wealth of the central trader diverges with increasing connectivity to eventual non-survivors (Color figure online)

We systematically investigate the effect of adding a finite number of non-local connections to a chosen central trader. In [10], it was shown that the growth or decay of the wealth of a trader is solely dictated by its relative rate of change versus the cumulative rate of change of its neighbours' wealth. The key to better survival should therefore lie in choosing to network with traders whose wealth is decaying strongly. We accordingly divide all possible non-local connections into two classes: class A comprises eventual non-survivors ($X < X_\star$), while class B comprises would-be survivors ($X > X_\star$). In the next subsection, we look at the outcome of networking with members of class A.

## *10.4.1 Non-local Connections with Eventual Non-survivors ($X_i < X_\star$)*

Recall that non-survivors ($X < X_\star$) die very early during Stage I. In connecting such traders to a given trader with $X > X_\star$, we can be sure that they will never be able to compete with him, much less run him out of business.

Let us consider a central trader who is an eventual survivor, as shown in Fig. 10.5. We now let him network with eventual non-survivors from all over the lattice, and record the growth of his wealth as a function of the number of traders in his network; the results are shown in Fig. 10.5(b). When all the neighbours go out of business, their contribution in (10.3) is zero, leading to the exponential solution shown in Fig. 10.5(b). The wealth of the central trader increases markedly as more and more small traders are connected to him, making him an even wealthier survivor asymptotically.

We now use this observation to return a 'dying' trader to life. Figure 10.6 shows our results: the central trader would eventually have gone bankrupt in his original environment, but on adding 100 small traders (whose wealth $X < X_\star$) to his network, he comes back to solvency: further additions, e.g. 200 or 364 such traders, evidently

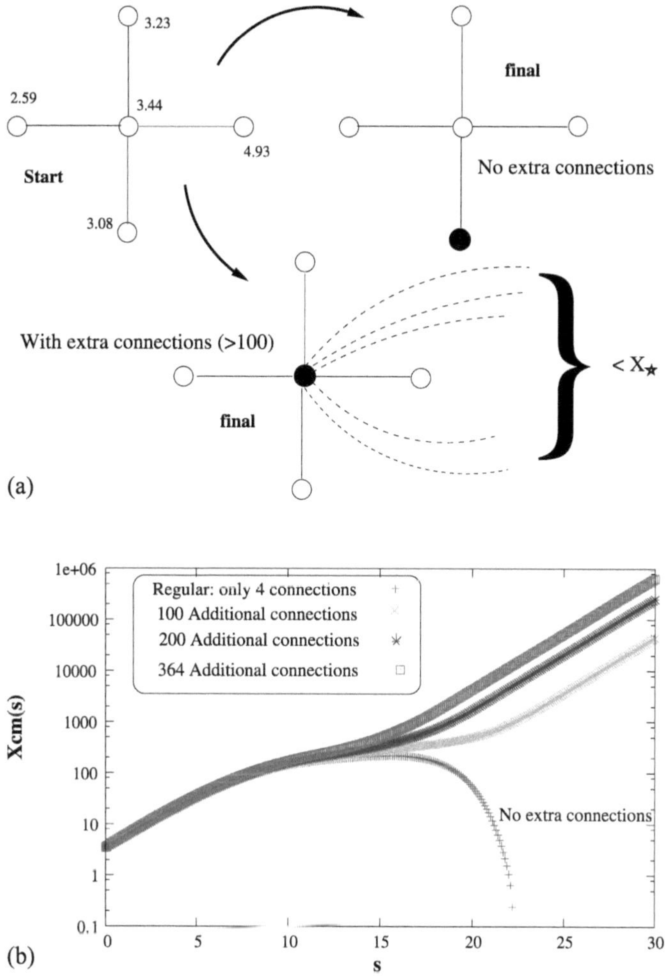


**Fig. 10.6** (**a**) The central trader goes bankrupt in his original configuration without extra connections. He becomes a survivor after linking up with more and more non-survivors. (**b**) A crossover is seen here as the central trader is returned to (financial) 'life', his wealth $X_{cm}$ increasing with increasing connections to non-surviving traders ($X < X_{\star}$) (Color figure online)

make him an even wealthier survivor (Fig. 10.6(b)). Thus, networking with eventual non-survivors is the surest way to invoke the Lazarus effect on a 'dying' trader.

### *10.4.2 Networking with Would-Be Survivors $X_i > X_{\star}$*

Choosing to network with traders whose intrinsic wealth is greater than $X_{\star}$ could turn out to be rather delicate. The financial lifespan of such traders will certainly exceed Stage I: and depending on their individual environments, they could either

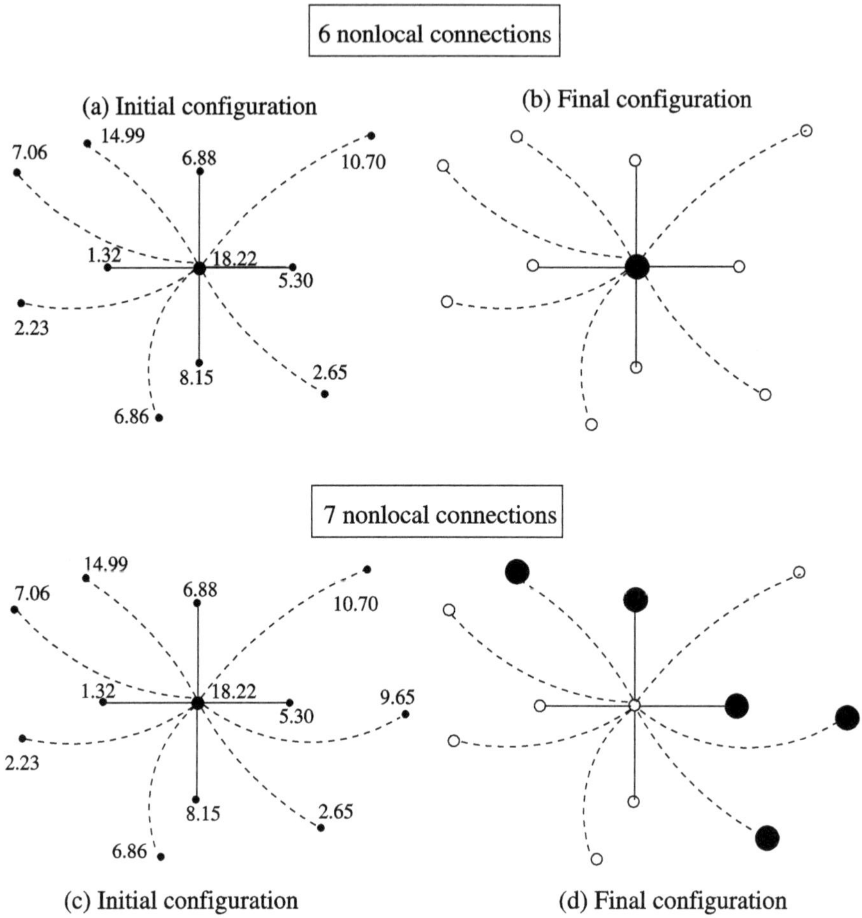


**Fig. 10.7** The central trader is networked with would-be survivors ($X_i > X_\star$) who are poorer than himself ($X_i < X_{cm}$). (**a**) The initial configuration, where the central trader has 6 non-local connections, (**b**) the asymptotic state with the only survivor being the central trader. (**c**) With the linkage of one more non-local trader to the existing configuration (**d**) the central trader goes bankrupt. In (**b**) and (**d**), the open *(dark) circles* represent asymptotic non-survivors (survivors)

survive through Stage II with a positive growth rate, or die as a result of a negative growth rate. Networking with such traders is like playing Russian roulette.

Consider first non-local connections with would-be survivors ($X_i > X_\star$) who are poorer than our chosen trader ($X_i < X_{cm}$). Such would-be survivors will live beyond Stage I, their wealth showing at least initially a positive growth rate (Fig. 10.1). From a mean field perspective, we would therefore expect to see a decrease in the chances of survival of the chosen trader as it networks with more and more would-be survivors.

Figure 10.7 shows a sample scenario, where the central trader is connected non-locally with would-be survivors who are poorer than himself. In Fig. 10.7(a) the cen-

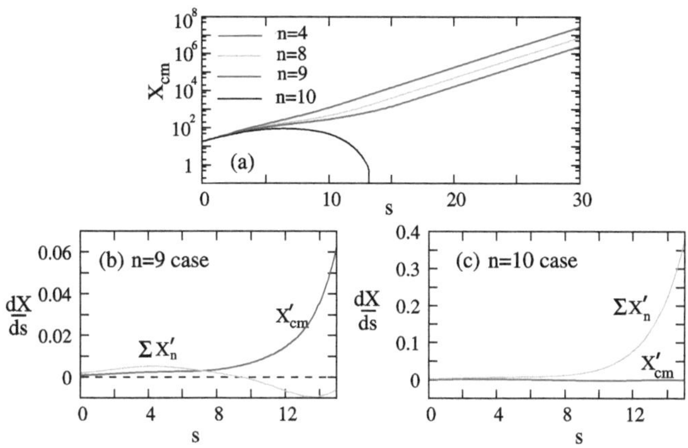


**Fig. 10.8** The central trader is networked with traders whose $X_n > X_\star$. In (**a**) the growth of $X_{cm}$ with an increasing number of connections is plotted—$n = 4$ corresponds to a regular lattice. There is a crossover seen when the number of connections increases from $n = 10$ to $n = 11$; for larger $n$ values, the central site goes bankrupt. This observation is supported by the rates of growth of $X_{cm}$ and its neighbours $X_n$ (**b**) when $n = 10$ and (**c**) when $n = 11$ (Color figure online)

tral trader networks with 6 would-be survivors and is able to survive asymptotically (Fig. 10.7(b)). On the other hand, adding one more would-be survivor (Fig. 10.7(c)) to the existing network of the central trader, causes him to go out of business at long times (Fig. 10.7(d)). As the central trader is made bankrupt by the arrival of the new connection, the fates of some of his other links are also changed (cf. Figs. 10.7(b) and (d)).

To understand the dynamics, we present the rates of growth for another sample scenario in Fig. 10.8. An increase in the number of non-local connections with would-be survivors, leads to a fall in the absolute value of $X_{cm}$ as well as its rate of growth $X'_{cm}$. Beyond a certain number of networked contacts, the wealth of the central trader begins to decay, and eventually vanishes. This crossover from life to death happens when the cumulative rate of the wealth growth of the neighbours $\Sigma X'_{i,j}$ is larger than that of the central trader $X'_{cm}$. Unfortunately, however, the intricate many-body nature of this problem precludes a prediction of when such crossovers might occur in general.

Finally, in the case where a given trader networks with would-be survivors who are richer than himself ($X_i > X_\star$ and $X_i > X_{cm}$), one would expect a speedier 'death'. One such sample scenario is depicted in Fig. 10.9 and the corresponding rates of evolution of the traders' wealth in Fig. 10.10. We notice that in his original configuration with four neighbours ($n = 4$), the central trader is a survivor. As we increase the number of networked connections, the growth rate of his wealth gets stunted; there is a substantial fall for 2 extra links ($n = 6$ in Fig. 10.10). Adding one more link ($n = 7$) does the final damage; the central trader goes bankrupt. The rates shown in Fig. 10.10(b) and (c) for $n = 6$ and $n = 7$ connections vividly capture the competition for survival, leading to solvency in one case and bankruptcy in the other.

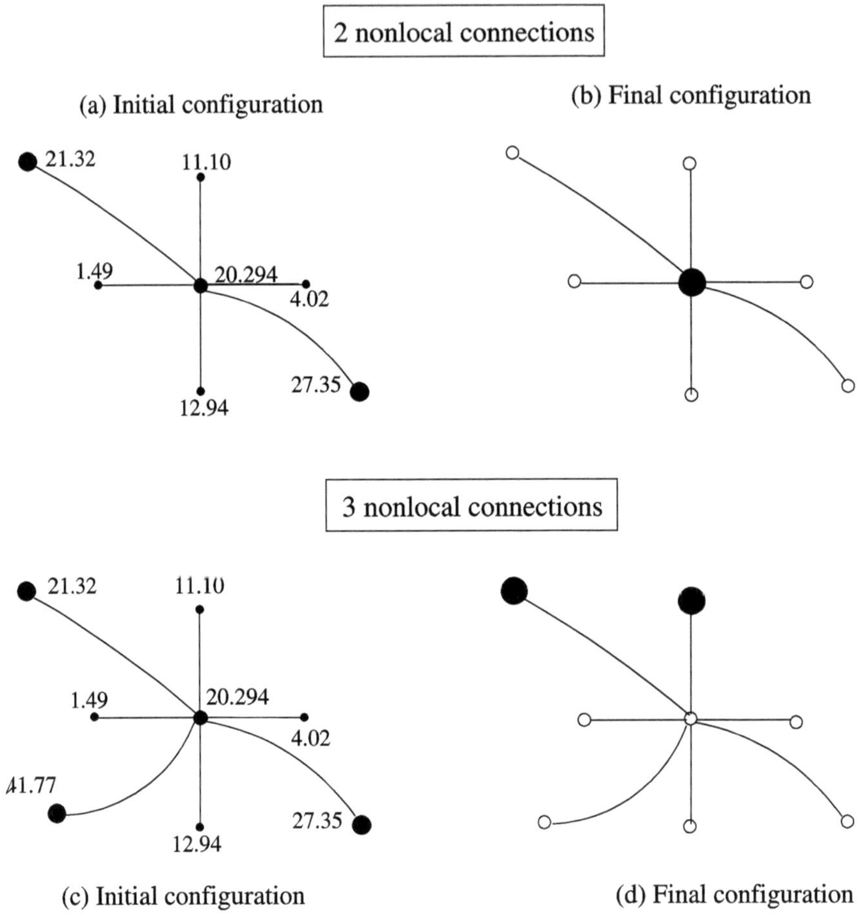


**Fig. 10.9** The central trader is networked with would-be survivors ($X_i > X_\star$) that are wealthier than himself ($X_i > X_{cm}$). (**a**) The initial configuration with 2 non-local connections, (**b**) and its asymptotic state; the only survivor here is the central trader. (**c**) With the addition of one more non-local trader to the existing configuration (**d**) the central trader goes bankrupt. In (**b**) and (**d**), the *open* (*dark*) *circles* represent asymptotic non-survivors (survivors)

As expected, we observe that fewer connections (here, $n = 7$) are needed, compared to the earlier case with smaller would-be survivors ($n = 11$), to eliminate the chosen trader. In closing, we should of course emphasise that the $n$ values mentioned here are illustrative.

## 10.5 Survivor Distributions and Rare Events

We have seen that the safest strategy for the Lazarus effect is to network with eventual non-survivors, i.e. those who will never get past Stage I. It is also relatively safe

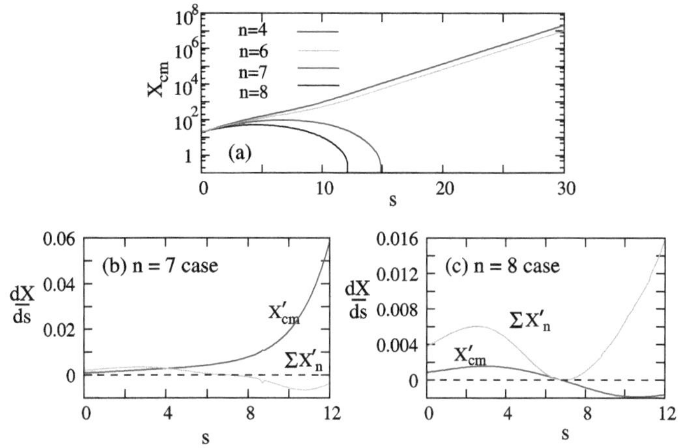


**Fig. 10.10** The central trader of a configuration given in Fig. 10.9(b) is connected with non-local traders with $X_n > X_{cm}$. In (**a**) the growth of $X_{cm}$ with an increasing number of connections is plotted—$n = 4$ corresponds to a regular lattice. There is a crossover seen when the number of connections increases from $n = 6$ to $n = 7$; for larger $n$ values, the central trader goes bankrupt. This observation is supported by the rates of growth of the wealth of the central trader, $X_{cm}$, and that of its neighbours $X_n$, (**b**) when $n = 6$ and (**c**) when $n = 7$ (Color figure online)

to network with would-be survivors, provided they are poorer than oneself. (This would explain why multinationals are not afraid to enter an arena where smaller-size retailers predominate, for example.) However, in this section we consider rare events, where the wealthiest trader in a given neighbourhood dies marginally, and a poorer one survives against the odds.

We look first at the four immediate neighbours of a given trader, and consider their pairwise interactions with him. Clearly, had such a pair been isolated, the larger trader would have won [4]. However, many-body interactions in the lattice mean that this is not always true. We therefore ask the question: what is the proportion of cases where the poorer trader wins?

Each survivor has four neighbours; we first calculate the probability distribution of the initial wealth differences in a pairwise fashion between a survivor and each of his neighbours. The initial wealth differences are given by $\delta X_i = X_{cm} - X_i$ ($i = 1, 2, 3, 4$) corresponding to the four neighbours—right, left, bottom and top—of a survivor. The distribution of $\delta X_i$ for all the survivors is shown in Fig. 10.11. Here, a negative $\delta X_i$ means that the survivor is poorer than his neighbour, and conversely for positive $\delta X_i$. All four distributions corresponding to four neighbouring pairs overlap due to isotropy; the resulting distributions are universal functions of *wealth differences*, depending only on $\mu$.

We also obtain the cumulative wealth difference between survivors and all of their four neighbours viz. $4X_{cm} - \sum_{i=1}^{4} X_i = \sum_{i=1}^{4} \delta X_i$ (see Fig. 10.12). The distributions of $\sum_{i=1}^{4} \delta_i$ are plotted in Fig. 10.12 for different values of $\mu$. For a positive cumulative wealth difference we know that the survivor is richer than his neighbours, matching our intuition based on the mean-field regime. The negative side of

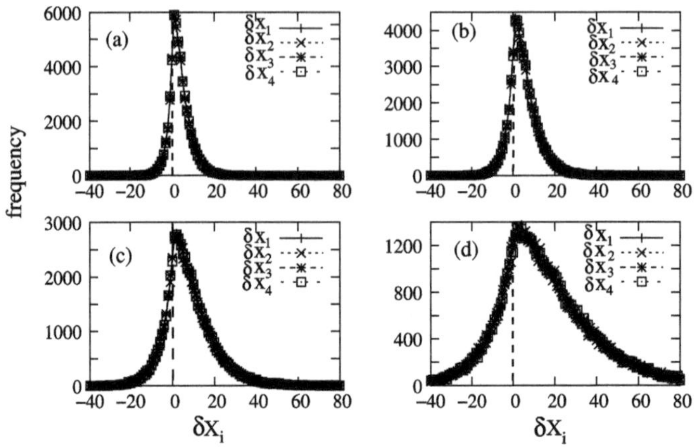


**Fig. 10.11** The plots show the distribution of pairwise wealth differences between survivors and their four neighbours. They are obtained for exponential distributions of initial wealth with different mean values $1/\mu$, where $\mu = -\log(S_1)/X_\star$. The plots are for (**a**) $1/\mu = 3.92$ ($S_1 = 0.6$) (**b**) 5.607 ($S_1 = 0.7$) (**c**) 8.963 ($S_1 = 0.8$) and (**d**) 18.982 ($S_1 = 0.9$). The system size is $400 \times 400$

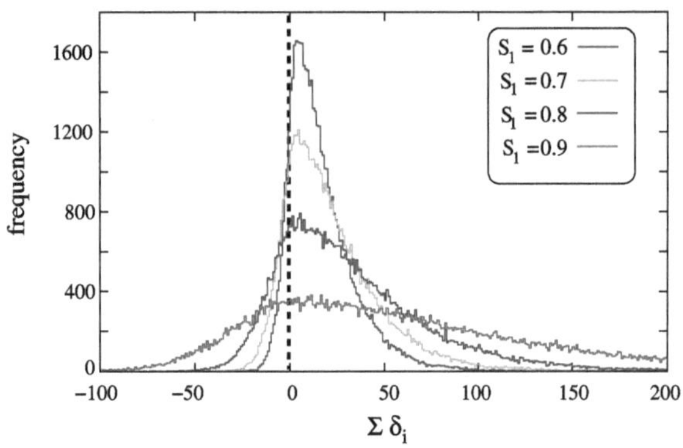


**Fig. 10.12** The plots show the distribution of wealth differences between survivors and all of their four neighbours. They are obtained for exponential distributions of initial wealth with different mean values $1/\mu$, where $\mu = -\log(S_1)/X_\star$. The plot in '*red*' represents the case for $1/\mu = 3.92$ ($S_1 = 0.6$), '*green*' represents $1/\mu = 5.607$ ($S_1 = 0.7$), '*blue*' represents $1/\mu = 8.963$ ($S_1 = 0.8$), and '*pink*' represents $1/\mu = 18.982$ ($S_1 = 0.9$). The system size is $400 \times 400$ (Color figure online)

the distribution is more interesting, *comprising traders who are poorer than their four neighbours combined, and who have won against the odds.*

Notice that both the survivor-neighbour pair distribution (Fig. 10.11), and the survivor—all neighbours cumulative distribution (Fig. 10.12) get broader with increasing $\mu$. This is because increasing $\mu = -\log(S_1)/X_\star$ [4] increases the number of potential survivors $S_1$ beyond Stage 1. In each case, the fraction of area under the negative side of the survivor pair-distribution gives an estimate of survivors against

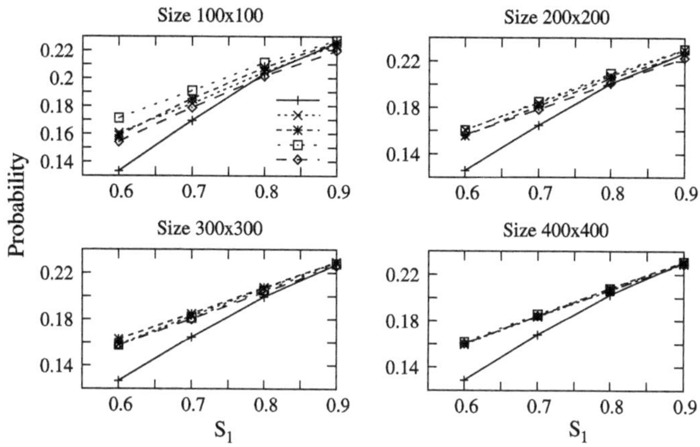


**Fig. 10.13** The plots show the probability of finding a poorer survivor, who has won against the odds. The probability increases with increase in $S_1$ (refer to Figs. 10.11 and 10.12). While the *solid line* shows the combined cumulative probabilities (CCP) (from Fig. 10.12), the rest of the lines represent the individual cumulative probabilities (from Fig. 10.11)

the odds—an example of some of the rare events alluded to at the beginning of this section.

Figure 10.13 shows this fraction, both in terms of individual survivor-pair distributions and cumulative distributions, as a function of the $\mu$ of the initial wealth distribution, for different system sizes. There are more survivors, hence more survivors *against* the odds, leading to an increase in the fraction plotted on the y-axis of Fig. 10.13 for both distributions. For the largest system size, there is full isotropy in the pairwise distributions; the probability of finding a survivor against the odds is now seen to be a *regular and universal function* of $\mu$ in both pairwise and cumulative cases, relying only on wealth differences rather than on wealth. Finally, the cumulative distribution gives a more stringent survival criterion than the pairwise one, as is to be expected from the global nature of the dynamics.

A major conclusion to be drawn from Fig. 10.13 is the following: there are traders who *are eliminated* against the odds (traders who are wealthier than the eventual survivor). These should be easier to revive (as they have failed marginally) by selective networking than those who have failed because they are indeed worse off. This question is of real economic relevance, and its mathematical resolution seems to us to be an important open problem.

## 10.6 Discussion

We have used the model of [4] to investigate two related issues in this paper on predatory trading [2]: first, that of systemic risk in the presence of increasing interactions, and next, the use of selective networking to prevent financial collapse. As long-range connections are introduced with probability $0 < p < 1$ to individual traders [10], we find that the qualitative features of the networked system remain the same as that of the regular case. The presence of two well-separated dynamical stages is retained, and the glassy dynamics and metastable states of [4] persist. However, the number of survivors decreases as expected with increasing $p$, quantitatively validating the thesis of [3]; and systemic risk is far greater as the complexity of interactions is increased. This view finds resonance with the present economic scenario, where it appears that some measure of insulation via economic firewalls, is needed to prevent individual, and hence eventually systemic, collapse.

Another central result of this paper is the use of smart networking strategies to modify the fate of an arbitrary trader. We find that it is safest to network with eventual non-survivors; their decay and eventual death lead to the transformation of the destiny of a given site, from bankruptcy to solvency, or from solvency to greater solvency. Networking with peers, or with those who are born richer, in general leads to the weakening of one's own finances, and an almost inevitable bankruptcy, given a predatory scenario.

However, the above is not immutable: the probability distributions in the last section of the paper indicate an interesting possibility of survival 'against the odds'. It would be interesting to find a predictive way of financial networking that would enable such a phenomenon to occur both at the individual, and at the societal, level.

**Acknowledgements** AM gratefully acknowledges the input of JM Luck, AS Majumdar and NN Thyagu to this manuscript.

## References

1. Brunnermeier MK, Pedersen LH (2005) Predatory trading. J Finance LX:1825–1864
2. Mehta A, Majumdar AS, Luck JM (2005) How the rich get richer. In: Chatterjee A et al (eds) Econophysics of wealth distributions. Springer-Verlag, Italia, pp 199–204
3. Haldane AG, May RM (2011) Systemic risk in banking ecosystems. Nature 469:351–355
4. Luck JM, Mehta A (2005) A deterministic model of competitive cluster growth: glassy dynamics, metastability and pattern formation. Eur Phys J B 44:79–92
5. Watts DJ, Strogatz SH (1998) Collective dynamics of small-world networks. Nature 393:440–442
6. Barabasi AL, Albert R (1999) Emergence of scaling in random networks. Science 286:509–512
7. Albert R, Barabasi AL (2002) Statistical mechanics of complex networks. Rev Mod Phys 74:47–97
8. Newman M, Barabasi AL, Watts DJ (2006) The structure and dynamics of networks. Princeton University Press, Princeton
9. Amaral LAN et al (2000) Classes of small-world networks. Proc Natl Acad Sci 97:11149–11152
10. Thyagu NN, Mehta A (2011) Competitive cluster growth on networks: complex dynamics and survival strategies. Physica A 390:1458–1473
11. Majumdar AS (2003) Domination of black hole accretion in brane cosmology. Phys Rev Lett 90:031303
12. Mezard M, Parisi G, Virasoro M (1987) Spin glass theory and beyond. World Scientific, Singapore
13. Lagarde C (2012) At the World Economic Forum, Davos
14. Luke 16:19–31

# Chapter 11
# Statistical Mechanics of Labor Markets

**He Chen and Jun-ichi Inoue**

**Abstract** We introduce a probabilistic model of labor markets for university graduates, in particular, in Japan. To make a model of the market efficiently, we take into account several hypotheses. Namely, each company fixes the (business year independent) number of opening positions for newcomers. The ability of gathering newcomers depends on the result of job matching process in past business years. This fact means that the ability of the company is weaken if the company did not make their quota or the company gathered applicants too much over the quota. All university graduates who are looking for their jobs can access the public information about the ranking of companies. Assuming the above essential key points, we construct the local energy function of each company and describe the probability that an arbitrary company gets students at each business year by a Boltzmann-Gibbs distribution. We evaluate the relevant physical quantities such as the employment rate. We find that the system undergoes a sort of 'phase transition' from the 'good employment phase' to 'poor employment phase' when one controls the degree of importance for the ranking.

## 11.1 Introduction

Deterioration of the employment rate is now one of the most serious problems in Japan and various attempts to overcome these difficulties have been done by central or local governments. Apparently, labor (work) is important not only for each of us to earn our daily bread, but also for our state to keep the revenues by collecting the taxes from labors. Especially, in recent Japan, the employment rate is getting worse and the government has over-issued quite a lot of government bonds to compensate

H. Chen · J. Inoue
Graduate School of Information Science and Technology, Hokkaido University, N14-W-9, Kita-ku, Sapporo 060-0814, Japan

H. Chen
e-mail: chen@complex.ist.hokudai.ac.jp

J. Inoue (✉)
e-mail: jinoue@cb4.so-net.ne.jp

F. Abergel et al. (eds.), *Econophysics of Systemic Risk and Network Dynamics*, New Economic Windows,
DOI 10.1007/978-88-470-2553-0_11, 

a lack of the tax revenues and the national debt (amount to $\sim 6 \times 10^{15}$ Japanese yen or more!) is now becoming a serious risk to cause a national-wide bankruptcy.

To make the matter worse, the earthquake and tsunami hit the northeast coast on 11th March 2011, and as the result, Fukushima nuclear power plant was seriously damaged and people living in that area has taken refuge from the nuclear radiation. These unpredictable disasters caused by nature and human error have made our country in financial difficulties that we have never encountered before. Many people lost their jobs and a lot of companies and plants could not be maintained. Hence, the reconstruction (improvement) of working condition for the labors and companies is now not something that can be ignored.

To consider the effective policy and to carry out it for sweeping away the unemployment uncertainty, it seems that we should investigate the labor markets scientifically and if it is possible, one should simulate artificial labor markets in personal computer to reveal the essential features of the problem. In fact, in macroeconomics (labor science), there exist a lot of effective attempts to discuss the macroscopic properties [1–6]. However, apparently, the macroscopic approaches lack of their microscopic view point, namely, in their arguments, the behavior of microscopic agents such as job seekers or companies are neglected.

Taking this fact in mind, in this paper, we shall propose a simple probabilistic model based on the concept of statistical mechanics for stochastic labor markets, in particular, Japanese labor markets for university (college) graduates.

## 11.2 Empirical Evidence: The Philips Curve

Let us first mention the relationship between the unemployment and the inflation rates in recent Japan (from 1970s to 2000s) in empirical evidence of the labor market. This relationship is generally called as *Philips curve*. In the original paper of Philips [7], he found the relation for the empirical data set from the middle of nineteen century to the beginning of twenty century (1861–1913) in UK. However, up to now, a lot of verifications have been done for various data sets in various countries. Therefore, we are confirmed that the Philips curve should be regarded as one of the 'universal properties' in labor markets.

In Fig. 11.1, we plot the Philips curves of our country in 1980s, 90s and 2000s and the age-dependence of the curve (the lower-right panel). In these panel, the fitting curve (the solid curve): $\pi + b \propto U^{-c}$ is actually obtained by the least square estimation for the parameter $b$ and $c$ from the $n$-data points $(X_i, Y_i) = (\log U_i, \log(\pi_i + b)), i = 1, \ldots, n$. From the lower-right panel in Fig. 11.1, we clearly find that the curve is slowly getting 'flat'.

From these empirical findings, we easily notice that the curve changes (evolves) in time and the behavior might be dependent on the situation of economy in the country. Therefore, it is important for us to reveal the dependence from the microscopic point of view and it is our motivation to deal with the problem in this paper.

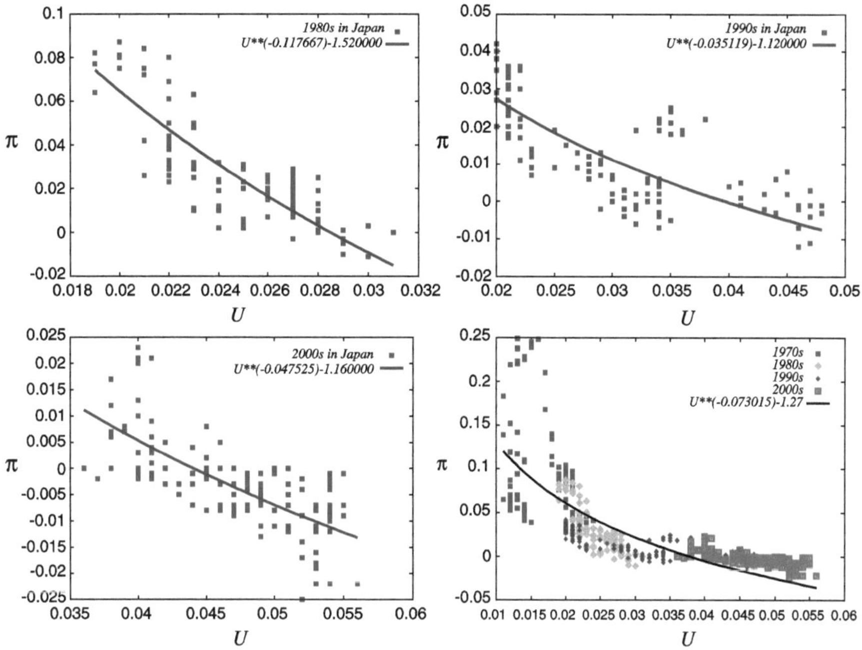


**Fig. 11.1** The Philips curves in Japan. From the *upper left* to the *lower right*, the curves in 1980s, 90s and 2000s are plotted. In the *lower-right panel*, Philips curve in Japan (from 1970s to 2000s) is shown by a *single plot*. We clearly find that the Philips curve is getting 'flat'

## 11.3 Hypotheses in Modeling

We first discuss several basic properties should be satisfied in our probabilistic modeling. As we already mentioned, in our modeling, the job seekers are restricted to university graduates and the other persons on the job searching are neglected for simplicity. Strictly speaking, this assumption should not be justified, however, as is well known, Japanese society is still somewhat conservative and once labors get their jobs, they stay in the company by their retirement age. In this sense, we omit the contribution from the on the job searching persons to the labor market because the fraction at the present time might be negligibly small.

Based on the above general 'hypotheses', we assume that the following three points (i)–(iii) should be taken into account to construct the effective labor markets.

(i) Each company recruits constant numbers of newcomers in each business year.
(ii) If the company takes too much or too less applications which are far beyond or far below the quota, the ability of the company to gather applicants in the next business year decreases.
(iii) Each company is apparently ranked according to various perspectives such as the stability, the wage-level, the location, the welfare of employees, etc. The ranking is useful information and it is available for all students.

In following, we shall attempt to construct labor markets by considering the above three essential points. Our model system is a variant of urn models [1, 8] or the so-called *Kolkata Paise Restaurant Problem (KPRP)* [9, 10].

## 11.4 System Parameters

First of all, in this section, we define the system parameters and their sizes such as the number of companies, open positions (vacancies) and applicants. Let us define the total number of companies as $K$ and each of them is distinguished by the label $k = 1, 2, \ldots, K$. Then, the number of the quota of the company $k$ is specified by $v_k^*$. In real labor markets, the quota $v_k^*$ itself fluctuates in time (business year) and it changes according to the gross margin in the previous year and some companies in financial difficulties might decrease the quota. However, in this paper, we fix the value and regard the quota as a 'uniform' and 'time-independent' variable.

Hence, the total job vacancy in society in each business year $V$ is given by

$$V = \sum_{k=1}^{K} v_k^*. \tag{11.1}$$

On the other hand, we define the number of new university graduates by $N$ and each of the students is specified by the index $i$ as $i = 1, 2, \ldots, N$. Then, we introduce

$$\alpha \equiv \frac{V}{N} \tag{11.2}$$

as *job offer ratio* and it is independent of system size for $\mathcal{O}(V) = \mathcal{O}(N)$. Apparently, for $\alpha = V/N > 1$, that is $V > N$, the labor market behaves as a 'seller's market', whereas for $\alpha < 1$, the market becomes a 'buyer's market'. For this model system, we might assume that each student post his/her single application (what we call 'entry sheet' or CV) to the company. In our analysis given below, the relevant system parameter is job offer ratio $\alpha$ rather than $V$ or $N$ because obviously the $\alpha = V/N$ is system size independent.

## 11.5 The Local Energy Function: A Link to Physics

Here we define a sort of 'local energy function' for each company which represents the ability of gathering applicants in each business year $t$. The energy function is a nice bridge to link the labor market to physics. Let us first define the following *local mismatch measurement*: $h_k(t)$ for each company $k(= 1, 2, \ldots, K)$ as

$$h_k(t) = \frac{1}{V}\left|v_k^* - v_k(t)\right| = \frac{1}{\alpha N}\left|v_k^* - v_k(t)\right| \tag{11.3}$$

where $v_k(t)$ denotes the number of students who seek for the position in the company $k$ at the business year $t$ (they will post their own 'entry sheet (CV)' to the company $k$). Hence, the local mismatch measurement $h_k(t)$ is the difference between the number of applicants $v_k(t)$ and the quota $v_k^*$. We should keep in mind that from the fact (i) mentioned before, the $v_k^*$ is a business year $t$-independent constant. Some analysts reported that the increase of unemployment rate in recent Japanese labor market is due to the mismatch between the university graduates and companies. Namely, most of students look for the positions in famous and already established large companies and they do not want to work for a small business in, say, like fostering venture businesses. Such a sort of 'local mismatch' could be quantified by $h_k(t)$.

On the other hand, we define the ranking of the company $k$ by $\epsilon_k (> 1)$ which is independent of the business year $t$. Here we assume that the ranking of the company $k$ is higher if the value of $\epsilon_k$ is larger. In this paper, we simply set the value as

$$\epsilon_k = 1 + \frac{k}{K}. \tag{11.4}$$

Namely, the company $k = K$ is the highest ranking company, whereas the company $k = 1$ is the lowest.

Form the above set-up and on the analogy of the Boltzmann-Gibbs distribution in conventional statistical mechanics, we define the probability $P_k(t)$ that the company $k$ gathers their applicants at time $t$ as

$$P_k(t) = \frac{\epsilon_k}{Z}\exp\big[-H_k(\boldsymbol{\beta}_k, \boldsymbol{h}_k)\big] \equiv \frac{\exp[-E(\epsilon_k, H_k(\boldsymbol{\beta}_k, \boldsymbol{h}_k))]}{Z}, \tag{11.5}$$

$$Z \equiv \sum_{k=1}^{K} \exp\big[-E\big(\epsilon_k, H_k(\boldsymbol{\beta}_k, \boldsymbol{h}_k)\big)\big] \tag{11.6}$$

where we defined $Z$ as the normalization constant for probability (a sort of partition function in statistical physics). We also defined two $\tau$-dimensional *market history vectors* with the length $\tau$: $\boldsymbol{\beta}_k \equiv (\beta_k(t-1), \ldots, \beta_k(t-\tau))$ and $\boldsymbol{h}_k \equiv (h_k(t-1), \ldots, h_k(t-\tau))$. Then, $H_k$ appearing in the probability (11.5) is defined by the inner product of these two vectors as

$$H_k(\boldsymbol{\beta}_k, \boldsymbol{h}_k) \equiv \boldsymbol{\beta}_k \cdot \boldsymbol{h}_k. \tag{11.7}$$

We should notice that the above inner product choice for the expression of $H_k$ is just an example and one can easily extend (modify) the functional form to much more generalized one: $H_k = f(\boldsymbol{\beta}_k, \boldsymbol{h}_k)$ including (11.7) as a special case.

With the above definitions, the local energy function is written explicitly by

$$E\big(\epsilon_k, H_k(\boldsymbol{\beta}_k, \boldsymbol{h}_k)\big) \equiv -\gamma \log \epsilon_k + \sum_{l=1}^{\tau} \beta_l h_k(t-l). \tag{11.8}$$

In this paper, we simply choose a particular market history vector as $\boldsymbol{\beta}_k = (\beta, 0, \ldots, 0)$.

Thus, the local energy function (11.8) is now simplified as

$$E\big(\epsilon_k, h_k(t-1)\big) = -\gamma \log \epsilon_k + \beta h_k(t-1). \tag{11.9}$$

The parameters $\gamma > 0$ and $\beta$ specify the probability from the macroscopic point of view. Namely, the company $k$ having relatively small $h_k(t)$ can gather a lot of applicants in the next business year and the ability is controlled by the parameter $\beta$ (we used the fact (ii) which was mentioned in the previous section). On the other hand, the high ranked company can gather lots of applicants and the degree of the ability is specified by the parameter $\gamma$ (we used the fact (iii) which was mentioned in the previous section).

Thus, local energy function $E(\epsilon_k, h_k(t-1))$ is written in terms of the sum of these two independent factors. Therefore, the result of the previous business year $h_k(t)$ is much more important factor for $\gamma > \beta$ and the ranking becomes more essential for $\gamma < \beta$ to decrease the energy.

We should bear in mind that even if the highest ranking company $k = K$ gathers a lot of applicants as over the quota $v_K(t) \gg v_K^*$ at some year $t$, however, the second term appearing in the energy function $\sim -\beta|v_K^* - v_K(t)| \ll 1$ acts as the 'negative feedback' on the first ranking preference to decrease the probability that the company $K$ gathers the applicants at the next business year $t+1$.

## 11.6 Job Matching Process: Microscopic Quantities

We should notice that for the probability $P_k(t)$, each student $i$ decides to post their entry sheet to the company $k$ at time $t$ as

$$a_{ik}(t) = \begin{cases} 1 & \text{(with prob. } P_k(t)), \\ 0 & \text{(with prob. } 1 - P_k(t)) \end{cases} \tag{11.10}$$

where $a_{ik} = 1$ means that the labor $i (= 1, \ldots, N)$ post their entry sheet to the company $k$ and $a_{ik} = 0$ denotes that he/she does not. In this paper, we assume that each labor post their entry sheets $a$-times on average. In other wards, the company $k$ takes $aNP_k(t)$-entry sheets on average.

We can now evaluate how many acceptances are obtained by a student and let us define the number by $s_i$ for each student $i (= 1, \ldots, N)$. Then, we should notice that the number of acceptances for the student $i$ is defined by $s_i(t) = \sum_{k=1}^{K} s_{ik}(t)$ with

$$s_{ik}(t) = \begin{cases} \Theta(v_k^* - v_k(t))\delta_{a_{ik}(t),1} & \text{(with prob. 1)}, \\ \Theta(v_k(t) - v_k^*)\delta_{a_{ik}(t),1} & \text{(with prob. } v_k^*/v_k(t)) \end{cases} \tag{11.11}$$

where $\Theta(\cdots)$ denotes the step function and $\delta_{a,b}$ stands for the Kronecker delta. Thus, (11.11) means that the $s_{ik}$ takes 1 when the student $i$ posts the sheet to the company $k$ and the total number of sheets the company $k$ gathers does not exceed the quota $v_k^*$. On the other hand, the variable $s_{ik}$ also takes 1 with probability $v_k^*/v_k(t)$ even if $v_k(t) > v_k^*$ holds. In other words, for $v_k(t) > v_k^*$, the $v_k^*$ students are randomly selected as winners' from $v_k(t)$ candidates.

In following, we investigate statistical properties of these microscopic quantities by means of their distribution.

### 11.6.1 The Distribution of Physical Quantities

For the model introduced in the previous section, we evaluate the distribution of several microscopic physical quantities. To calculate the distribution numerically, we define the distribution of such quantity $A(t)$ by

$$P(A) = \lim_{T\to\infty} \frac{1}{T} \sum_{t=0}^{T-1} \delta_{A(t),A}. \tag{11.12}$$

Using the above definition, the distribution of the number of entry sheets $v_k(t)$ which the company $k(=1,\ldots,K)$ obtains is evaluated by substituting $A(t) = v_k(t)$ into the above definition (11.12) for finite system size $K, N(\gg 1)$. After recursively updating the (11.5), (11.6) for $T(\gg 1)$-times, we obtain the distribution of any micro- or macroscopic quantities through (11.12).

In Fig. 11.2(left), we show the distribution of the microscopic quantity $a_{ik}$. The system size is chosen as $N = 10000$, $K = 1000$ and fix the quota for each company so as to be a company-independent variable as $v_k^* = v \equiv 30$ (homogeneous quota) for simplicity. This choice leads to $\alpha = 3$. The time step to evaluate (11.12) is set to $T = 10^4$ and the ranking factor $\gamma$ is fixed as $\gamma = 1$. From this figure, we find that $a_{ik}$ is distributed around the average $a$. In our computer simulations to be given in the next sections, we use the $P(a_{ik})$ to determine the number of entry sheets posted by each student.

We next consider the distribution of the number of entry sheets obtained by company $k$. We plot the result in Fig. 11.2(right). From this figure, we clearly find that in the regime of market history actually works (the market history is switched on) $\beta = \gamma = 1$, the distribution has a single peak around the relatively large value of $v_k$. However, as one increases the ranking effect $\gamma$ as $\gamma \gg \beta$, the peak eventually moves to zero. The fraction (probability) of companies getting no entry sheet is easily evaluated as follows.

Obviously, the probability $P(v_k = \bar{v})$ follows a binomial distribution $P(v_k = \bar{v}) = {}_{aN}C_{\bar{v}} P_k^{\bar{v}}(1-P_k)^{aN-\bar{v}}$, where we neglected the time-dependent part in $P_k(t)$, that is $\sim -\beta|v_k^* - v_k(t)|$ for $\gamma \gg \beta$. Then, $P(v_k = 0)$ is written as $P(v_k = 0) = (1-P_k)^{aN} \simeq \exp(-aNP_k)$, where $P_k$ is roughly estimated in the thermodynamic limit as

$$P_k = \frac{(1+k/K)^\gamma}{\sum_{k=1}^{K}(1+k/K)^\gamma} \simeq \frac{(1+k/K)^\gamma}{\int_1^K (1+k/K)^\gamma dk} = \frac{(\gamma+1)}{K}\frac{(1+k/K)^\gamma}{2^{\gamma+1}-1}. \tag{11.13}$$

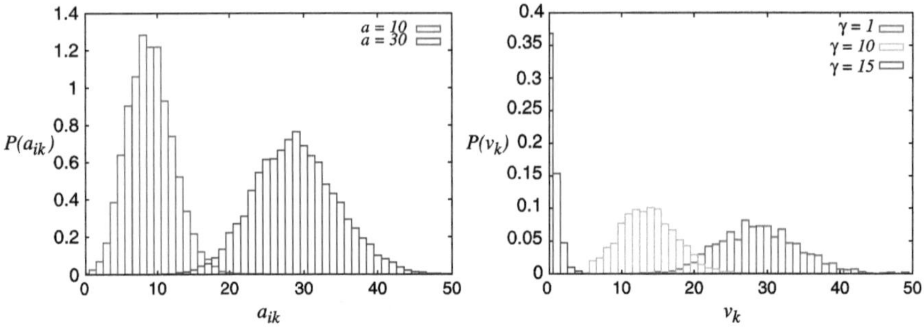


**Fig. 11.2** The distribution of $a_{ik}$ (*left*). In our simulations, we set $N = 10000$, $K = 1000$ and for simplicity $v_k^* = v = 30$, which reads $\alpha = 3$. The time step to evaluate (11.12) is set to $T = 10^4$. $\beta$ and $\gamma$ are fixed as $\beta = \gamma = 1$. We plot the $P(a_{ik})$ for $a = 10$ and 30. The *right panel* shows the distribution of the number of entry sheets the company $k$ obtains. In our simulations, we set $N = 10000$, $K = 1000$ and $v_k^* = v = 30$ leading to $\alpha = 3$. The time step to evaluate (11.12) is set to $T = 10^4$. We set $\beta = 1$ and change $\gamma$ as $\gamma = 1, 10$ and 15

Hence, in the limit of $K \propto N \to \infty$ (in other words, $K, N \to \infty$ keeping $\rho \equiv K/N = \mathcal{O}(1)$), the highest ranking company has $P_{k=K} \simeq (\gamma + 1)/2K$, whereas the lowest ranking company gets $P_{k=1} \simeq (\gamma + 1)/2^{\gamma+1}K$. Substituting these results into $P(v_k = 0)$, we immediately have

$$P(v_K = 0) = \exp\bigl[-aN(\gamma + 1)/2K\bigr], \tag{11.14}$$

$$P(v_1 = 0) = \exp\bigl[-aN(\gamma + 1)/2^{\gamma+1}K\bigr]. \tag{11.15}$$

These results imply that in the limit of $\gamma \to \infty$, the lowest ranking company gets no entry sheet with probability $P(v_1 = 0) = \exp(0) = 1$ from (11.14), whereas the highest ranking company always can get macroscopic order of entry sheet as $P(v_k = 0) = 0$ from (11.15). We easily notice that the argument for (11.15) should be valid for any companies which satisfy $k \ll K$. Therefore, for large $\gamma \gg 1$, macroscopic number of companies completely lose their applicants (the entry sheets) by this probabilistic nature. We can actually confirm this result indirectly from Fig. 11.2(right) as $P(v_k) = 0$ for $\gamma = 15$ which is relatively a large value in our simulations.

In following, we investigate the macroscopic behavior of this system through several physical quantities.

## 11.7 Unemployment Rate: Macroscopic Quantity

In the previous section, we modeled the microscopic matching process between students and companies by means of the probability distribution $P_k(t)$ of Boltzmann-Gibbs-type. As our main purpose is to reconstruct the macroscopic behavior of labor markets from the microscopic description, we should calculate the macroscopic quantities by means of the microscopic variables. To reconsider the results in the previous section from the macroscopic viewpoint, here we can calculate the unemployment rate $U$ as a function of $t$ as follows.

$$U_t = \frac{1}{N} \sum_{i=1}^{N} \delta_{s_i(t),0}. \tag{11.16}$$

Namely, the unemployment rate at business year $t$ is defined as a ratio of students who could not get any job ($s_i(t) = 0$) to the total number of students $N$.

### 11.7.1 The Order Parameter

To discuss the macroscopic quantity, we consider the long time average of $U_t$ as an 'order parameter' $U$ as usually used in statistical physics. Namely, we define the order parameter as follows.

$$U = \lim_{T \to \infty} \frac{1}{T} \sum_{t=0}^{T-1} U_t. \tag{11.17}$$

Here it should be noted that the above time average should be identical to the ensemble average $\langle U \rangle$, where the bracket $\langle \ldots \rangle$ stands for the average over the joint probability for the microscopic quantities $P(\boldsymbol{a}_1, \ldots, \boldsymbol{a}_K; s_1, \ldots, s_N)$ with $\boldsymbol{a}_k \equiv (a_{1k}, \ldots, a_{Nk})$, $a_{ik} \in \{0, 1\}$, $s_i \in \{0, 1, 2, \ldots, a\}$, $i = 1, \ldots, N$, $k = 1, \ldots, K$ when the system can reach the equilibrium state.

### 11.7.2 The Beveridge Curve

In Fig. 11.3(left), we plot the employment rate $1 - U$ as a function of $\alpha$ for several choices of $(\gamma, \beta) = (1, 1), (1, 5), (5, 1)$. From this figure, we certainly find that the employment rate is lower than $\sim 0.7$. In these numerical simulations, we assumed that each student posts their entry sheet just only once on average. However, if the number of posting increases the situation might be changed. Thus, we next consider the case in which each student posts their applications $a$ times on average. Therefore, we increases the number of entry sheets $a$ for each student to post to the market. We check the case $a = 1, 2$ and $a = 3$. The result is shown in Fig. 11.3(right). From this figure, as we expected, we find that the employment rate increases up to near 1 (perfect employment) when we increase the number $a$.

We should notice that the plots shown in Fig. 11.3(left) and Fig. 11.3(right) correspond to the so-called *Beveridge curve* in economics (labor science). Usually, the Beveridge curve is defined as the behavior of the number of vacancy $V$ against the

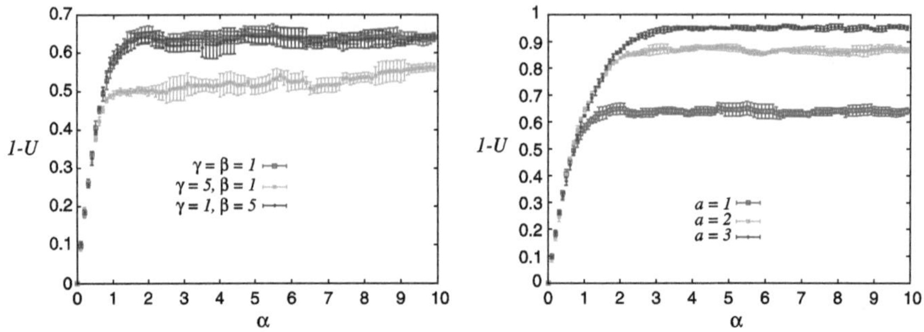


**Fig. 11.3** The employment rate $1-U$ as a function of $\alpha$ (*left*). We evaluated the rate for the cases $(\gamma, \beta) = (1, 1), (1, 5), (5, 1)$. In our simulations, we set the system sizes as $N = 500$, $K = 50$. The errorbars are calculated by five independent trials. The *right panel* shows the employment rate $1-U$ as a function of $\alpha$ for the case of $a = 1, 2$ and $a = 3$ keeping $\gamma = \beta = 1$

unemployment rate $U$ (hence, it is sometimes referred to as *UV curve*). However, as we already mentioned, in the thermodynamic limit $V, N \to \infty$, the relevant system parameter is job offer rate $\alpha$ rather than the number of vacancies $V$. Thus, we might regard the $\alpha$ as the effective number of vacancies. Hence, the $U$–$\alpha$ curves shown in these figures correspond to the conventional Beveridge curves.

## 11.8 Phase Transitions in Labor Markets

We next consider the $\gamma$-dependence of the employment rate for several values of $a$. In Fig. 11.4, we show the employment rate $1-U$ as a function of $\gamma$ for a fixed $\beta$ value ($\beta = 1$). The results are plotted for $a = 1, 2, 3$ and $a = 10$. The left panel is given for $\alpha = 1$, whereas the right panel is obtained for $\alpha = 10$. From this panel, as we expected, we find that the employment rate for a relatively high job offer ratio $\alpha = 10$ increases up to near 1 when we increase the number $a$.

One of the remarkable features of the results is existing a sort of 'phase transitions' in our probabilistic labor market. Namely, in the lower panel in Fig. 11.4, we clearly find that there exist two distinct phases, namely, 'perfect employment phase' ($1-U \simeq 1$) and 'perfect unemployment phase' ($1-U \simeq 0$, that is, $U \simeq 1$), and the system changes gradually from the perfect employment phase to the poor employment phase around $\gamma \simeq 10(\equiv \gamma_c)$. To evaluate the residual employment rate, we consider the extreme limit $\gamma \to \infty$ for the simplest case $\alpha = a = 1$. In this limit, all students want to post their own single entry sheet to the highest-ranking company $k = K$. As the result, only $v_K^* = v = 10$ students get their jobs and the residual employment rate is approximately evaluated as $1-U \simeq v/N = 10/500 = 0.02$.

From the viewpoint of labor markets, the phase transition might be understood as follows. For the case of a society with quite high job offer ratio such as $\alpha = 10$, the employment rate can be almost 1 when each labors can post $a = 10$ entry sheets on almost randomly to companies. However, as labors start to take into account

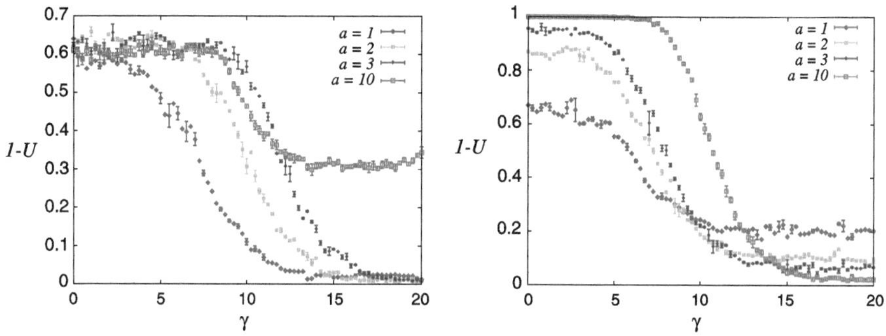


**Fig. 11.4** The employment rate $1 - U$ as a function of $\gamma$ for a fixed $\beta$ value ($\beta = 1$). The results are plotted for $a = 1, 2, 3$ and $a = 10$. The *left panel* is given for $\alpha = 1$, whereas the *right panel* is obtained for $\alpha = 10$

the ranking of company, namely, for $\gamma \simeq \gamma_c$, the employment rate is gradually decreased to the zero-level. Apparently, this result is induced due to the global mismatch between students and companies as observed in recent Japanese labor market for university graduates.

## 11.9 The Philips Curve

In the previous section, we have made a simple probabilistic model for job matching process between university graduates and companies in Japan. We evaluated the unemployment rate $U$ as an order parameter and found that the system undergoes a phase transition when one changes the system parameters.

We next consider the Philips curve for our labor market. To obtain the Philips curve, we should evaluate the inflation rate separately and it needs some information about the production process of the companies and consumption procedure by consumers. Moreover, the bargaining process of wages of labors for each company also should be taken into account to determine the inflation rate. In this paper, we shall use the macroscopic formula for the inflation rate given by Neugart [6]. Then, by making coupled equations with our result of unemployment rate, we attempt to draw the Philips curve.

### *11.9.1 Macroscopic Neugart Model*

In this subsection, we shall briefly explain the derivation of non-linear maps with respect to the unemployment and inflation rates according to Neugart [6].

We first defined the update rule for the unemployment rate as follows.

$$U_{t+1} = U_t + \xi(1 - U_t) - o_t U_t \tag{11.18}$$

where the second term in the right hand side denotes the contribution of labors who are fired (lost) their jobs at time $t$ and the ratio is controlled by a single parameter $\xi$. The third term means the contribution of labors who get their job at time $t$ and $o_t$ is a time-dependent rate. The $o_t$ is explicitly given by

$$o_t = \frac{J_s + \Gamma(m-\pi_t)}{U_t + d(1-U_t)} \tag{11.19}$$

where the denominator denotes the total amount of labors who seek for the jobs at time $t$ and $d(1-U_t)$ corresponds to the labors who are 'on the job searching'. On the other hand, in the numerator, $J_s$ denotes the time-independent number of job openings and the time-dependent part of job openings comes from the second term $\Gamma(m-\pi_t)$. In this term, $\pi_t$ denotes the inflation rate at time $t$ and $m$ stands for a constant growth rate for the value of money. Hence, (11.19) means that the probability getting jobs $o_t$ decreases when the number of labors who seek for jobs increases, and increases when the inflation rate is smaller than the growth rate for the value of money.

On the other hand, the inflation rate at time $t$ is written in terms of the 'expected inflation rate' $\pi_{e,t}$ as follows.

$$\pi_t = \frac{1}{\delta}\left(\pi_{e,t} + \frac{w_{b,t}-w_p}{w_p}\right) = \frac{1}{\delta}\left(\pi_{e,t} + \frac{\mu-(1-c_2)U_t}{1-\mu}\right) \tag{11.20}$$

where $\delta$ stands for a scaling factor and $w_{b,t}$ is bargaining wages and we naturally set $w_{b,t} = 1-(1-c_2)U_t$ with a constant $0 \le c_2 \le 1$. The justification of this choice depends on the validity of our assumption that the bargaining should go well when the unemployment rate is low. The union having enough number of labors can negotiate with management for wage increases well. The $w_p$ is a base wage and it is controlled a single parameter $\mu$ as $w_p = 1-\mu$.

Then, the expected inflation rate is updated by means of a linear combination of inflation rate $\pi_t$ and the expected inflation rate $\pi_{e,t}$ with $0 \le c_1 \le 1$ as follows.

$$\pi_{e,t+1} = c_1\pi_t + (1-c_1)\pi_{e,t}. \tag{11.21}$$

From (11.18), (11.19), (11.20) and (11.21), we obtain the following non-linear maps with respect to $U$ and $\pi$:

$$U_{t+1} = U_t + \xi(1-U_t) - U_t\frac{J_s+\Gamma(m-\pi_t)}{U_t+d(1-U_t)}, \tag{11.22}$$

$$\begin{aligned}\pi_{t+1} = {} & \frac{1}{\delta}\left(\frac{\mu}{1-\mu} + c_1\pi_t + (1-c_1)\left(\delta\pi_t - \frac{\mu-(1-c_2)U_t}{1-\mu}\right)\right) \\ & - \frac{1}{\delta}\left(\frac{1-c_2}{1-\mu}\left(U_t + \xi(1-U_t) - U_t\frac{J_s+\Gamma(m-\pi_t)}{U_t+d(1-U_t)}\right)\right).\end{aligned} \tag{11.23}$$

The fixed point of the above non-linear maps is easily obtained as $(U^*, \pi^*) = (\{\mu - m(\delta-1)(1-\mu)\}/(1-b), m)$. Then, according to Neugart [6], we set the value of $J_s$

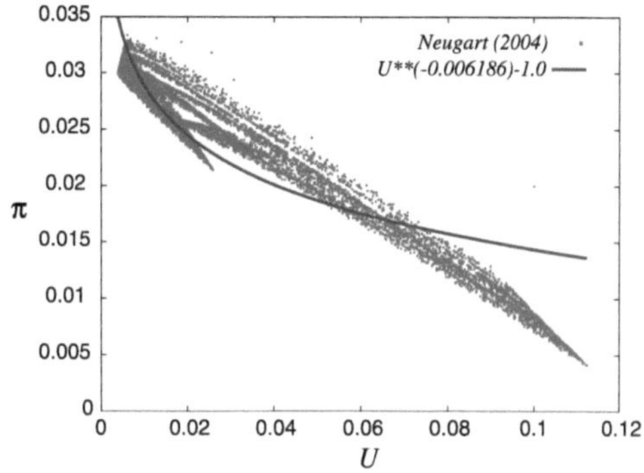


**Fig. 11.5** Philips curve obtained by the Neugart model [6]

in terms of the above fixed point, namely, by inserting the fixed point $U_{t+1} = U_t = U^*$, $\pi_{t+1} = \pi_t = \pi^*$ into (11.22), we obtain $J_s = J_s^* \equiv \xi(1 - U_*)(U^* + d(1 - U^*))/U^*$.

The chaotic attractor $(U_t, \pi_y)$ gives the Philips curve. In Fig. 11.5, we plot the Philips curve obtained by the set of parameters: $\xi = 0.18$, $d = 0.01$, $c_1 = c_2 = 0.5$, $\mu = 0.04$, $\Gamma = 0.5$, $\delta = 2$ and $m = 0.03$. From this figure, we observe that the chaotic attractor follows the scaling form: $\pi + 1.0 \propto U^{-0.006186}$.

## 11.9.2 Coupling with Our Probabilistic Model

As is shown in the previous sections, we can make a model for the probabilistic labor market which described by $a_{ik}(t)$: (11.10) and $s_i(t)$: (11.11) with $s_i(t) = \sum_k s_{ik}(t)$ microscopically. The behavior of the system is macroscopically written in terms of the time-dependence of unemployment rate $U_t$ or the order parameter $U$ which is defined as long-time average of the $U_t$ as $U = \lim_{T\to\infty}(1/T)\sum_{t=0}^{T-1} U_t$. To draw the Philips curve, here we consider the coupled equations for $U_t$ obtained in our model and $\pi_t$ which is one of the no-linear map in the Neugart model. Namely, here we use (11.23) with $J_s = J_s^*$ for the update rule for the inflation rate. The other parameters in the Neugart model are set to $\xi = 0.18$, $d = 0.01$, $c_1 = c_2 = 0.5$, $\mu = 0.04$, $\Gamma = 0.5$, $\delta = 2$ and $m = 0.03$.

### 11.9.2.1 Typical Dynamics

We first show the typical dynamics of the unemployment rate $U_t$ and inflation rate $\pi_t$ in Fig. 11.6. From this figure we find that both $U_t$ and $\pi_t$ are 'clustered' during the interval $100 \sim 200$. Within each interval, the these quantities behave periodically (oscillate).

Finally we plot the Philips curve as a trajectory $(U_t, \pi_t)$ of the dynamics. The results are shown in Fig. 11.7. We find that the negative correlation between $U_t$ and $\pi_t$ is actually observed for $\beta = 10$, $\gamma = 1$ and the curve is 'well-fitted' by the form: $\pi + 1.49 \propto U^{-0.54}$.

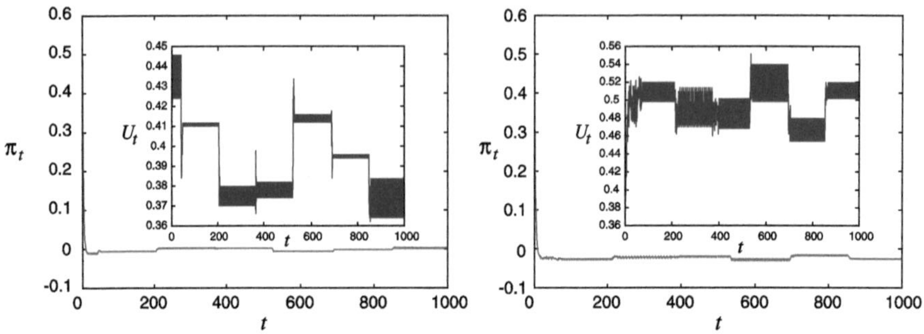


**Fig. 11.6** Typical dynamics of the unemployment rate $U_t$ and inflation rate $\pi_t$. The *upper panel* is given for $\beta = \gamma = 1$, whereas the *lower panel* is plotted for $\beta = 10$, $\gamma = 1$. We set $N = 500$, $K = 50$, $v_k^* = 10 (\alpha = 1)$ and $a = 10$. We find that both $U_t$ and $\pi_t$ are 'clustered' during the interval $100 \sim 200$. Within each interval of 'clustering', the these quantities behave periodically (oscillate)

**Fig. 11.7** The Philips curve for our probabilistic model for $\beta = 10$, $\gamma = 1$. We set $N = 500$, $K = 50$, $v = 10$ ($\alpha = 1$) and $a = 10$. The curve is 'well-fitted' the form: $\pi + 1.49 \propto U^{-0.54}$

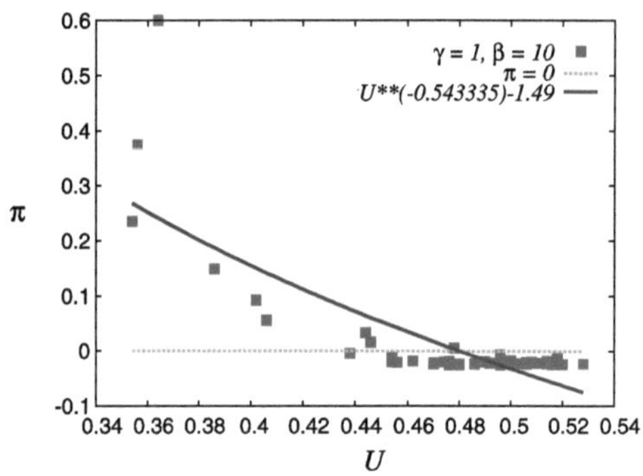


## 11.10 Summary

In this paper, on the basis of statistical physics, we proposed a minimal model to describe Japanese labor markets from the microscopic point of view. The model is definitely the simplest one and it might be possible for us to consider various extensions.

**Acknowledgements** We thank organizers of *Econophysics-Kolkata VI*, in particular, Frederic Abergel, Anirban Chakraborti, Asim K. Ghosh and Bikas K. Chakrabarti. We also thank Enrico Scalas, Giacomo Livan, Nobuyasu Ito, Koji Oishi, Takero Ibuki and Yu Chen for valuable discussion. This work was financially supported by Grant-in-Aid for Scientific Research (C) of Japan Society for the Promotion of Science, No. 22500195.

## References

1. Aoki M, Yoshikawa H (2006) Reconstructing macroeconomics: a perspective from statistical physics and combinatorial stochastic processes. Cambridge University Press, Cambridge

2. Boeri T, van Ours J (2008) The economics of imperfect labor markets. Princeton University Press, Princeton
3. Gabriele R (2002) Labor market dynamics and institution: an evolutionary approach. Working Paper in Laboratory of Economics and Management Sant'Anna School of Advances Studies. Pisa, Italy
4. Fagiolo G, Dosi G, Gabriele R (2004) Adv Complex Syst 7(2):157–186
5. Casares M, Moreno A, Vázquez J. (2010) An estimated new-Keynesian model with unemployment as excess supply of labor. Working paper. Universidad Pública de Navarra
6. Neugart M (2004) J Econ Behav Optim 53:193–213
7. Philips AW (1958) Economica, New Ser 25(100):283–299
8. Garibaldi U, Scalas E (2010) Finitary probabilistic methods in econophysics. Cambridge University Press, Cambridge
9. Chakrabarti BK (2007) Econophysics of markets and business networks. New economic windows series. Springer, Milan, pp 220–227
10. Ghosh A, Chatterjee A, Mitra M, Chakrabarti BK (2010) New J Phys 12:077033

# Chapter 12
# Kolkata Paise Restaurant Problem: An Introduction

**Asim Ghosh, Soumyajyoti Biswas, Arnab Chatterjee, Anindya Sundar Chakrabarti, Tapan Naskar, Manipushpak Mitra, and Bikas K. Chakrabarti**

**Abstract** We discuss several stochastic optimization strategies in games with many players having large number of choices (Kolkata Paise Restaurant Problem) and two choices (minority game problem). It is seen that a stochastic crowd avoiding strategy gives very efficient utilization in KPR problem. A slightly modified strategy in the minority game problem gives full utilization but the dynamics stops after reaching full efficiency, thereby making the utilization helpful for only about half of the population (those in minority). We further discuss the ways in which the dynamics may be continued and the utilization becomes effective for all the agents keeping fluctuation arbitrarily small.

A. Ghosh (✉) · S. Biswas · B.K. Chakrabarti
Theoretical Condensed Matter Physics Division, Saha Institute of Nuclear Physics, 1/AF Bidhannagar, Kolkata 700 064, India
e-mail: asimghosh066@gmail.com

S. Biswas
e-mail: soumyajyotibiswas@gmail.com

B.K. Chakrabarti
e-mail: bikask.chakrabarti@saha.ac.in

A. Chatterjee
Department of Biomedical Engineering and Computational Science (BECS), Aalto University School of Science, FI-00076 AALTO, Espoo, Finland
e-mail: arnab.chatterjee@aalto.fi

A.S. Chakrabarti
Economics Department, Boston University, 270 Bay State Road, Boston, MA, USA
e-mail: aschakrabarti@gmail.com

T. Naskar
Theory Division, Saha Institute of Nuclear Physics, 1/AF Bidhannagar, Kolkata 700 064, India
e-mail: tnaskar@gmail.com

M. Mitra · B.K. Chakrabarti
Economic Research Unit, Indian Statistical Institute, 203 B. T. Road, Kolkata 700 018, India

M. Mitra
e-mail: mmitra@isical.ac.in

F. Abergel et al. (eds.), *Econophysics of Systemic Risk and Network Dynamics*, New Economic Windows,
DOI 10.1007/978-88-470-2553-0_12, 

## 12.1 Introduction

The Kolkata Paise Restaurant (KPR) problem [1–5] is a repeated game, played between a large number ($N$) of agents having no interaction amongst themselves. In KPR problem, prospective customers (agents) choose from $n$ restaurants each evening simultaneously (in parallel); $N$ and $n$ are both large and fixed (typically $n = N$). Each restaurant has the same price for a meal (hence no budget constraint for the agents). It is assumed that each can serve only one customer any evening (generalization to a larger value is trivial). The information regarding the customer distributions for earlier evenings is available to everyone. If more than one customer arrives at any restaurant on any evening, one of them is randomly chosen (each of them are anonymously treated) and is served, while the rest do not get dinner that evening. An alternative visualization can be one in which multiple customers arriving in a single restaurant have to share the food meant for one customer, keeping all of them unhappy. The utilization fraction $\bar{f}$ in the problem is defined as the average fraction of restaurants which were visited by people any evening in the steady state. Each agent develops its own (parallel) algorithm to choose the restaurant every evening such that he/she is alone there. Also, the times required to converge/settle to such a solution (if exists), should be low (faster than, say, $\log N$). If the restaurants have different ranks which are agreed upon by all the agents, additional complications may arise.

*Paisa* is the smallest monetary unit in Indian currency, and the use of the word would essentially be synonymous with anything that is very cheap. In Kolkata, there used to be very cheap and fixed rate "Paise Restaurant" which were popular among the daily labourers. During lunch hours, the labourers used to walk (to save the transport costs) to one of these restaurants and would miss lunch if they got to a restaurant where there were too many customers. Walking down to the next restaurant would mean failing to report back to work on time! There were indeed some well-known rankings of these restaurants, as some of them would offer tastier items compared to the others. A more general example of such a problem would be when the society provides hospitals (and beds) in every locality but the local patients go to hospitals of better rank (commonly perceived) elsewhere, thereby competing with the local patients of those hospitals. Unavailability of treatment in time may be considered as a lack of service for those people and consequently as (social) wastage of service by those unvisited hospitals.

A dictator's solution to the KPR problem is the following: everyone is asked to form a queue and is assigned a restaurant with rank matching the sequence of the person in the queue on the first evening. Then each person is asked to go to the next ranked restaurant in the following evening, thus for the person in the last ranked restaurant this means going to the first ranked restaurant. This shift process continues for successive evenings, thus providing clearly the most efficient solution (with utilization fraction $\bar{f}$ of the services by the restaurants equal to unity) and the system arriving at this solution trivially and immediately (from the first evening itself). However, in reality this cannot be the true solution of the KPR problem, where each agent decides on his own (in parallel and democratically) every evening, based on complete information about past events. In this game, the customers try to evolve a learning strategy to eventually get dinners at the best possible ranked restaurant, avoiding the crowd. It is seen that these strategies take considerable time to converge and even after that the eventual utilization fraction $\bar{f}$ is far below unity.

## 12.2 Kolkata Paise Restaurant Problem

In this review, we will talk about the KPR problem where $N$ agents are parallelly visiting $n$ restaurants on every day [$n$, $N \to \infty$; keeping $n/N$ finite]. Each agent has been trying to get food from the best rank restaurants every day. But, each day, one agent can visit one restaurant and every restaurant has the capacity to serve food for one customer per evening. Therefore, as mentioned before, many agents go to a particular restaurant then one of the agents will be randomly chosen and will be served and the rest of the agents will not get dinner for that day, thus satisfying one of them. An alternative picture is one in which many customers have to share the food served for one customer, leaving all of them unsatisfied. Gencrally one can scc that a few of the restaurants are not visited by any of the agents on a particular evening and that many agents crowd in other restaurants and do not get dinner for the evening. The utilization fraction $\bar{f}$ in the problem is therefore given by the average fraction of restaurants which were visited by customers on any evening in the steady state.

We discuss the case where instead of deterministic strategies, if everyone follows stochastic strategies, then one gets not only to higher values of the utilization fraction, but also attains it in very small convergence time (usually of order $\log N$ or smaller).

In general in the KPR problem $n = gN$ and $N \to \infty$ and in its primitive version, $g = 1$ ($n = N$), while for general phase transition studies (see Sect. 12.3) one considers $g \leq 1$. For the Minority Game (see Sect. 12.4) $n = 2$ (with $N \to \infty$ as usual).

### 12.2.1 Random-Choice Case (Stochastic)

Suppose there are $N$ agents and $n$ restaurants. Any agent can select any restaurant with equal probability. Therefore, the probability that a single restaurant is chosen by $m$ agents is a Poisson distribution in the limit $N \to \infty$, $n \to \infty$:

$$\begin{aligned} \Delta(m) &= \binom{N}{m} p^m (1-p)^{N-m}; \quad p = \frac{1}{n} \\ &= \frac{(N/n)^m}{m!} \exp(-N/n) \quad \text{as } N \to \infty,\ n \to \infty. \end{aligned} \tag{12.1}$$

Therefore the fraction of restaurants not chosen by any agent is given by $\Delta(m=0) = \exp(-(N/n))$ and that implies that average fraction of restaurants occupied on any evening is given by [2]

$$\bar{f} = 1 - \exp(-N/n) \simeq 0.63 \tag{12.2}$$

for $n = N$ in the KPR problem.

### 12.2.2 Rank Dependent Strategies (Stochastic)

Let us now consider that all restaurants have a well defined rank (agreed by every agent) depending upon quality of food, services, etc. although price of a meal is same for all restaurants. Thus, all agents will try to get food from best rank restaurants. But since a restaurant can serve only one customer, it means that many of the agents in crowded restaurants will remain unsatisfied. Now, assume that any $k$th restaurant have rank $k$ and any agent choses that restaurant with probability $p_k(t) = k^\zeta / \sum k^\zeta$ (here $\zeta$ is any natural number). Here we discuss the results for such kind of strategy.

If an agent selects any restaurant with uniform probability $p$ then the probability that a single restaurant is chosen by $m$ agents is given by

$$\Delta(m) = \binom{N}{m} p^m (1-p)^{N-m}. \tag{12.3}$$

Therefore, the probability that a restaurant with rank $k$ is not chosen by any of the agents will be given by

$$\begin{aligned} \Delta_k(m=0) &= \binom{N}{0} (1-p_k)^N; \quad p_k = \frac{k^\zeta}{\sum k^\zeta} \\ &\simeq \exp\left(\frac{-k^\zeta N}{\widetilde{N}}\right) \quad \text{as } N \to \infty, \end{aligned} \tag{12.4}$$

where $\widetilde{N} = \sum_{k=1}^N k^\zeta \simeq \int_0^N k^\zeta \, dk = \frac{N^{\zeta+1}}{(\zeta+1)}$. Hence

$$\Delta_k(m=0) = \exp\left(-\frac{k^\zeta(\zeta+1)}{N^\zeta}\right). \tag{12.5}$$

Therefore the average fraction of agents getting dinner in the $k$th ranked restaurant is given by

$$\bar{f}_k = 1 - \Delta_k(m=0) \tag{12.6}$$

and the numerical estimates of $\bar{f}_k$ is shown in Fig. 12.1. Naturally for $\zeta = 0$, the problem corresponding to random choice $\bar{f}_k = 1 - e^{-1}$, giving $\bar{f} = \sum \bar{f}_k/N \simeq 0.63$ and for $\zeta = 1$, $\bar{f}_k = 1 - e^{-2k/N}$ giving $\bar{f} = \sum \bar{f}_k/N \simeq 0.57$.

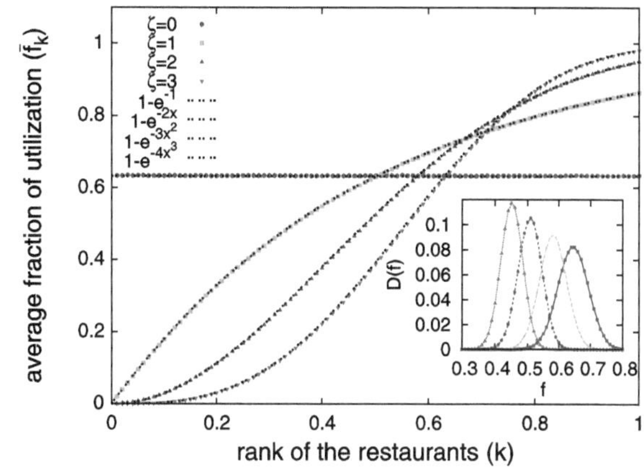


**Fig. 12.1** The *main figure* shows average fraction of utilization ($\bar{f}_k$) versus rank of the restaurants ($k$) for different $\zeta$ values. The *inset* shows the distribution $D(f = \sum \bar{f}_k/N)$ of the fraction $f$ agent getting dinner any evening for different $\zeta$ values. The simulations are done for $N = 10^4$ and $n = 10^4$. From [5]

### *12.2.3 Strict Crowd-Avoiding Case (Mixed)*

We consider the case (see [4, 5]) where each agent chooses on any evening ($t$) randomly among the restaurants in which nobody had gone in the last evening ($t - 1$). It was observed [5] that the distribution $D(f)$ of the fraction $f$ of utilized restaurants is again Gaussian with a most probable value at $\bar{f} \simeq 0.46$. The explanation was given in the following way: As the fraction $\bar{f}$ of restaurants visited by the agents in the last evening is avoided by the agents this evening, the number of available restaurants is $N(1 - \bar{f})$ for this evening and is chosen randomly by all the $N$ agents. Hence, it fits with (12.1) by considering $(N/n) = 1/(1 - \bar{f})$. Therefore, following (12.1),

$$(1 - \bar{f})\left[1 - \exp\left(-\frac{1}{1 - \bar{f}}\right)\right] = \bar{f}. \tag{12.7}$$

The solution of this equation gives $\bar{f} \simeq 0.46$.

### *12.2.4 Stochastic Crowd Avoiding Case*

Up to this point it is seen that indeed the random choice gives best utilization. Following a rank or strictly avoiding the crowd do not improve this fraction. While following a rank inherently prefers some restaurants and thereby making those crowed, the strict crowd avoidance on the other hand eliminates the possibility of a high utilization by not allowing repeated (successful) visits to a given restaurant.

However, in this section, we describe the following stochastic strategy: [5] If an agent goes to restaurant $k$ on an evening ($t - 1$) then the agent goes to the same restaurant next evening with probability $p_k(t) = \frac{1}{n_k(t-1)}$ where $n_k(t - 1)$ is the number of customers in $k$th restaurant on $t - 1$ day's evening or otherwise choose any

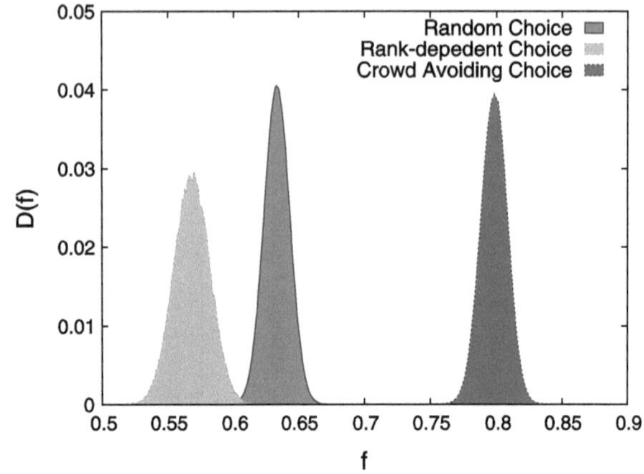


**Fig. 12.2** The figure shows that distribution of utilization fraction in different condition of the KPR problem. All simulation data are shown for $N = 10^4$ and $n = 10^4$

other restaurant $k'(\neq k)$ with uniform probability. In this process, the average utilization fraction is $\bar{f} \simeq 0.8$ in the steady state and the distribution $D(f)$ is a Gaussian with peak at $f \simeq 0.8$ (see Fig. 12.2).

An approximate estimate of $\bar{f}$ can be made using the following argument: Let $a_i$ denote the fraction of restaurants where exactly $i$ agents ($i = 0, \ldots, N$) appeared on any evening and assume that $a_i = 0$ for $i \geq 3$. Therefore, $a_0 + a_1 + a_2 = 1$, $a_1 + 2a_2 = 1$ and hence $a_0 = a_2$. Given this strategy, $a_2$ fraction of agents will attempt to leave their respective restaurants in the next evening $(t + 1)$, while no intrinsic activity will occur at the restaurants where, nobody came $(a_0)$ or only one came $(a_1)$ in the previous evening $(t)$. These $a_2$ fraction of agents will now get equally divided (each in the remaining $N - 1$ restaurants). Of these $a_2$, the fraction going to the vacant restaurants ($a_0$ in the earlier evening) is $a_0 a_2$. Hence the new fraction of vacant restaurants is now $a_0 - a_0 a_2$. In restaurants having exactly two agents ($a_2$ fraction in the last evening), some vacancy will be created due to this process, and this is equal to $\frac{a_2}{4} - a_2 \frac{a_2}{4}$. Steady state implies that $a_0 - a_0 a_2 + \frac{a_2}{4} - a_2 \frac{a_2}{4} = a_0$ and hence using $a_0 = a_2$ we get $a_0 = a_2 = 0.2$, giving $a_1 = 0.6$ and $\bar{f} = a_1 + a_2 = 0.8$. Of course, the above calculation is approximate as none of the restaurant is assumed to get more than two customers on any evening ($a_i = 0$ for $i \geq 3$). The advantage in assuming only $a_0$, $a_1$ and $a_2$ to be non vanishing on any evening is that the activity of redistribution on the next evening starts from this $a_2$ fraction of the restaurants. This of course affects $a_0$ and $a_1$ for the next evening and for steady state these changes must balance. The computer simulation results also conform that $a_i \leq 0.03$ for $i \geq 3$ and hence the above approximation does not lead to a serious error.

### *12.2.5 A General Study for Crowd Avoiding Case*

The stochastic crowd avoiding case can be generalized by modifying the probability of an agent to choose the same restaurant as the previous evening as $p_i(t) =$

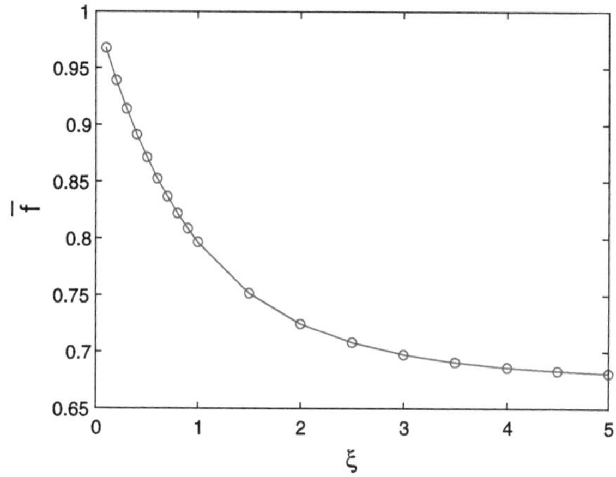


**Fig. 12.3** The figure shows the average utilization fraction ($\bar{f}$) for different values of $\xi$. All simulation data are shown for $N = 10^4$ and $n = 10^4$

$1/n_i^{\xi}(t-1)$ where $\xi$ is positive real number. Of course $\xi = 1$ is the case discussed in the previous section. It is observed (numerically) that the utilization fraction increases with decreasing $\xi$. However, the time to reach steady state value also increases. So, in this method we can reach a better utilization fraction as $\xi \to 0$ (Fig. 12.3). We observe, trivially, that the $\xi = 0$ case does not have any dynamics. On the other hand, the utilization fraction decreases to a limiting value ($\bar{f} \simeq 0.676$) for $\xi \to \infty$. The details of the critical behavior of this model will be reported elsewhere [6].

## 12.3 KPR and Phase Transition

Recently Ghosh et al. applied a stochastic crowd avoiding strategy in the KPR problem with considering $gN$ agents and $N$ number of restaurants [7]. It was observed that if the stochastic crowd avoiding strategy is applied to the problem then one can find out a particular value of $g = g_c$ below which all the agents are satisfied (and the state is called an absorbing or frozen state) and above the value of $g_c$, some of the agents will not be satisfied (and the state is called an active state). Therefore there is a phase transition between the an absorbing state and an active state with variation of $g$. The exponents of the transition in this process is well fitted with stochastic sandpile model.

### *12.3.1 The Models*

Consider $gN$ ($g < 1$) agents and the $N$ restaurants. It is reminded that a restaurant can serve only one agent in an evening. Suppose in any evening a particular restaurant ($i$th restaurant) is visited by $n_i$ agents and then one of the agents is chosen randomly and is served and rest $(n_i - 1)$ agents do not get any dinner for that day. Suppose all the agents are following the stochastic crowd avoiding dynamics mentioned before. Here two cases of the model are discussed (Model A & Model B). In model A, if any ($i$th) restaurant is visited by $n_i$ agents in any evening then in the next evening each of the $n_i$ agents will independently choose the same restaurant with probability $p = 1/n_i$ or a different restaurant otherwise with uniform probability. But in model B, if any ($i$th) restaurant is visited by $n_i$ $(n_i > 1)$ agents in any evening then in the next evening all agents will independently choose any of the restaurants with uniformly probability $(p = 1/N)$. If, however, $n_i = 1$ then the agent will stick to his/her choice in the next evening. In both the models, one can find a value of $g = g_c$ below which all the agents will be getting food and when $g > g_c$, some of agents will not be satisfied. The order parameter is given by the steady state density of active sites $\rho_a$ (density of sites having $n > 1$). So the absorbing phase corresponds to $\rho_a = 0$ $(g < g_c)$ whereas, for $g > g_c$ the steady state gives a non-zero value of the order parameter $(\rho_a > 0)$. Here the lattice versions (1D & 2D) models are also discussed.

### 12.3.2 Numerical Results

In this model one can see that below $g_c$ the order parameter $\rho_a$ goes to zero with time and above $g_c$, $\rho_a$ goes to a stationary non zero value with time. Now, it is known that the evolution of order parameter is an exponential form and can be expressed as

$$\rho_a(t) = \rho_a^0\left[1 - e^{-t/\tau}\right] \qquad (12.8)$$

for $g > g_c$, and

$$\rho_a(t) = \rho_a^0 e^{-t/\tau} \qquad (12.9)$$

for $g < g_c$, where $\tau$ in the above expressions represents the relaxation time in the system. Therefore, the order parameter asymptotically goes to steady state value with time. Now, near critical point the order parameter can be scaled as $\rho_a \sim (g - g_c)^\beta$ where $\beta$ is the order parameter exponent, similarly $\tau$ also scales as $\tau \sim (g - g_c)^{-\nu}$. A scaling form for $\rho_a$ can be written as

$$\rho_a \sim t^{-\alpha} F\left(\frac{t}{\tau}\right); \qquad \tau \sim (g - g_c)^{-\nu} \sim L^z, \qquad (12.10)$$

where $L$ denotes size of the system and $\alpha$, $z$ are dynamic exponents near critical point. For time $t \to \infty$, and using (12.8), (12.9) and (12.10) we get a scaling relation $\beta = \nu\alpha$. The exponents have been obtained by numerical simulations and the scaling relations are also discussed.

## 12.3.3 Model A

### 12.3.3.1 Mean Field Case

The model in its original form (as discussed so far) is mean-field (i.e. infinite range) type, in the sense that the excess agents from a restaurant can choose from all the remaining restaurant in the next evening and the geometrical distance was not an issue. In the mean filed case, the simulations are done by taking system size $L = 10^6$ and different scaling exponents are estimated (see Fig. 12.4). The simulation results suggest that $g_c = 0.7502 \pm 0.002$ and $\beta = 0.98 \pm 0.02$. Also doing the data collapse it has been shown $z = 0.50 \pm 0.01$, $\nu = 1.00 \pm 0.01$ and $\alpha = 1.00 \pm 0.01$. Therefore, the scaling relation $\beta = \nu\alpha$ is satisfied by the estimated exponents for this case.

### 12.3.3.2 Lattice Cases

This model was also studied for 1-d and 2-d lattices. In 1-d, by studying the dynamics in the lattice it is meant that the excess agents can only go to the nearest neighbor sites in the next step. For 1-d, lattice size $N = L = 10^4$ have been taken and averaging over $10^3$ initial conditions were permormed. For 2-d, a square lattice ($N = L^2$) with $L = 1000$ and averaging over $10^3$ initial conditions were considered. Periodic boundary condition have been employed in both cases.

1. The model is defined for 1-d as follows: The agents are allowed to hop only to their nearest neighbor restaurants, and each agent can choose either left or right neighbor randomly. It is found that $g_c = 1$ and hence the phase transition is not very interesting.
2. In the 2-d version of the model, a square lattice is considered and the agents are to choose one of the 4 nearest neighbors randomly in next evening. For $N = 1000 \times 1000$, $g_c = 0.88 \pm 0.01$, $\beta = 0.68 \pm 0.01$, $z = 1.65 \pm 0.02$, $\nu = 1.24 \pm 0.01$ and $\alpha = 0.42 \pm 0.01$. It was observed that these independently estimated exponent values do not fit with the scaling relation $\beta = \nu\alpha$. However, this type of scaling violation was also observed previously in many active-absorbing transition cases [8].

## 12.3.4 Model B

### 12.3.4.1 Mean Field Case

For the mean field case, $N = 10^6$, averaging over $10^3$ initial condition were taken. The phase diagram and the universality classes of the transition has been numerically investigated. In the mean field case, the phase boundary seems to be linear starting $g_c = 1/2$ for $p = 0$ and ending at $g_c = 1$ for $p = 1$ (Fig. 12.5), obeying $g_c = \frac{1}{2}(1 + p)$. In this case, for $p = 0$, $g_c = 1/2$, and this is similar to the fixed energy sandpiles [9–11]. Again the critical exponents are the same along the phase boundary and they match with those of model A.

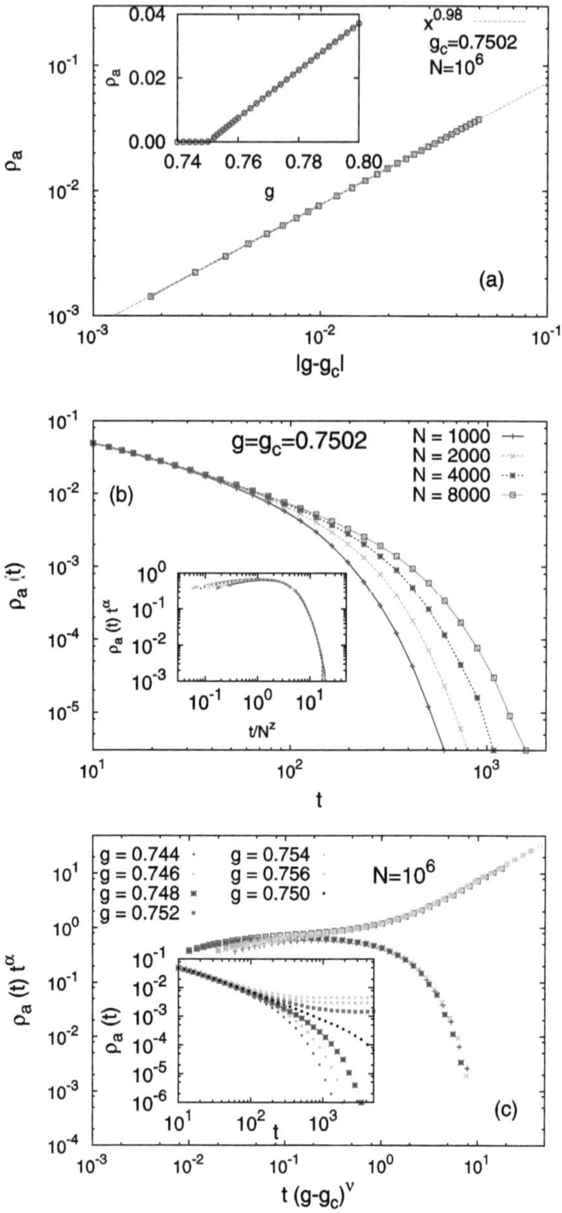


**Fig. 12.4** Simulation results for mean field case, $g_c = 0.7502 \pm 0.0002$. (**a**) Variation of steady state density $\rho_a$ of active sites versus $g - g_c$, fitting to $\beta = 0.98 \pm 0.02$. The *inset* shows the variation of $\rho_a$ with density $g$. (**b**) Relaxation to absorbing state near critical point for different system sizes, the *inset* showing the scaling collapse giving estimates of critical exponents $\alpha = 1.00 \pm 0.01$ and $z = 0.50 \pm 0.01$. (**c**) Scaling collapse of $\rho_a(t)$. The *inset* shows the variation of $\rho_a(t)$ versus time $t$ for different densities $g$. The estimated critical exponent is $\nu = 1.00 \perp 0.01$. The system sizes $N$ are mentioned. From [7]

#### 12.3.4.2 Lattice Cases

This model was also studied for 1-d and 2-d lattices. For a linear chain in 1-d, $N = L = 10^4$ and average over $10^3$ initial condition were considered. For 2-d, square

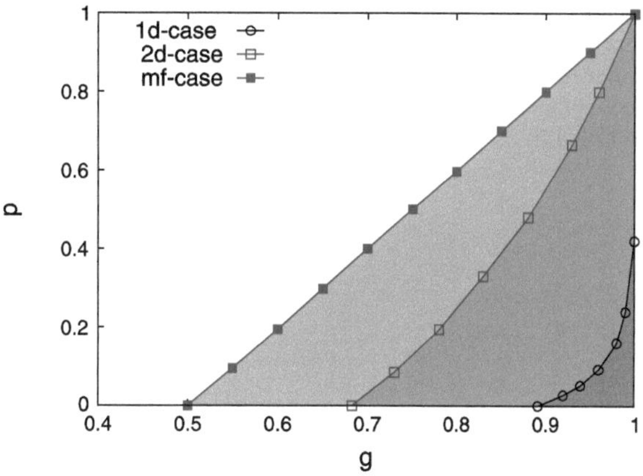


**Fig. 12.5** Phase diagram for the generalized model in the $(g, p)$ plane, showing the phase boundaries separating the active and absorbing phases in 1-d, 2-d and mean field cases. The active phases are on the right of the phase boundaries while the absorbing phases are on the left in the respective cases. The system sizes are $N = 10^5$ for mean field, $1000 \times 1000$ for 2-d, and $10^4$ for 1-d. From [7]

restaurants (lattice) with $L = 1000$ and averaging over $10^3$ initial conditions were considered.

1. For 1-d, for the case $p = 0$, $g_c = 0.89 \pm 0.01$, with $\beta = 0.42 \pm 0.01$, $z = 1.55 \pm 0.02$, $\nu = 1.90 \pm 0.02$ and $\alpha = 0.16 \pm 0.01$. The phase boundary in $(g, p)$ is nonlinear: it starts from $g_c = 0.89 \pm 0.01$ at $p = 0$ to $p = 0.43 \pm 0.03$ at $g = 1$ (Fig. 12.5). Thus, one can independently define a model at unit density ($g = 1$) and calculate the critical probability $p_c$ for which the system goes from an active to an absorbing phase.
2. For 2-d, for the case $p = 0$, $g_c = 0.683 \pm 0.002$, with $\beta = 0.67 \pm 0.02$, $z = 1.55 \pm 0.02$, $\nu = 1.20 \pm 0.03$ and $\alpha = 0.42 \pm 0.01$. The phase boundary seems nonlinear, from $g_c = 0.683 \pm 0.002$ for $p = 0$ (Fig. 12.5) extending to $g_c = 1$ at $p = 1$.

*In summary*, it is shown how a crowd dynamics in a resources allocation game gives rise to a phase transition between an active and a frozen phase, as the density varies. In this respect, a class of models has been defined and studied, where $gN$ agents compete among themselves to get the best service from $N$ restaurants of same rank, generalizing the 'Kolkata Paise Restaurant' problem. In the original problem, where density $g = 1$, the model was far from its critical value $g_c$, the relaxation time $\tau$, given by (12.10) never showed any $L = N^{1/d}$ dependence. As long as $g \leq g_c$, absorbing frozen configurations are present, and whether that can be reachable or not, depends on the underlying dynamics. The existence of a critical point $g_c$ above which the agents are unable to find frozen configurations was found. In the case in which the agents are moving if and only if they are unsatisfied (model B) with $p = 0$, they fail to reach satisfactory configurations if the density is above $g_c = 1/2$. Strategies where agents wait longer (higher $p$) speed up the convergence, increasing $g_c$ and decreasing the time to reach saturation configurations (faster-is-slower effect). The exponent values of the phase transitions in finite dimensions are in good agreement with the exponents of stochastic fixed-energy sandpile (Table. 12.1) [9–13]. Thus, it is a simple model for resource allocation, which is solvable (the MF

**Table 12.1** The table shows that comparison of the critical exponents of this model with those of the conserved Manna model [13]

| | | Model A | Model B | Manna |
|---|---|---|---|---|
| $\beta$ | 1D | | $0.42 \pm 0.01$ | $0.382 \pm 0.019$ |
| | 2D | $0.68 \pm 0.01$ | $0.67 \pm 0.02$ | $0.639 \pm 0.009$ |
| | MF | $0.98 \pm 0.02$ | $0.99 \pm 0.01$ | 1 |
| $z$ | 1D | | $1.55 \pm 0.02$ | $1.393 \pm 0.037$ |
| | 2D | $1.65 \pm 0.02$ | $1.55 \pm 0.02$ | $1.533 \pm 0.024$ |
| | MF | $0.50 \pm 0.01$ | $0.50 \pm 0.01$ | 2 |
| $\alpha$ | 1D | | $0.16 \pm 0.01$ | $0.141 \pm 0.024$ |
| | 2D | $0.42 \pm 0.01$ | $0.42 \pm 0.01$ | $0.419\pm 0.015$ |
| | MF | $1.00 \pm 0.01$ | $1.00 \pm 0.01$ | 1 |
| $\nu$ | 1D | | $1.90 \pm 0.02$ | $1.876 \pm 0.135$ |
| | 2D | $1.24 \pm 0.01$ | $1.20 \pm 0.03$ | $1.225 \pm 0.029$ |
| | MF | $1.00 \pm 0.01$ | $1.00 \pm 0.01$ | 1 |

limit), and shows a variety of interesting features including phase transitions as in well known models.

## 12.4 KPR and Its Application on MG

So far we have dealt with the cases where the number of choices and the number of agents making those choices are of comparable magnitudes (KPR problem). However, there is another very well studied limit where the number of agents remain large but the number of choices is only two. A pay-off is given to the agents belonging to the minority group. Given there is no dictator and the agents do not communicate among themselves, how to device a strategy to extract maximum gain for maximum number of people, has been a long standing question. This problem goes by the name Minority Game (MG). This is, in fact, a particular version of the El Farol bar problem introduced by Brian Arthur [14].

In MG, the total number of agents ($N$) being odd, the maximum possible utilization can come when $(N-1)/2$ agents are in the minority. However, if the agents choose randomly, the utilization is far from the maximum value, in fact the deviation is of the order of $\sqrt{N}$. However, there can be deterministic strategies, where agents learn from their past experiences and in those cases this fluctuation can be considerably reduced, giving a self-organized, efficient market [15–20]. But in all those cases, the fluctuations (deviation from maximum utilization) scales with system size as $\sqrt{N}$. Only the pre-factor, depending upon the particulars of the strategy, can be reduced.

Recently, Dhar et al. [21] applied a stochastic strategy, inspired by the stochastic strategy used in KPR [2, 4, 5], to show that the fluctuations, or deviation from maximum utilization, can be reduced to be of the order of $N^{\epsilon}$ for any $\epsilon > 0$ in $\log\log N$

time. Stochastic strategy was used in MG before [22], where the fluctuation could be made of the order 1, but the time to reach that state scaled with $\sqrt{N}$. The strategy taken by Dhar et al., is the first of its kind that gives smallest fluctuation in very short time. In the following sections we discuss the main results of this strategy and its subsequent modifications.

## 12.5 KPR Strategy in MG: Results

As mentioned before, Minority Game deals with $N$ (odd) agents selecting between two choices, when an incentive is associated with people belonging to minority. For example, consider the situation where there are only two restaurants in a locality and $N = 2M + 1$ agents select between these two restaurants for dinner. An agent is happy if he or she goes to the less crowded restaurant. But they cannot communicate among themselves and cannot change their choices once they fix it for a given evening. The agents, however, have in their possession the entire history of which restaurant was more crowded. This is a classic example to the MG problem. Other examples can be buying or selling of stocks and so on.

For any configuration at time (day) $t$, one can write the populations in the two restaurants as, $M - \Delta(t)$ and $M + \Delta(t) + 1$. In the this strategy, a deviation from the classic MG problem was made that the knowledge of $\Delta(t)$ was also available to the agents, while originally only its sign was known. In that sense, agents have more information than usual.

The strategy of the agents is as follows: At $t = 0$ the agents select randomly. Then the agents belonging to the minority stick to their choice in the next day. But the agents in the majority change their choice with a probability

$$p = \frac{\Delta(t)}{(M + \Delta(t) + 1)} \tag{12.11}$$

for $\Delta(t) > 0$ and stick to their choice with probability $1 - p$. As it is a probabilistic strategy, the number of people shifting will also have a fluctuation of the order $\sqrt{\Delta(t)}$, which is the new difference between the two populations; which leads us to the recurrence relation $\Delta(t + 1) = \sqrt{\Delta(t)}$. This shows that after $\log \log N$ time $\Delta(t)$ becomes of the order 1 and remains there.

Therefore, by following the same stochastic strategy, the difference between the populations in the two restaurants can be minimized in a very short time. This is in contrast with standard MG strategies, where the agents indeed try to differ in their strategies to maximize individual gain. However, the difference being in general the strategies were deterministic, i.e. given a history, all the subsequent steps are known. The stochasticity itself makes the agents differ. Furthermore, that the agents follow the same stochastic strategy and do not deviate from it, can be justified if it can be shown that a single individual does not gain by deviating from this strategy. Indeed it was shown that for this strategy, an individual will not gain by deviating from this strategy.

## 12.5.1 Stability Against Individual Deviator

In the above discussions it is discussed that if the agents in the MG problem follow a simple stochastic strategy, the difference between the two choices can be made of the order 1 in $\log\log N$ time. However, it is not always expected that all the agents will follow the same strategy, until it is shown that no one will gain by deviating from the strategy.

### 12.5.1.1 Game with One Cheater

Defining *cheater* as one who will not follow the strategy followed by rest present in the majority. Now suppose there is a cheater, say $X_1$ in the majority, say in restaurant A. If he want to stay, then the number of agents in the restaurant A, who will follow the conventional strategy is $M + \Delta(t)$. The probability that $\tilde{r}$ agents from $M + \Delta(t)$ agents in A will shift from A to minority, say restaurant B is

$$P(\tilde{r}) = \binom{M + \Delta(t)}{\tilde{r}} p^{\tilde{r}} (1 - p)^{M+\Delta(t)\ \tilde{r}}. \tag{12.12}$$

For $M \to \infty$ the probability distribution will become Poisson with $\lambda = p(M + 1)$. So this distribution will be

$$P(r) = \frac{\lambda^r}{r!} \exp(-\lambda)(1 + Br), \tag{12.13}$$

where

$$B = \frac{\lambda}{M} - \left(\frac{\lambda^3}{2} + \lambda^2 \Delta - \frac{\lambda^2}{2}\right) \frac{1}{M^2}. \tag{12.14}$$

Using the above probability distribution, it can be shown that [21] there exist a value of $\lambda$ for a given $\Delta(t)$ such that existence of cheater does not effect the dynamics of the game. This $\lambda$ is given by

$$\lambda - \Delta = \frac{1}{6} + \frac{\lambda^2}{2M} \sqrt{\frac{\lambda}{\Delta}} \left(1 + \frac{\Delta}{\lambda}\right). \tag{12.15}$$

Or restoring the inequality given that $X_1$ will gain switching as he is in majority then we get

$$\lambda < \Delta + \frac{1}{6} + \frac{\lambda^2}{2M} \sqrt{\frac{\lambda}{\Delta}} \left(1 + \frac{\Delta}{\lambda}\right). \tag{12.16}$$

As $\lambda \propto \Delta$, this means for a large difference $\Delta$ we can increase the noise safely up to $\frac{1}{6}+$ without letting the cheater to win. We have seen in Fig. 12.6 that (12.15) match the simulation result. In the simulation we took $p = \frac{\Delta + c}{M + \Delta + 1}$, with vary-

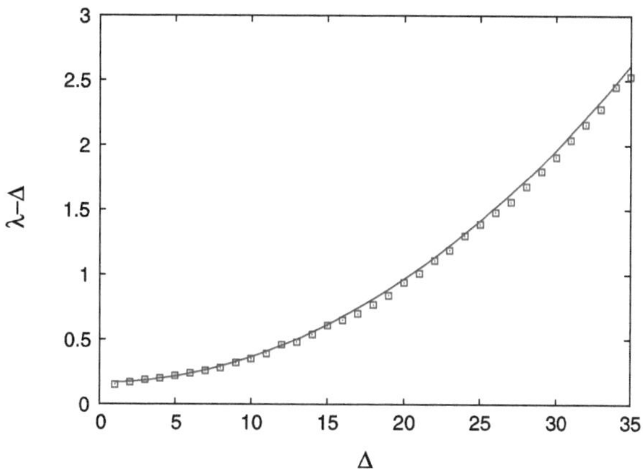


**Fig. 12.6** The data points are the simulation data and the line is (12.15). The total number of player is 2001

ing the noise parameter $c$. Below this optimal value of $\lambda$, cheater will gain if he shift from majority to minority, above this optimal value a cheater will gain if he shift from minority to majority.

#### 12.5.1.2 Majority Stay or Minority Flip

For a situation when one agent will stay if he finds himself in majority (in A) and will shift if he finds himself in minority (in B). Then he will win by staying in majority if $r$ number of agent shift from majority A to minority B, given $r \geq \Delta + 1$. The total probability $P(win \,|\, stay\ in\ majority)$ that he will win, which is same as expected payoff is

$$EP(majority \,|\, stay) = \sum_{r=\Delta+1}^{\infty} P(r). \tag{12.17}$$

Now, if he is in B having total number of agent $M - \Delta$ and shift to A having total number of agent $M + \Delta + 1$, he will win if $r$ number of agent come from A to B, with $r \geq \Delta + 2$. The total probability of his win if he flip is $P(win \,|\, flip\ from\ minority)$, which is same as his payoff given by

$$EP(minority \,|\, flip) = \sum_{r=\Delta+2}^{\infty} P(r) \tag{12.18}$$

where $P(r)$ is given by (12.13). Total probability of win or expected payoff, if he stay at majority and flip if he is in minority is sum of (12.18) and (12.17), which after little algebra is given by

$$EP_I = 2\left(1 - \frac{\Gamma(\Delta + 1, \lambda)}{\Delta!}\right) - \frac{\lambda^{\Delta+1}}{(\Delta + 1)!} \tag{12.19}$$

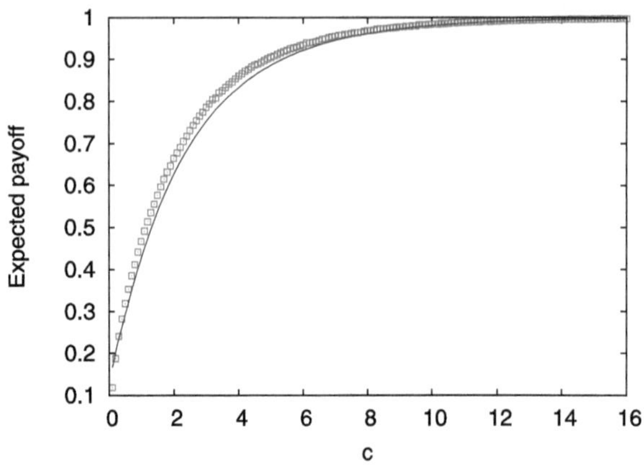


**Fig. 12.7** The simulation data is compared with the solid line which is (12.19) with given in (12.21) neglecting the small correction term in the bracket, thus $\lambda = 2c$. The total number of player is 2001

where $\Gamma(s, x)$ is a *incomplete gamma function*. To get more accurate result we need to average the expected payoff ($EP_I$). From numerical experiment we know that fluctuation in $\Delta$ is very small. So this gives very little error when fitted with the simulated result. This error can be minimized by little adjustment of the constant terms. The best fit will come for the first argument $\Delta + a$ where $a \neq 1$, but $a = 0.65$. and the second argument $\lambda = 2\Delta$ in the $\Gamma(s, x)$. From Fig. 12.7 we find that the noise $c$ can not be increased to very large value else a cheater will always gain the game.

We have seen that if we take $\Delta = 0$ so that $\lambda = c$, the noise parameter, then the curve have same features as the simulated curve, this is due to the fact that $\Delta$ does not become zero in the presence of non zero noise. So we need to know the average $\Delta$ in this case, which is given by

$$\langle \Delta \rangle = \frac{1}{2}\lambda\left(1 - \frac{\Gamma(\Delta_0, \lambda)}{(\Delta_0 - 1)!}\right). \tag{12.20}$$

This is the average difference if A become minority after a shift of agents. So we get

$$\lambda = 2\langle \Delta \rangle\left(1 - \frac{\Gamma(\Delta_0, \lambda)}{(\Delta_0 - 1)!}\right)^{-1}. \tag{12.21}$$

#### 12.5.1.3 Minority Stay or Majority Flip

Let the cheater is in A having $M + \Delta + 1$ agents who will shift to B there are $M - \Delta$ agents. He will shift to B making $n_A = M + \Delta$ and $n_B = M - \Delta + 1$. Now he will win if $r$ number of people from A shift to B with $r \leq \Delta - 1$. Then the probability that he will win is given by $P(win \,|\, \textit{flip from majority})$. Now if he is in B, then he will stay. If $r \leq \Delta$ number of people shift from A to B he will win. The probability that he will win staying in B is $P(win \,|\, \textit{stay in minority})$. The total probability of

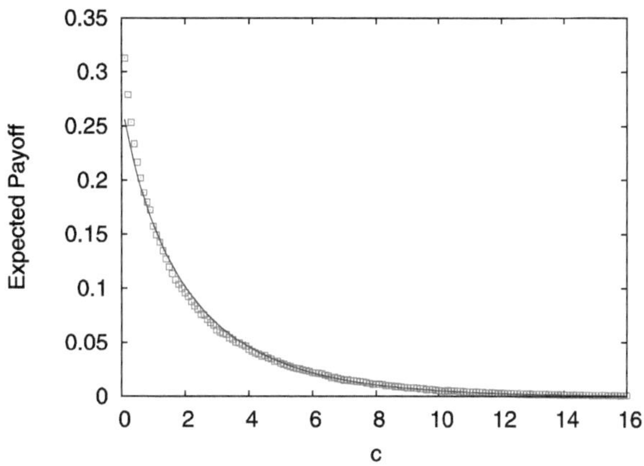


**Fig. 12.8** The symbols represent the simulation data and the line is (12.22) with $\lambda$ given in (12.23). In the plot average $\Delta$ is not the noise parameter c but a little less, so instead of $\lambda = 2(c+1)$, $\lambda = 2c + 1.85$ is plotted in the theoretical curve. The total number of player is 2001

winning if he always stays in minority, which is same as expected payoff $E_{II}$ in the same case

$$EP_{II} = \sum_{r=0}^{\Delta-1} P(r) + \sum_{r=0}^{\Delta} P(r) \approx 2\frac{\Gamma(\Delta, \lambda)}{(\Delta-1)!} + \frac{\lambda^{\Delta}}{\Delta!}\exp(-\lambda). \tag{12.22}$$

We have seen that if we take $\Delta = 0$ so that $\lambda = c$, the noise parameter, then the curve have same features as the simulated curve (see Fig. 12.8), this is because $\Delta$ does not become zero in the presence of non zero noise. So we need to know the average $\Delta$ in this case which is

$$\lambda = 2\big(\langle\Delta\rangle + 1\big). \tag{12.23}$$

### 12.5.2 *Freezing of Dynamics and Escape Routes from It*

It is clear from the strategy discussed above, that once $\Delta(t) = 0$ i.e., the difference of population in the two restaurants is 1 (which is the minimum possible value as the total number is odd), the dynamics stops. This leaves the system highly asymmetric in the sense that the people in the majority (minority) will remain in the majority (minority) forever. This situation is of course socially unacceptable, although this is the most efficient division.

#### 12.5.2.1 Resetting After a Given Time

To resolve this status quo, Dhar et al. [21] suggested that once $\Delta(t) = 0$ a major reshuffle can take place if all the agents (whether in majority or in minority) shift after waiting $T$ time steps. This time period needs to be much smaller than the life-time of the agents. If an inefficiency parameter is defined as follows

$$\eta = \lim_{N\to\infty} \frac{4}{N}\langle (r - N/2)^2\rangle, \tag{12.24}$$

then for this case this would be

$$\eta \simeq \frac{K_1 N^{\epsilon-1}}{T + K_2 \log\log N} \tag{12.25}$$

where $K_1$, $K_2$ and $\epsilon$ are constant. This means that efficiency increases with $T$. However, large $T$ would mean longer wait in the majority. Clearly, other parameters like overall social welfare and equality needs to be considered here. Also, as indicated in the Dhar et al. paper, it will be interesting to see what if the agents try to maximize their pay-offs for next $n > 1$ days.

### 12.5.2.2 Continuous Transition of Social Efficiency

In the above method, the system becomes efficient only when the agents act for overall social welfare or have a long-term gain strategy. Even then, efficiency depends upon time waiting time $T$, which gives rise to a competition regarding its magnitude.

Biswas et al. [23] suggested a subsequent modification in the strategy such that the fluctuation could be reduced to any arbitrarily small value by tuning a parameter. This, therefore, gives a continuous phase transition and as long as a finite fluctuation is kept in the system, the frozen condition can be avoided.

The modified strategy is the following: The agents in the majority in a given day shifts to the other choice with a probability

$$p_+(t) = \frac{\Delta'(t)}{M + \Delta'(t) + 1}, \tag{12.26}$$

(where $\Delta'(t) = q\Delta(t)$ and $q$ is a constant) and people in the minority stick to their choices ($p_- = 0$).

Regarding the steady-state behavior, consider the following: Suppose the populations in the majority and minority are $M + \Delta(t)$ and $M - \Delta(t)$ respectively, at time $t$. Now, if $2\Delta(t)$ number of people can be shifted from majority to minority, then the population difference will remain same and the same process can be repeated, sustaining a steady state. Of course, this possibility can only arise when $q > 1$. If $\Delta_s$ is the steady state value for fluctuation, then

$$(M + \Delta_s + 1)\frac{q\Delta_s}{M + q\Delta_s + 1} = 2\Delta_s. \tag{12.27}$$

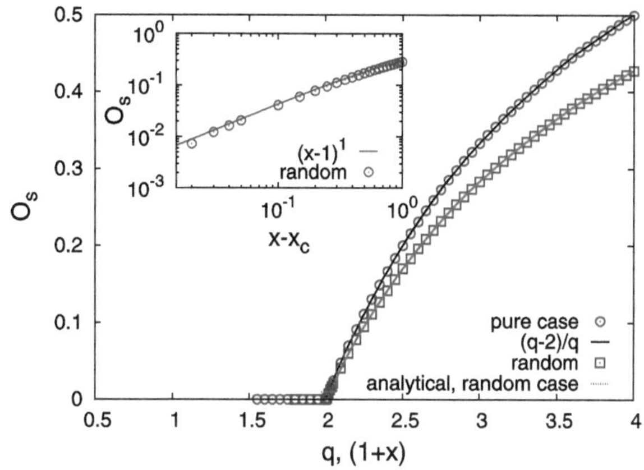


**Fig. 12.9** Steady state values of the order parameter $O_s$ for different values of $q$ and $x$. The *solid lines* show the analytical results for the pure and annealed disordered cases. Both match very well with the simulation points. *Inset* shows the log–log plot near the critical point for the disordered case, confirming $\beta = 1.00 \pm 0.01$. All simulation data are shown for $M = 10^5$. From [23]

The two solutions of this quadratic equation are

$$\Delta_s = 0 \quad \text{or} \quad \frac{q-2}{q}(M+1). \tag{12.28}$$

This means that for $q < q_c = 2$, the system will reach the zero fluctuation state (although the dynamics of the system will be very much different for $q < 1$ and $q > 1$), and for $q > 2$ there will remain a residual fluctuation in the system signifying an active-absorbing type phase transition around $q = q_c = 2$.

Formally, one can define an order parameter like $O(t) = \Delta(t)/M$ and in the steady state the saturation value is $O_s = 0$ when $q < 2$ and $O_s = (q - q_c)/q$ for $q > 2$ both for $M \gg 1$, giving the order parameter exponent $\beta = 1$. Figure 12.9 shows the numerical results and its comparison with the above calculations.

Regarding the dynamics of the system in approaching this steady state, assume that at time $t$ the populations at the two restaurants are $N_A(t)$ and $N_B(t)$ and $N_A(t) > N_B(t)$. Therefore,

$$\Delta(t) = \frac{N_A(t) - N_B(t) - 1}{2}. \tag{12.29}$$

Now, according to the strategy in (12.26), the number of people shifted from choice A to choice B will be

$$\begin{aligned} S(t) &= \frac{q\Delta(t)}{M + q\Delta(t) + 1}\big(M + \Delta(t) + 1\big) \\ &\approx q\Delta(t) \end{aligned} \tag{12.30}$$

up to leading order term, when $\Delta(t) \ll M$, i.e., when $q$ is close to $q_c$, or in the long time limit if $q < q_c$ and not too close to it. With this transfer amount, in the next step $N_A(t+1) = N_A(t) - S(t)$ and $N_B(t+1) = N_B(t) + S(t)$. For $q > 1$, majority will become minority, so

$$\begin{aligned}\Delta(t+1) &= \frac{N_B(t+1) - N_A(t+1) - 1}{2} \\ &\approx q\Delta(t) - \Delta(t) - 1.\end{aligned} \tag{12.31}$$

Subtracting $\Delta(t)$ from both sides and dividing by $M$, one arrives at

$$\frac{dO(t)}{dt} = -(2-q)O(t) - \frac{1}{M}. \tag{12.32}$$

The last term can be neglected for large $M$. The it follows

$$O(t) = O(0)\exp\bigl[-(2-q)t\bigr]. \tag{12.33}$$

So this exponential decay in the region $1 < q < 2$ gives a time scale $\tau \sim (q_c - q)^{-1}$, diverging at the critical point with exponent 1.

In (12.30), if one keeps the second order term, one gets

$$S(t) \approx q\Delta(t) - \frac{1}{M}\bigl(q^2\Delta^2(t) - q\Delta^2(t)\bigr). \tag{12.34}$$

The time evolution equation becomes

$$\frac{dO(t)}{dt} = -(2-q)O(t) - q(q-1)O^2(t). \tag{12.35}$$

Now, exactly at the critical point $q = 2$, the solution is

$$O(t) = \frac{O(0)}{2O(0)t + 1}, \tag{12.36}$$

which, in the long time limit gives $O(t) \sim t^{-1}$, giving the critical exponent value $\alpha = 1$.

A more general solution of (12.35) can be obtained (for any $q$) as follow: Consider the auxiliary variable $u(t) = |q-1|^t/O(t)$ and substitute it in (12.35). This gives after simplifications

$$u(t+1) = u(t) + q|q-1|^t. \tag{12.37}$$

Using this recursion relation, one can write $u(t)$ in a GP series and can perform the sum to get the following:

$$O(t) = \frac{1-|q-1|}{q}\,\frac{1}{(\frac{1-|q-1|}{qO(0)} + 1)|q-1|^{-t} - 1}. \tag{12.38}$$

Putting $q = 2$ in the above equation, one gets back (12.36). Also, a time scale is obtained from the above equation in the form

$$\tau \sim \frac{1}{|\ln(|q-1|)|}. \tag{12.39}$$

As $q \to q_c$, the power law divergence $(q_c - q)^{-1}$ is recovered.

Furthermore, for $q < 1$ the dynamical equation (12.38) will reduce to

$$O(t) \sim \frac{O(0)}{O(0)+1}(1-q)^t. \tag{12.40}$$

## 12.5.3 *Reducing Fluctuation with Less Informed Agents*

As is clear from the strategies discussed above, the agents in those versions of the game, posses more information than the usual minority game problem. Particularly, the agents are aware of the amount of excess population in the majority, while in the usual case then only know whether they were in the majority or minority. This extra information is important. Although it is logical that the would agents eventually come to know about this excess population, there have been studies to confirm if this extra information is essential in obtaining the maximum efficient state. It is found that this information is not essential. The system can indeed reach the maximum efficient state even when this knowledge is partially or even fully absent.

### 12.5.3.1 Non-uniform Guessing of the Excess Crowd: Phase Transition

It has been argued in Ref. [23] that in considering less informed agent a natural step would be the agents with different guessing abilities. This means that although the agents do not know the exact value of the excess population, they can make a guess about the value. This acts as an annealed disorder. Formally, the $i$th agent at time $t$ makes a guess about $\Delta(t)$ which is

$$\Delta_i(t) = \Delta(t)(1+\epsilon_i), \tag{12.41}$$

where $\epsilon_i$ is an annealed random variable taken from a uniform distribution in the range $[0:2x]$. This means,

$$\langle \Delta_i(t)\rangle = \Delta(t)\bigl(1+\langle\epsilon_i\rangle\bigr) = \Delta(t)(1+x), \tag{12.42}$$

where the angular brackets denote average over disorder. One can generally write

$$\Delta(t+1) = \bigl|\Delta(t) - S(t)\bigr|, \tag{12.43}$$

where

$$S(t) = \left|\left\langle \frac{\Delta(t)(1+\epsilon)}{M+\Delta(t)(1+\epsilon)}\right\rangle\right|. \tag{12.44}$$

This leads to

$$O(t+1) = O(t)\left|\left\langle \frac{\epsilon}{1+(1+\epsilon)O(t)}\right\rangle\right|. \tag{12.45}$$

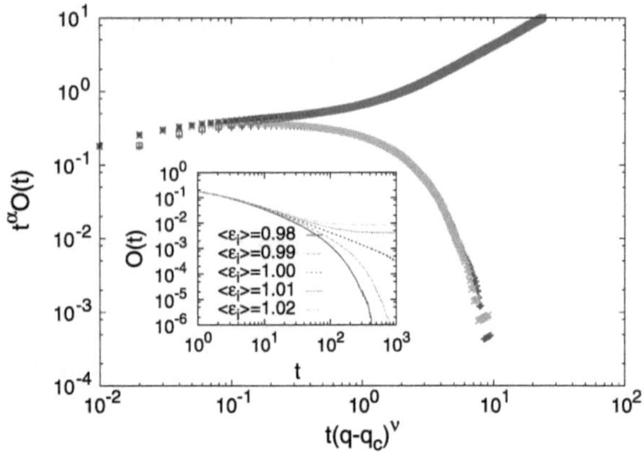


**Fig. 12.10** Data collapse for finding $\nu$ in the disordered case for different $x$ values. The estimate is $\nu = 1.00 \pm 0.01$. *Inset* shows the uncollapsed data. The *straight line* at the critical point gives $\alpha = 1.00 \pm 0.01$. Simulation data is shown for $M = 10^6$. From [23]

In the steady state $O(t+1) = O(t) = O^*$, leading to

$$\frac{(1-O^*)2xO^*}{(1+O^*)} = \ln\left[1 + \frac{2xO^*}{1+O^*}\right]. \tag{12.46}$$

A numerical solution of this self-consistent equation was found to agree with the simulation results (see Fig. 12.9). For small $O^*$, $O^* \sim (x-1)$ giving $\beta = 1$. Also, for small $O(t)$, the dynamical equation can be written as

$$\frac{dO(t)}{dt} = (x-1)O(t) - xO^2(t). \tag{12.47}$$

The critical point is at $x_c = 1$. So at the critical point, $O(t) \sim t^{-1}$, giving $\alpha = 1$ and above the critical point the exponential decay would give a time scale, diverging at $x = x_c$ with an exponent $\nu = 1$.

The above results were also verified using numerical simulations. A finite size scaling form was considered

$$O(t) \approx t^{-\alpha}\mathcal{F}\big(t^{1/\nu}(q-q_c), t^{d/z}/N\big), \tag{12.48}$$

where $d$ is the spatial dimension, which was taken as 4 in this mean-field scenario. This form suggests that at the critical point the order parameter decays in a power-law, with exponent $\alpha$, which was numerically found to be $1.00 \pm 0.01$ (see inset of Fig. 12.10). One can also plot (see Fig. 12.10) $O(t)t^\alpha$ against $t(q-q_c)^\nu$, where by knowing $\alpha$, $\nu$ can be tuned to get best data collapse, giving $\nu = 1.00 \pm 0.01$. Also, $O(t)t^\alpha$ can be plotted against $t/N^{z/d}$, where $z/d$ can be obtained from the data collapse (Fig. 12.11) to be $0.50 \pm 0.01$. Therefore, it was concluded that the analytical estimates were verified and the scaling relation $\alpha = \beta/\nu$ was satisfied.

In the above mentioned case, the non-uniform guessing power acts as an annealed disorder. When this disorder is quenched, the case slightly complicated. It is no longer possible to tackle analytically as done above. It was seen that the agents with higher $\epsilon$ are more likely to change side and be in the majority. So, if the average pay-offs are plotted against $\epsilon$, a monotonic decay is observed (Fig. 12.12).

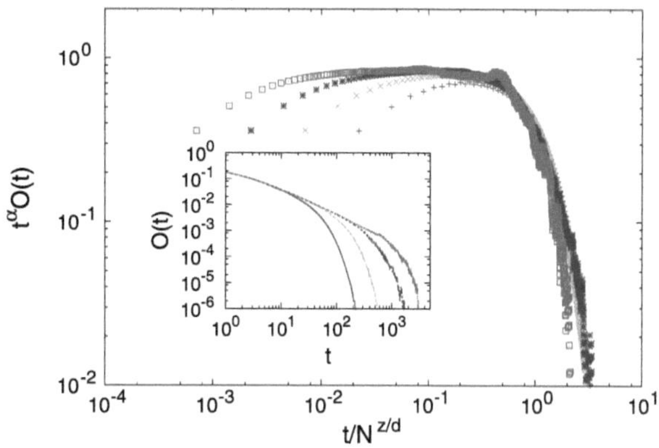


**Fig. 12.11** Data collapse for finding $z$ in the disordered case for different system sizes ($M = 10^3, 10^4, 10^5, 10^6$) at $x = 1.0$. The estimate is $z/d = 0.50 \pm 0.01$. *Inset* shows the uncollapsed data. The *linear part* in the *inset* confirms $\alpha = 1.00 \pm 0.01$. From [23]

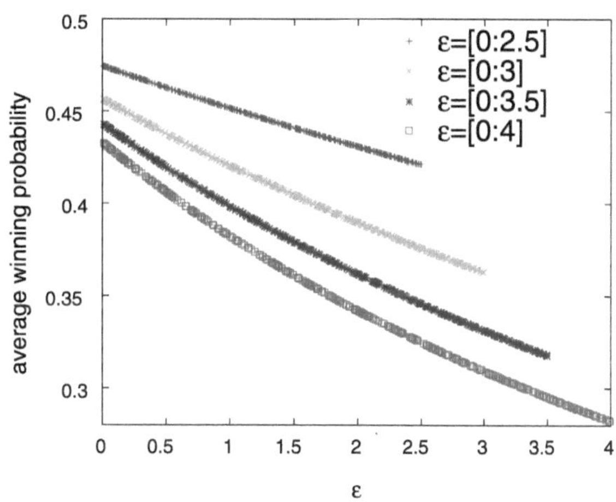


**Fig. 12.12** For quenched $\epsilon_i$ the average pay-offs of the agents are plotted for different $\epsilon$ values having different ranges as indicated. The monotonic decay with increasing $\epsilon$ clearly indicates that agents with higher $\epsilon$ are more likely to be in the majority. From [23]

#### 12.5.3.2 Following an Annealing Schedule

Usually in minority game, agents do not have any information about the amount of excess population in the majority. They are only aware whether they are in the minority or majority. All the strategies mentioned above require this information in some form (fully or partially). However, it was studied in Ref. [23] that even without this information, the system can reach the fully efficient state in $\ln N$ time.

In this case of least informed agents, the agents assume a simple time evolution for the excess population. An example can be

$$\Delta^T(t) = \Delta^T(0)\exp(-t/\tau), \tag{12.49}$$

where $\Delta^T(0)$ is close to $\sqrt{M}$, corresponding to the initial random choice. Assuming this form, one can plot the actual $\Delta(t)$ along with this trial function with time. They have a simple relation as follows:

$$2\Delta(t) = \Delta^T(t). \tag{12.50}$$

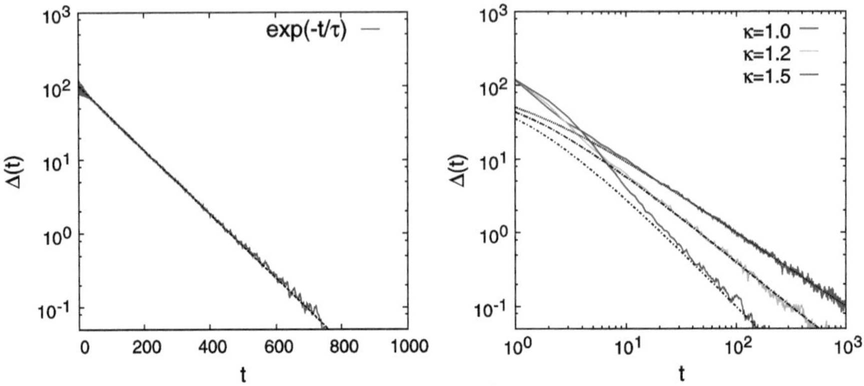


**Fig. 12.13** Time variation of the excess population $\Delta(t)$ are plotted for different functional forms of $\Delta^T(t)$. *Left*: In log-linear scale the excess population are plotted for exponential decay. *Right*: For power law ($\Delta^T(0)/(1+t)^\kappa$ decay, with different values of $\kappa$). $M = 5 \times 10^3$ for the simulations. From [23]

This implies that even when the agents are completely unaware of the excess population, they can reach an efficient state ($\Delta(t) \sim 1$) in $\ln N$ time.

It was also checked in Ref. [23] if this process is specific to the functional form considered for the trial function. For this purpose a power-law decay was also considered

$$\Delta^T(t) = \frac{\Delta^T(0)}{(1+t)^\kappa}. \tag{12.51}$$

Again it was found that for different $\kappa$ values, the relation in (12.50) is satisfied. It was therefore concluded that this relation is true for a wide range of the functional form (the restrictions in the functional form is discussed later).

The behavior of the order parameter when a trial function is considered, can be verified as follows: The dynamical evolution of $O(t)$ would be

$$O(t+1) = \frac{|\eta(t) - O(t)|}{1+\eta(t)}, \tag{12.52}$$

where $\eta(t) = \Delta^T(t)/M$. When $\eta(t) > O(t)$, one can obtain (by Taylor series expansion)

$$\frac{dO(t)}{dt} - \big[\eta(t) - 2\big]O(t) = \eta(t)\big[1 - \eta(t)\big]. \tag{12.53}$$

A general solution of the above equation will be of the form

$$O(t) = \frac{\int_0^t dt_1\, \eta(t_1)(1-\eta(t_1))e^{\int_0^{t_1}(2-\eta(t_2))\,dt_2}}{e^{\int_0^t (2-\eta(t_1))\,dt_1}} + C_1 e^{-\int_0^t (2-\eta(t_1))\,dt_1}, \tag{12.54}$$

where $C_1$ is a constant. This is valid only when $\eta(t)$ is not a fast decaying function. When $\eta(t) < 2$, the dominant terms in the above equation is

$$O(t) \approx \frac{\eta(t)(1-\eta(t))}{2-\eta(t)} \approx \frac{\eta(t)}{2}, \tag{12.55}$$

which was the numerical observation (see Fig. 12.13). If one evaluates (12.54) using $\eta(t) = \eta_0 \exp(-t/\tau)$ for $\tau > 1/2$, one gets

$$O(t) \sim \frac{\tau}{2\tau - 1}\eta(t). \tag{12.56}$$

Therefore, $O(t) \approx \eta(t)/2$ is only valid when $\tau \gg 1/2$, which limits the fastness in the trial function.

When one considers a fast decaying trail function, one would simply have

$$O(t) \sim O(t-1) - \eta(t-1) \sim O(0) - \sum_{k=0}^{t-1} \eta(k). \tag{12.57}$$

So, $O(t)$ will saturate to a finite value in this case.

## 12.5.4 *Effect of Random Traders*

The above mentioned strategies concern with agents following a given strategy (this does not remove their heterogeneity, since these are stochastic strategies that involve uncorrelated random numbers). However, it is often the case in real markets that there exist agents who do not follow the market signals (fluctuations) in deciding their trade strategies. Whatever might be their logic, it terms of market signals, they can be treated as random traders who decide completely randomly as opposed to the chartists who follow given strategies (deterministic or stochastic). Following discussions deal with effect of such random traders in minority games.

### 12.5.4.1 Single Random Trader

Consider the scenario when there is only one random trader in the system. The other agents follow some strategy mentioned before, and reach the minimum fluctuation state. After that $\Delta(t) = 0$, so no chartist will shift from his or her choice. However, the single random trader will continue to shift on average in a 2 days time period. The majority will be determined by this random trader. Therefore, that random trader will always be the loser. Although the resource utilization will be perfect in this case, it will be at the cost of one player being in the majority for ever.

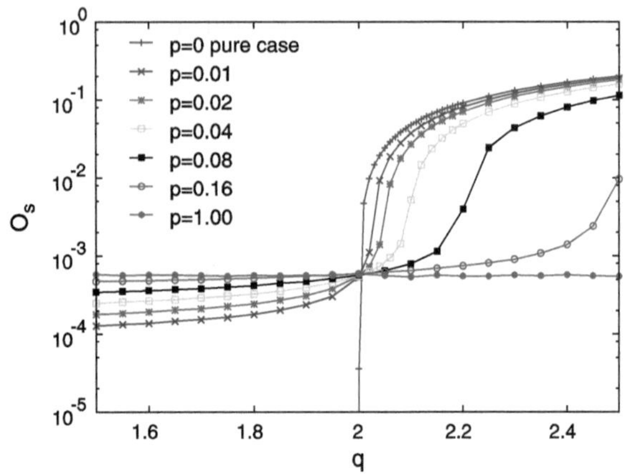


**Fig. 12.14** The saturation values of $O_s$ are plotted against $q$ for different fractions $p$ of the random traders. $M = 10^6$ for the simulations. From [23]

#### 12.5.4.2 More than One Random Trader

The case of the single random trader has the problem that the random trader is always a loser. This makes the system unstable in the sense that resource allocation is unfair for that agent as long as he or she follows that strategy (random in this case). However, this problem can be avoided by considering more than one random player. In this case it is not always possible to keep all the random players in the majority, since the majority is no longer determined by a single random player. Also, as the average time period is 2 days for the random players, both the choices will become majority and minority in this time period (due to symmetry of the choices). It is true that random players would make the fluctuations to grow. If the number of random player is $pN$, then the fluctuation would scale as $\sqrt{N}$ (see Fig. 12.14). However, one can always keep the number of random players at a minimum value. If this number is 2, then the fluctuation would be minimum and uniform resource allocation is guaranteed.

## 12.6 Summary

We consider a repetitive game performed by $N$ agents choosing every time (parallelly) one among the $n(\leq N)$ choices, such that each agent can be in minority: no one else made the same choice in the KPR case (typically $n = N$) and $N_k < N/2$ for the Minority Game ($n = 2$; $k = 1, 2$). The strategies to achieve this objective evolve with time bounded by $N$. Acceptable strategies are which evolve quickly (say within $\log N$ time). Also the effectiveness of a strategy is measured by the resulting utilization factor $\bar{f}$ giving the (steady state) number of occupied restaurants in any evening for the KPR, by the value of fluctuation $\Delta$ in the minority game case ($\Delta = 0$ corresponds to maximum efficiency).

The study of the KPR problem shows that a dictated solution leads to one of the best possible solution to the problem, with each agent getting his dinner at the best ranked restaurant with a period of $N$ evenings, and with best possible value of $\bar{f}$ (=1) starting from the first evening on itself. For a democratic situation (for parallel decision strategies), the agents employ stochastic algorithms based on past occupation informations (e.g., of $N_k(t)$)). These strategies are of course less efficient ($\bar{f} \ll 1$; the best one discussed in [5], giving $\bar{f} \simeq 0.8$ only). Here the time required is very weakly dependent on $N$, if any. We also note that most of the "smarter" strategies lead to much lower efficiency.

Finally we note that the stochastic strategy Minority Game [21], a very efficient one: The strategy is described by (12.11), where the agents very quickly (in $\log\log N$ time; $N = 2M + 1$) get divided almost equally ($M$ and $M + 1$) between the two choices. This strategy guarantees that a single cheater, who does not follow this strategy, will always be a loser. However, the dynamics in the system stops very quickly (leading to the absorbing state), making the resource distribution highly asymmetric (people in the majority stays there for ever) thereby making this strategy socially unacceptable. To rectify this, we noted that the presence of a single random trader (who picks between the two choices completely randomly) will avoid this absorbing state and the asymmetric distribution. However, this will always make that random trader a loser. But the presence of more than one random trader will avoid that situation too, making the average time period of switching between majority and minority for all the traders (irrespective of whether they are chartists or random traders) to be 2. Hence, the system will always evolve collectively such that only two agents will make random choices between the binary choices, while the rest $N - 2$ will follow the probabilities given by (12.11).

## References


1. Chakrabarti BK, Chakraborti A, Chatterjee A (eds) (2006) Econophysics and sociophysics. Wiley-VCH, Berlin
2. Chakrabarti AS, Chakrabarti BK, Chatterjee A, Mitra M (2009) Physica A 388:2420
3. Ghosh A, Chakrabarti BK (2009) Kolkata Paise Restaurant (KPR) problem. http://demonstrations.wolfram.com/KolkataPaiseRestaurantKPRProblem
4. Ghosh A, Chakrabarti AS, Charabarti BK (2010) In: Basu B, Charabarti BK, Ghosh A, Gangopadhyay K (eds) Econophysics and economics of games. Social choices and quantitative techniques Springer, Milan
5. Ghosh A, Chatterjee A, Mitra M, Chakrabarti BK (2010) New J Phys 12:075033
6. Ghosh A, Chatterjee A, Chakrabarti BK (in preparation)
7. Ghosh A, Martino DD, Chatterjee A, Marsili M, Chakrabarti BK (2012) Phys Rev E 85:021116
8. Rossi M, Pastor-Satorras R, Vespignani A (2000) Phys Rev Lett 85:1803
9. Dickman R, Muñoz MA, Vespignani A, Zapperi S (2000) Braz J Phys 30:27
10. Vespignani A, Dickman R, Muñoz MA, Zapperi S (2000) Phys Rev E 62:4564
11. Dickman R, Alava M, Muñoz MA, Peltola J, Vespignani A, Zapperi S (2001) Phys Rev E 64:056104
12. Manna SS (1991) J Phys A 24:L363
13. Lübeck S (2004) Int J Mod Phys B 18:3977
14. Arthur BW (1994) Am Econ Assoc Pap & Proc 84:406
15. Challet D, Zhang YC (1997) Physica A 246:407

16. Challet D, Zhang YC (1998) Physica A 256:514
17. Challet D, Marsili M, Zhang Y-C (2005) Minority games: interacting agents in financial markets. Oxford University Press, Oxford
18. Moro E (2004) In: Korutcheva E, Cuerno R (eds) Advances in condensed matter and statistical mechanics. Nova Science Publishers, New York. arXiv:cond-mat/0402651v1
19. De Martino A, Marsili M (2006) J Phys A 39:R465
20. Kets W (2007) Preprint. arXiv:0706.4432v1 [q-fin.GN]
21. Dhar D, Sasidevan V, Chakrabarti BK (2011) Physica A 390:3477
22. Reents G, Metzler R, Kinzel W (2001) Physica A 299:253
23. Biswas S, Ghosh A, Chatterjee A, Naskar T, Chakrabarti BK (2012) Phys Rev E 85:031104
24. Evans MR, Hanney T (2005) J Phys A, Math Gen 38:R195

# Chapter 13
# Kolkata Paise Restaurant Problem and the Cyclically Fair Norm

**Priyodorshi Banerjee, Manipushpak Mitra, and Conan Mukherjee**

**Abstract** In this paper we revisit the Kolkata Paise Restaurant problem by allowing for a more general (but common) preference of the $n$ customers defined over the set of $n$ restaurants. This generalization allows for the possibility that each pure strategy Nash equilibrium differs from the Pareto efficient allocation. By assuming that $n$ is small and by allowing for mutual interaction across all customers we design strategies to sustain cyclically fair norm as a sub-game perfect equilibrium of the Kolkata Paise Restaurant problem. We have a cyclically fair norm if $n$ strategically different Pareto efficient strategies are sequentially sustained in a way such that each customer gets serviced in all the $n$ restaurants exactly once between periods 1 and $n$ and then again the same process is repeated between periods $(n+1)$ and $2n$ and so on.

## 13.1 Introduction

The Kolkata Paise Restaurant problem [2, 3, 5–7] is a repeated game with identical stage (or one-shot) games and with the same set of $n$ customers (or agents or players). In each stage these $n$ customers have to simultaneously choose between $n$ restaurants to get served. All the customers have a common and rational preference ordering over the service of these $n$ restaurants and, to each customer, the least preferred outcome is not getting the service. Without loss of generality, we assume that the first restaurant is the most preferred followed by the second restaurant and so on and that getting served in the last restaurant is preferred to not getting the service. The price of getting the service from each restaurant is identical. Each restaurant

The authors would like to thank Tarun Kabiraj for his comments and suggestions.

P. Banerjee (✉) · M. Mitra · C. Mukherjee
Economic Research Unit, Indian Statistical Institute, Kolkata, India
e-mail: banpriyo@gmail.com

M. Mitra
e-mail: mmitra@isical.ac.in

C. Mukherjee
e-mail: conanmukherjee@gmail.com

F. Abergel et al. (eds.), *Econophysics of Systemic Risk and Network Dynamics*, New Economic Windows,
DOI 10.1007/978-88-470-2553-0_13, © Springer-Verlag Italia 2013

can serve only one customer so that if more than one customer arrives at the same restaurant, the restaurant randomly chooses one customer to serve and the others do not get the service in that stage. Thus, given the common preferences of the customers over the set of restaurants, the stage game of the Kolkata Paise Restaurant problem is a symmetric one. Moreover, as long as the first restaurant is strictly preferred to the last restaurant, the stage game of the Kolkata Paise Restaurant problem is non-trivial. Given the restrictions on the preferences, Pareto efficiency means that each customer goes to a different restaurant and each restaurant gets exactly one customer to serve.

In the very first work on the Kolkata Paise Restaurant problem [2], it was assumed that the common preferences of the customers is such that going to any unoccupied restaurant is strictly preferred to going to any other restaurant where at least another customer is present. This restriction implied that the set of pure strategy Nash equilibria of the stage game were all Pareto efficient. Hence there are exactly $n!(= n(n-1)\dots 2.1)$ pure strategy Nash equilibria of this version of the stage game of the Kolkata Paise Restaurant problem. If customers are rational, $n$ is small and if customers can mutually interact, then, given the fact that the set of pure strategy Nash equilibrium are also Pareto efficient, one can show that it is easy to sustain any pure strategy Nash equilibrium of the stage game of the Kolkata Paise Restaurant problem as a sub-game perfect equilibrium outcome of the Kolkata Paise Restaurant problem without designing any punishment strategy. This is because, in this context, unilateral deviation means going to a restaurant where there is already another customer which is payoff reducing. In this context it seems quite unfair to sustain exactly one pure strategy Nash equilibrium of the stage game repeatedly as a sub-game perfect Nash equilibrium of the Kolkata Paise Restaurant problem. This is because in any pure strategy Nash equilibrium of the stage game, the customer going to the first restaurant derives a strictly higher payoff than the customer going to the last restaurant. Instead it seems more natural to sustain the *cyclically fair norm* where $n$ strategically different Pareto efficient allocations are sequentially sustained in a way such that each customer gets serviced in all the $n$ restaurants exactly once between periods 1 and $n$ and then again the same process is repeated from the $(n+1)$th period to period $2n$ and so on. A variant of the cyclically fair norm was proposed in [7] under the large player assumption. However, this type of cyclically fair norm can also be sustained as a sub-game perfect Nash equilibrium because unilateral deviation at any stage means going to a restaurant already occupied by another customer which is always payoff reducing. Therefore, the existing structure of the Kolkata Paise Restaurant problem is such that if the number of customers $n$ is small and if the customers can coordinate their action then the problem becomes uninteresting as there is no need to design punishment strategies to induce customers to remain in the equilibrium path. Thus it is natural that the existing literature on Kolkata Paise Restaurant problem [2, 5–7] deals with situations where $n$ is macroscopically large so that the agents cannot rely on the other agents' actions and therefore what matters to each agent is the past collective configuration of actions and the resulting average utilization of the restaurants.

In this paper we revisit the Kolkata Paise Restaurant problem by relaxing the assumption on preferences that ensures the Pareto efficiency of all the pure strategy Nash equilibria of the stage game. Therefore, we analyze the Kolkata Paise Restaurant problem by looking at a more general (but common across agents) rational preference structure over the restaurants such that the stage game allows for the possibility of inefficient pure strategy Nash equilibria. In this scenario we assume that $n$ is small and that customers can take coordinated actions and then analyze the possibility of sustaining the cyclically fair norm as a sub-game perfect equilibrium of the Kolkata Paise Restaurant problem. Clearly, in this context, there is a need for designing punishment schemes in order to sustain the cyclically fair norm as a sub-game perfect equilibrium since unilateral deviation from the proposed norm can be payoff enhancing as the configurations under the cyclically fair norm may not be pure strategy Nash equilibria of the stage game.

## 13.2 The Stage Game

We start by formally defining and analyzing the stage game associated with the Kolkata Paise Restaurant (or KPR) problem. Let $N=\{1,\dots,n\}$ be the finite set of agents, $S=\{R1,\dots,Rn\}$ be the set of restaurants and vector $u=(u_1,\dots,u_n)\in\Re^n$ represent the utility (in terms of money) associated with each restaurant which is common to all customers or agents. Assume w.l.o.g. that $0<u_n\leq\cdots<u_2\leq u_1$ with $u_1\neq u_n$. Formally, the one shot KPR game is $G(u)\equiv(N,S,\pi)$, where $S=\{R1,\dots,Rn\}$ is the common *action* space and $\pi_i:S^n\mapsto\Re$ is the payoff function of agent $i$. For any agent $i$, $s_i=k\in S$ implies that agent $i$ chooses the strategy of going to restaurant $k$. It may so happen that more than one agents goes to the same restaurant. In that case, service is provided to only one of them and this selection is completely random. Therefore, for any strategy profile $s\in S^n$, the *expected* payoff to agent $i$, $\pi_i(s)=\frac{u_{s_i}}{\eta_i(s)}$ where $\eta_i(s)-1+|\{j\in N: j\neq i, s_i=s_j\}|$ is the number of agents that have selected the same restaurant as agent $i$. We call a strategy profile $s=(s_1,\dots,s_n)\in S^n$ Pareto efficient, if the sum of payoffs of the agents is maximized, that is,

$$s\in\arg\max_{s'\in S^n}\sum_{i\in N}\pi_i(s').$$

Given the current setting, a strategy combination leads to Pareto efficiency if and only if the strategies of the agents are such that they end up in different restaurants, that is, $\forall i,j\in N$, $s_i\neq s_j$. A strategy combination $s^*=(s_1^*,\dots,s_n^*)$ is a pure strategy *Nash equilibrium* (NE) if no agent $i$ has incentive to deviate from the existing strategy $s_i^*$ given the strategies $s_{-i}^*=(s_1^*,\dots,s_{i-1}^*,s_{i+1}^*,\dots,s_n^*)$ of the other players, that is, for each agent $i\in N$,

$$\pi_i(s^*)\geq\pi_i(s_i,s_{-i}^*)\quad\forall s_i\in S.$$

*Remark 13.1* It was proved in [2] that if $u_1 < 2u_n$ then the set of all pure strategy Nash equilibria of the one-shot KPR problem coincides with the set of all Pareto efficient strategies. This result is quite intuitive since the restriction $u_1 < 2u_n$ means that going to any unoccupied restaurant is strictly preferred to going to any other restaurant where at least another agent is present. Hence for any agent $i \in N$, given the strategy of all other agents, it is always optimum for agent $i$ to select the most preferred unoccupied restaurant. Since the number of restaurant is the same as the number of agents, it is always possible for agent $i$ to find an unoccupied restaurant. Hence in any pure strategy Nash equilibrium all agents end up in different restaurants which is also a Pareto optimal strategy combination.

Before concluding this section we provide a discussion on symmetric mixed strategy equilibria in the following remark.

*Remark 13.2* (Symmetric mixed strategy equilibria) Let $u_1 < 2u_n$ and let $A(S)$ denote the set of all mixed strategies defined over $S$.[1] A symmetric mixed strategy Nash equilibrium $\underline{p}^* = \underbrace{(p^*, \dots, p^*)}_{n} \in A(S)^N$ where $p^* = (p_1^*, \dots, p_n^*) \in [0,1]^n$ with $\sum_{i=1}^n p_i^* = 1$ is a solution to the following sets of equation:

For each $i \in N$, $\sum_{k=0}^{n-1}(1-p_i^*)^k = \frac{nc(n)}{u_i}$ for some constant $c(n)$ which is positive real.[2]

(i) For $N = \{1, 2\}$, the symmetric mixed strategy Nash equilibrium is $\underline{p}^* = (p^*, p^*)$ where $p^* = (p_1^* = \frac{2u_1-u_2}{u_1+u_2}, p_2^* = \frac{2u_2-u_1}{u_1+u_2})$ and $c(2) = \frac{3u_1u_2}{2(u_1+u_2)}$.

(ii) For $N = \{1, 2, 3\}$, there are two symmetric mixed strategy Nash equilibria. These equilibria are characterized by $p^* = (p_1^*, p_2^*, p_3^*)$ and $c(3)$ where $p_i^* = \frac{3}{2} - \frac{1}{2}\sqrt{\frac{12c(3)}{u_i} - 3}$ for all $i \in \{1, 2, 3\}$, the constant $c(3)$ takes two values given by $c(3) = \sqrt{E_1E_2E_3}\left(\frac{3(E_1+E_2+E_3)\pm\sqrt{9(E_1+E_2+E_3)^2-20(E_1^2+E_2^2+E_3^2)}}{(E_1^2+E_2^2+E_3^2)}\right)$ and $E_i = u_ju_l$ for all $i \neq j \neq l \neq i$. It can be verified that given $u_3 < 2u_1$, $9(E_1 + E_2 + E_3)^2 - 20(E_1^2 + E_2^2 + E_3^2) > 0$ and hence $c(3)$ is always positive real.

(iii) In general, for $n > 3$ such symmetric mixed strategy equilibria always exists [1]. A general feature of the symmetric mixed strategy equilibria is that $0 < p_n^* \leq \dots \leq p_1^* < 1$ and $p_1^* \neq p_n^*$.

It is quite clear from Remark 13.2 that working out the mixed strategy equilibria, in general, is difficult. Therefore, in this paper, we concentrate only on pure strategy equilibria of the stage game.

---

[1] A mixed strategy is a probability distribution defined on the strategy set. Therefore, in the present context, $A(S)$ is the set of all probability distributions on the set of restaurants $S$.

[2] For mixed strategy equilibria the required condition is $\sum_{r=0}^{n-1}\{\binom{n-1}{r}(p_i^*)^r(1-p_i^*)^{n-r-1}\frac{u_i}{r+1}\} = c(n)$ for all $i \in N$ and after simplification we get $\sum_{k=0}^{n-1}(1-p_i^*)^k = \frac{nc(n)}{u_i}$ for all $i \in N$.

## 13.3 The KPR Problem

The KPR problem is an infinitely repeated game where in each stage the same set of $N = \{1, \dots, n\}$ agents play the one shot KPR game $G(u)$ defined in the previous section.[3] We represent the KPR problem as $G^{\infty}(u) = (N, (\Sigma_i)_{i \in N}, (\Pi_i)_{i \in N})$ where $N$ is the set of agents and for any agent $i$, $\Sigma_i$ is the set of *strategies* available to $i$, while $\Pi_i$ is the payoff function of $i$. However, the concepts of strategy and payoff, have now become more complex, due to this repeated interaction setting.[4]

Let us start with the concept of strategy in $G^{\infty}(u)$. In each period $t$, the play of the one-shot KPR game would result in some action profile $s^t = (s_1^t, \dots, s_n^t) \in S^n$. Given any period $t$, define history $h_t = (s^1, s^2, \dots, s^{t-1})$ as the description of past play. That is, $h_t$ is a sequence of action profiles realized through times 1 to $t-1$. For any $t$, $h_t$ is assumed to be common knowledge. Let $H^t$ denote the set of all possible histories at time $t$. Strategy of $i$ in $G^{\infty}(u)$, specifies an action, that is, the restaurant that $i$ goes to, in each period $t$, for each possible history $h^t$. Therefore, $\forall\, i \in N, \forall\, \sigma_i \in \Sigma_i, \sigma_i : H^t \mapsto S$.

For each possible sequence of action profiles over time, we get a sequence of payoffs, for each agent. To calculate the payoff of an agent we define the concept of the discount factor $\delta \in (0, 1)$. It is presumed that agents are impatient, and hence, discount future payoffs, so that *present discounted value* of a dollar to be received one period later is $\delta$, two periods later is $\delta^2$, and so on. In general, any payoff $x$ to be received $\tau$ periods later, is valued at the present period as $\delta^{\tau} x$. Therefore, present discounted value of the infinite sequence of payoffs corresponding to any infinite sequence of action profiles $\{s^1, s^2, s^3, \dots\}$, for agent $i$, is $\sum_{t=1}^{\infty} \delta^{t-1} \pi_i(s^t)$. We assume that each agent discounts the future payoffs at same rate.[5]

*Remark 13.3* In this remark we provide two interpretations of the discount factor.

(i) The popular interpretation of discount factor $\delta$ is that it is the *time-value* of money. Suppose a person puts an amount of money $x$ in a bank at the beginning of present period. If the bank pays interest $r$ per period, upon withdrawal the person gets $x(1+r)$ money at the beginning of the next period. Therefore, we can say that amount $x$ to be received in the beginning of the next period is worth only $\frac{1}{1+r}x$ money in the present period. Setting $\delta = \frac{1}{1+r}$ we get that; at present, the next period payoff $x$ is worth $\delta x$ and the next to next period payoff $x$ is worth $\delta^2 x$. Therefore, a sequence of future payoffs $\{x^1, x^2, x^3, \dots\}$ is worth $\sum_{t=1}^{\infty} \delta^{t-1} x^t$ at present.

(ii) The concept of $\delta$, can also be used to view the infinitely repeated game as a *finite period* repeated game that ends after a *random* number of periods. Suppose that

---

[3] An infinitely repeated game like the KPR problem, where the same stage game is played repeatedly, is also referred to as a *supergame* [4].

[4] The analysis of the concepts of repeated game theory is taken from [8] and [9].

[5] It can be easily verified that the conclusions of this paper remain qualitatively same if we allow for unequal discount factors across agents.

after each period is played, a (possible weighted) coin is flipped to determine whether the game will end. If the probability that the game ends immediately is $p$ and then, with probability $1-p$, the game continues for a least one more period and then the payoff $x$, to be obtained in the next stage (if it occurs), is worth only $\frac{(1-p)x}{(1+r)}$. Similarly, a payoff $x$ to be received two periods from now (if both periods are played) is worth only $\frac{(1-p)^2x}{(1+r)^2}$ before this stage's coin flip occurs. Therefore, the sum $x+\delta x+\delta^2x+\cdots$ with $\delta=\frac{1-p}{1+r}$ reflects both the time value of money and the possibility that the game may end.

For different values of $\delta$, we get different KPR problems $G_\delta^\infty(u)$. Therefore, for any agent $i$, payoff function $\Pi_i$ in $G_\delta^\infty(u)$ is a mapping $\Pi_i:\Sigma_1\times\ldots\Sigma_n\mapsto\Re$ such that for any strategy profile $\sigma=(\sigma_1,\ldots,\sigma_n)$, $\Pi_i(\sigma)=\sum_{t=1}^\infty\delta^{t-1}\pi_i(\sigma_1^t,\sigma_2^t,\ldots,\sigma_n^t)$. A strategy profile $\sigma^*=(\sigma_1^*,\ldots,\sigma_n^*)$ is a Nash equilibrium (NE) of $G_\delta^\infty(u)$, if no agent $i$ finds it profitable to deviate unilaterally from $\sigma^*$, that is for each $i\in N$,

$$\Pi_i\big(\sigma^*\big)\geq\Pi_i\big(\sigma_i,\sigma_{-i}^*\big)\quad\forall\sigma_i\in\Sigma_i.$$

We focus on a particular strategy profile $\bar{\sigma}$ satisfying the following conditions.

(i) Without loss of generality, in period $t=1$ each agent $i(\in N)$ goes to restau rant $i$.
(ii) For any period $t>1$, if agent $i$ went to restaurant 1 in the last period $t-1$, then $i$ goes to restaurant $n$ at period $t$.
(iii) For any period $t>1$, if agent $i$ went to restaurant $k>1$ in the last period $t-1$, then $i$ goes to restaurant $k-1$ at period $t$.

Note that strategy $\bar{\sigma}$ requires that action of any agent $i$ at any period $t$ depend only on $i$'s action at period $t-1$ and *not on other agents' actions* in the past. If all agents play $\bar{\sigma}$ at $G_\delta^\infty(u)$, we get the *cyclically fair norm.*

**Proposition 13.1** *If $u_1<2u_n$, then for all $\delta\in(0,1)$, $\bar{\sigma}$ is a Nash equilibrium of $G_\delta^\infty(u)$.*

*Proof* If $u_1<2u_n$ then we know that in any period $t$, going to any unoccupied restaurant is strictly preferred to going to any other restaurant where at least another agent is present. Hence it is always optimum for any agent $i\in N$, in any period $t$, to select the most preferred unoccupied restaurant. Since the number of restaurant is the same as the number of agents, it is always possible for agent $i$ in any period $t$ to find an unoccupied restaurant.

Given $\bar{\sigma}$ it is clear that any unilateral deviation from $\bar{\sigma}$ by any agent $i$, at any time $t$, would lead to $i$ being tied with another agent at some restaurant thereby ensuring a strict reduction in payoff in that period.

Depending on the deviation strategy $\sigma_i$, in all periods after $t$, agent $i$ can face a tie or he may not face a tie. If agent $i$ faces a tie then he is strictly worse off in that period in comparison to $\bar{\sigma}_i$ and if he does not face a tie then he gets the same payoff in that period in comparison to $\bar{\sigma}_i$. The reason is that given that all other agents $j \in N \setminus \{i\}$ are continuing with the strategy $\bar{\sigma}_j$, in each period after $t$, for agent $i$, there is exactly one restaurant which is not occupied and hence, given the preference of agent $i$ we get the result. Thus, in either case $\Pi_i(\sigma_i, \bar{\sigma}_{-i}) < \Pi_i(\bar{\sigma})$ implying that $\bar{\sigma}$ is a Nash equilibrium of $G_\delta^\infty(u)$. □

Our objective is to sustain $\bar{\sigma}$ as a sub-game perfect equilibrium in order to implement the cyclically fair norm. That is, we need to show that $\bar{\sigma}$ constitutes a sub-game perfect equilibrium of $G_\delta^\infty(u)$. Before defining the sub-game perfect equilibrium we need to define a sub-game. We call any 'piece' of game $G_\delta^\infty(u)$ following any history $h^t$, at any period $t$, a *subgame* of $G_\delta^\infty(u)$. Therefore, a sub-game is that piece of the game that remains to be played beginning at any point at which the complete history of the game thus far is common knowledge. The definition of a strategy in any infinitely repeated game is closely related to the definition of a sub-game. In particular, an agent's strategy specifies the actions the agent will take in the first period of the repeated game and the first stage of each of its sub-game. There are infinite number of sub-games of $G_\delta^\infty(u)$. Since $G_\delta^\infty(u)$ is an infinitely repeated game, each of its sub-games, beginning at period $t+1$ of $G_\delta^\infty(u)$ is identical to $G_\delta^\infty(u)$. Note that the $t$th period of a repeated game taken in isolation is *not* a sub-game of the repeated game. Therefore, a sub-game is a piece of the original game that not only starts at a point where the history of the game thus far is common knowledge among the agents, but also includes all the moves that follow this point in the original game. A Nash equilibrium strategy profile $\sigma^{**} = (\sigma_1^{**}, \ldots, \sigma_n^{**})$ is a *sub-game perfect equilibrium* if these strategies constitute a Nash equilibrium in every sub-game.

**Corollary 13.1** *If $u_1 < 2u_n$, then for any $\delta \in (0, 1)$, $\bar{\sigma}$ is a sub-game perfect equilibrium of $G_\delta^\infty(u)$.*

*Proof* Given particular property of $\bar{\sigma}$ where $i$'s behavior depends only on his own past behavior, no deviation by $i$ in any period $t$ (and hence in the sub-game starting from period $t$) can induce a change in future actions of other agents in $N \setminus \{i\}$. Thus, using the arguments from the last paragraph of the proof of Proposition 13.1, it follows that $\bar{\sigma}$ continues to be a Nash equilibrium in every sub-game of $G_\delta^\infty(u)$. □

*Remark 13.4* If $u_1 = 2u_n$ then, by making minor alterations in the arguments in the proofs of Proposition 13.1 and Corollary 13.1, one can implement the cyclically fair norm as a sub-game perfect equilibrium with the same strategy $\bar{\sigma}$ and for any $\delta \in (0, 1)$. The proof is left to the reader.

Observe that the strategy profile $\bar{\sigma}$, that implements the cyclically fair norm as a sub-game perfect equilibrium when $u_1 \leq 2u_n$, is such that there is no specification of punishment in the sense that it is silent about what happens if one agent deviates from the existing strategy. However, if $u_1 > 2u_n$ then implementing the cyclically fair norm becomes non-trivial and one has to design appropriate punishment

**Table 13.1** The two-agent payoff matrix

| $G(u_1, u_2)$ | $R1$ | $R2$ |
|---|---|---|
| $R1$ | $(\frac{u_1}{2}, \frac{u_1}{2})$ | $(u_1, u_2)$ |
| $R2$ | $(u_2, u_1)$ | $(\frac{u_2}{2}, \frac{u_2}{2})$ |

schemes. To see this, consider the simple stage game with $n = 2$. The two-agent payoff matrix with 1 as the row player and 2 as the column player is presented in Table 13.1.

If $u_1 < 2u_2$ then there are only two pure strategy Nash equilibria $(R1, R2)$ and $(R2, R1)$ of the stage game. Even when $u_1 = 2u_2$, $(R1, R2)$ and $(R2, R1)$ continue to be pure strategy Nash equilibria of the stage game. Therefore, the Pareto efficient strategies where both agents go to different restaurants are pure strategy Nash equilibria for $u_2 \geq \frac{u_1}{2}$. It is this strong result that drives Proposition 13.1 and we can easily implement the cyclically fair norm. Note that if $u_2 = \frac{u_1}{2}$, then there are three pure strategy Nash equilibria $(R1, R2)$, $(R2, R1)$ and $(R1, R1)$ of the stage game. Therefore, for $u_1 = 2u_2$, there exists a sub-game prefect equilibrium that leads to inefficiency in every period. Specifically, the strategy that specifies that each agent should go to the first restaurant in all periods is a sub-game perfect equilibrium, where sum of the stage game payoffs of the two agents is $u_1$ which is strictly less that sum $u_1 + u_2$ that results under any Pareto optimal strategy. The problem gets only worse if $2u_2 < u_1$, because now there is only one pure strategy Nash equilibrium $(R1, R1)$ of the stage game which is not Pareto efficient. How to design strategies to implement the cyclically fair norm as a sub-game perfect equilibrium when $n = 2$ and $2u_2 < u_1$ is discussed in the next section.

## 13.4 The Two Agent Problem

In this section we show that for $N = \{1, 2\}$ and for $u_2 < \frac{u_1}{2}$, if agents are sufficiently patient (that is, if $\delta$ is sufficiently high), then, by designing an appropriate strategy one can implement the cyclically fair norm as a sub-game perfect equilibrium of $G_\delta^\infty(u_1, u_2)$. The strategy profile we propose, to implement the cyclically fair norm, is $\sigma^c = (\sigma_1^c, \sigma_2^c)$, that specifies the following.

(i) Without loss of generality, if $t$ is odd, then agent 1 plays $R1$ and agent 2 plays $R2$.
(ii) If $t$ is even, then agent 2 plays $R1$ and agent 1 plays $R2$.
(iii) If in any period $t$ both agents end up in the same restaurant, then from $t + 1$ onwards both agents play $R1$.

Strategy profiles of the type $\bar{\sigma}^c$ above, are called *trigger strategies* because agents cooperate until someone fails to cooperate, which triggers a switch to non-cooperation forever. In other words, each agent is willing to settle for lower payoffs under the expectation that the other agent would do the same. If some agent breaks this cooperative arrangement, the other agent *punishes* the deal-breaker by playing certain actions (for all periods in future) that ensure lower present discounted payoffs. Thus, a deviation *triggers* a punishment play by the non-deviating agents. Such trigger strategies are sub-game perfect only if the punishment play for all future periods, induced by these strategies, are *credible*. This credibility, in turn, requires that the punishment play be the Nash equilibrium of $G_\delta^\infty(u_1, u_2)$ as a whole.

**Proposition 13.2** *For all* $\delta \in (\frac{u_1-2u_2}{u_1}, 1)$, *the strategy profile* $\bar{\sigma}^c = (\sigma_1^c, \sigma_2^c)$ *is a Nash equilibrium of* $G_\delta^\infty(u_1, u_2)$.

*Proof* We first assume that agent 1 plays strategy $\bar{\sigma}_1^c$. Given $\bar{\sigma}_1^c$, we show that if $\delta \in (\frac{u_1-2u_2}{u_1}, 1)$ then $\bar{\sigma}_2^c$ is the best response of agent 2.

It is clear that at any history, if agent 1 decides to play $R1$ in all future periods, then, given $2u_2 < u_1$, the best response of agent 2 is to play $R1$ in all future periods.

Consider the other possibility, that is, agent 1 decides to alternate between restaurants 1 and 2 at each odd and even period, respectively. Then it is obvious that the best response of agent 2 at any even period is to play $R1$ (since agent 1 is playing $R2$ and $u_1 > u_2$). However, finding the best response of agent 2, at *odd* periods (when agent 1 plays $R1$), is a little more complicated. If agent 2 chooses $R1$, then as per $\bar{\sigma}^c$, agent 1 plays $R1$ at all future periods, giving 2 a present discounted payoff $\frac{u_1}{2} + \delta\frac{u_1}{2} + \delta^2\frac{u_1}{2} + \cdots = \frac{u_1}{2(1-\delta)}$. Define $P$ to be the present discounted payoff that agent 2 gets by making the *optimal action choice* at any such odd period. Therefore, if the optimal choice of agent 2 is $R1$, then $P = \frac{u_1}{2(1-\delta)}$. If the optimal choice of agent 2 is $R2$, then in the next period, that is in period $t+1$ which is even, agent 1 plays $R2$. As mentioned before, the best response of 2 at $t+1$ is $R1$, and so we have the following: (i) agent 2 gets payoff $u_1$ at $t+1$ and (ii) agent 2 faces the same choice problem in period $t+2$ as in period $t$ and, since all sub-games of $G_\delta^\infty(u_1, u_2)$ is $G_\delta^\infty(u_1, u_2)$ itself, agent 2 selects $P$. Therefore, if $R2$ has to be the optimal choice of agent 2 at all odd periods $t$ then agent 2 gets $u_2 + \delta u_1 + \delta^2 P$ and hence, by definition of $P$ and using observations (i) and (ii), it follows that $P$ has to satisfy the condition that $P = \max\{\frac{u_1}{2(1-\delta)}, u_2 + \delta u_1 + \delta^2 P\} = u_2 + \delta u_1 + \delta^2 P$. If $P = u_2 + \delta u_1 + \delta^2 P$ then we get $P = \frac{u_1\delta + u_2}{1-\delta^2}$. Finally, for $\bar{\sigma}_2^c$ to be the best response it is both necessary and sufficient that $\frac{u_1\delta+u_2}{1-\delta^2} > \frac{u_1}{2(1-\delta)}$ which holds for all $\delta \in (\frac{u_1-2u_2}{u_1}, 1)$. Therefore, at any odd period the best response of agent 2 is $R2$ implying that for all $\delta \in (\frac{u_1-2u_2}{u_1}, 1)$, the strategy $\bar{\sigma}_2^c$ is the best response of agent 2 when agent 1 plays $\bar{\sigma}_1^c$. Using very similar arguments it is now quite easy to show that for all $\delta \in (\frac{u_1-2u_2}{u_1}, 1)$, the strategy $\bar{\sigma}_1^c$ is the best response of agent 1 when agent 2 plays $\bar{\sigma}_2^c$. Hence, $\bar{\sigma}^c = (\bar{\sigma}_1^c, \bar{\sigma}_2^c)$ is a Nash equilibrium of $G_\delta^\infty(u_1, u_2)$ for all $\delta \in (\frac{u_1-2u_2}{u_1}, 1)$. □

**Corollary 13.2** *For all* $\delta \in (\frac{u_1-2u_2}{u_1}, 1)$, *the strategy profile* $\bar{\sigma}^c = (\sigma_1^c, \sigma_2^c)$ *is a sub-game perfect equilibrium of* $G_\delta^\infty(u_1, u_2)$.

*Proof* The set of sub-games of $G_\delta^\infty(u_1, u_2)$ can be partitioned into two classes. One class following those histories where each agent followed the cyclically fair norm and alternated between restaurants 1 and 2 in a way such that Pareto efficiency is achieved in every period. The other class following those histories where there has been a tie at some restaurant and agents have shifted to $(R1, R1)$ from the next period onwards. Recall that every sub-game of $G_\delta^\infty(u_1, u_2)$ is $G_\delta^\infty(u_1, u_2)$ itself. If agents adopt strategy $\bar{\sigma}^c$ for the game as a whole, then they end up playing strategy $\bar{\sigma}^c$ in sub-games of the first class and (ii) the punishment play $(R1, R1)$ in each period of sub-games of the second class. By Proposition 13.2, for sub-games of the first type, strategies $\bar{\sigma}^c$ constitute a Nash equilibrium. For sub-games of the second type, the punishment play of $R1$ by both agents at all periods constitutes a Nash equilibrium of $G_\delta^\infty(u_1, u_2)$ since $(R1, R1)$ is the unique Nash equilibrium of $G(u_1, u_2)$ when $2u_2 < u_1$. Hence, the punishment play is always credible and the result follows. □

From Corollary 13.2 it follows that as long as agents are sufficiently patient, the strategy profile $\bar{\sigma}^c$ implements the cyclically fair norm. Therefore, the bound on $\delta$, obtained in Proposition 13.2 above, signifies the need for sufficiently patient agents to implement cyclically fair norm that calls for cooperative behavior. If agents feel the need to obtain high payoffs in the future (or equivalently if $\delta$ is high enough) then they are willing to make a sacrifice by going to the inferior restaurant in alternate periods in order to maximize long term individual payoff. In the next section we analyze the KPR problem with three agents. We show how using different strategy profiles one can implement the cyclically fair norm.

## 13.5 The Three Agent Problem

We depict the payoff matrices of $G(u_1, u_2, u_3)$ in Tables 13.2, 13.3 and 13.4.

Recall that if the one shot game $G(u_1, u_2, u_3)$ represents the one shot game of the KPR problem then $u_1 \geq u_2 \geq u_3 > 0$ and $u_1 \neq u_3$. With different types of additional conditions on $u_1$, $u_2$ and $u_3$, we identify and discuss the associated set of pure strategy Nash equilibria in the following cases.

(N1) If $u_2 < \frac{u_1}{3}$ then there is a unique pure strategy Nash equilibrium $(R1, R1, R1)$ of $G(u_1, u_2, u_3)$. This equilibrium is inefficient.

(N2) If $u_3 < u_2 = \frac{u_1}{3}$ then the four pure strategy Nash equilibria of $G(u_1, u_2, u_3)$ are $(R1, R1, R1)$, $(R1, R1, R2)$, $(R1, R2, R1)$ and $(R2, R1, R1)$. All these equilibria are inefficient. The equilibrium $(R1, R1, R1)$ is Pareto dominated by all the remaining equilibria since the aggregate payoff under $(R1, R1, R1)$ is $u_1$ which is strictly less than the aggregate payoff $(u_1 + u_2)$ that results from each of the remaining equilibria.

(N3) If we have $u_3 = u_2 = \frac{u_1}{3}$, then $(R1, R1, R3)$, $(R1, R3, R1)$ and $(R3, R1, R1)$ are also pure strategy Nash equilibria along with the other equilibria specified in (N2) and hence we have seven pure strategy Nash equilibria. Again,

**Table 13.2** The payoff matrix when agent 3 plays $R1$

| $G(u_1, u_2, u_3)$ | $R1$ | $R2$ | $R3$ |
|---|---|---|---|
| $R1$ | $(\frac{u_1}{3}, \frac{u_1}{3}, \frac{u_1}{3})$ | $(\frac{u_1}{2}, u_2, \frac{u_1}{2})$ | $(\frac{u_1}{2}, u_3, \frac{u_1}{2})$ |
| $R2$ | $(u_2, \frac{u_1}{2}, \frac{u_1}{2})$ | $(\frac{u_2}{2}, \frac{u_2}{2}, u_1)$ | $(u_2, u_3, u_1)$ |
| $R3$ | $(u_3, \frac{u_1}{2}, \frac{u_1}{2})$ | $(u_3, u_2, u_1)$ | $(\frac{u_3}{2}, \frac{u_3}{2}, u_1)$ |

**Table 13.3** The payoff matrix when agent 3 plays $R2$

| $G(u_1, u_2, u_3)$ | $R1$ | $R2$ | $R3$ |
|---|---|---|---|
| $R1$ | $(\frac{u_1}{2}, \frac{u_1}{2}, u_2)$ | $(u_1, \frac{u_2}{2}, \frac{u_2}{2})$ | $(u_1, u_3, u_2)$ |
| $R2$ | $(\frac{u_2}{2}, u_1, \frac{u_2}{2})$ | $(\frac{u_2}{3}, \frac{u_2}{3}, \frac{u_2}{3})$ | $(\frac{u_2}{2}, u_3, \frac{u_2}{2})$ |
| $R3$ | $(u_3, u_1, u_2)$ | $(u_3, \frac{u_2}{2}, \frac{u_2}{2})$ | $(\frac{u_3}{2}, \frac{u_3}{2}, u_2)$ |

**Table 13.4** The payoff matrix when agent 3 plays $R3$

| $G(u_1, u_2, u_3)$ | $R1$ | $R2$ | $R3$ |
|---|---|---|---|
| $R1$ | $(\frac{u_1}{2}, \frac{u_1}{2}, u_3)$ | $(u_1, u_2, u_3)$ | $(u_1, \frac{u_3}{2}, \frac{u_3}{2})$ |
| $R2$ | $(u_2, u_1, u_3)$ | $(\frac{u_2}{2}, \frac{u_2}{2}, u_3)$ | $(u_2, \frac{u_3}{2}, \frac{u_3}{2})$ |
| $R3$ | $(\frac{u_3}{2}, u_1, \frac{u_3}{2})$ | $(\frac{u_3}{2}, u_2, \frac{u_3}{2})$ | $(\frac{u_3}{3}, \frac{u_3}{3}, \frac{u_3}{3})$ |

the equilibrium $(R1, R1, R1)$ is Pareto dominated by the other six non-comparable equilibria. The equilibria are inefficient.

(N4) If $\max\{u_3, \frac{u_1}{3}\} < u_2 < \frac{u_1}{2}$ then the three pure Nash strategy equilibria of the game $G(u_1, u_2, u_3)$ are $(R1, R1, R2)$, $(R1, R2, R1)$ and $(R2, R1, R1)$. All these equilibria lead to the same aggregate payoff and hence, are Pareto non-comparable. The equilibria are inefficient.

(N5) If $u_3 = u_2 < \frac{u_1}{2}$ then $(R1, R1, R3)$, $(R1, R3, R1)$ and $(R3, R1, R1)$ are also pure strategy Nash equilibria along with the other equilibria specified in (N4) and hence we have six pure strategy Nash equilibria of $G(u_1, u_2, u_3)$. The equilibria are Pareto non comparable and inefficient.

(N6) If $u_3 < \frac{u_1}{2} \leq u_2 \leq u_1$ then the three pure strategy equilibria of $G(u_1, u_2, u_3)$ are $(R1, R1, R2)$, $(R1, R2, R1)$ and $(R2, R1, R1)$. All these equilibria are inefficient and Pareto non-comparable.

(N7) If $\frac{u_1}{2} \leq u_3$ then we do not identify all possible pure strategy Nash equilibria. However, what is important is that the Pareto efficient strategies, characterized by all agents going to different restaurants, are all included in the set of all pure strategy Nash equilibria.

The equilibria in case (N7) above are uninteresting as implementation of the cyclically fair norm as a sub-game perfect equilibrium of $G^{\infty}(u_1, u_2, u_3)$ is quite easy (see Corollary 13.1 and Remark 13.4). For cases (N1)–(N3), all agents going to the best restaurant, that is, the strategy profile $(R1, R1, R1)$ constitutes a pure strategy Nash equilibrium and is Pareto inefficient. Therefore, as in Proposition 13.2 and Corollary 13.2, for cases (N1)–(N3), we can implement the cyclically fair norm as a sub-game perfect equilibrium of $G_\delta^\infty(u_1, u_2, u_3)$ for $\delta$ sufficiently close to one. It can be easily shown that, by designing a strategy profile which is a natural extension of $\sigma^c = (\sigma_1^c, \sigma_2^c)$ to the three agent case and that specifies that the non-deviating agents punish the deviating agent by going to the best restaurant for all future periods, one can implement the cyclically fair norm. For the rest of the cases, that is, (N4)–(N6), strategy profile $(R1, R1, R1)$ fails to be a pure strategy Nash equilibrium of the stage game, and so, implementing the cyclically fair norm becomes more subtle. This is because the threat of punishment embodied in strategy of type $\sigma^c$, that is, going to the best restaurant for all future periods, is no longer a credible one as it is not a Nash equilibrium of the stage game.

In the rest of this section, we focus on the interesting cases (N4), (N5) and (N6). Since there is no qualitative difference between (N4) and (N5), we analyze only cases (N4) and (N6) in detail. As long as agents are sufficiently patient, we can show that for both cases we can implement the cyclically fair norm as a sub-game perfect equilibrium of the KPR problem $G_\delta^\infty(u_1, u_2, u_3)$. Interesting to note here is that, for cases (N4) and (N6), the set of pure strategy Nash equilibria is $\{(R1, R1, R2), (R1, R2, R1), (R2, R1, R1)\}$. However, while for (N4), each agent playing $R1$ gets *more expected payoff* than the agent playing $R2$, for (N6), each agent playing $R1$ gets an expected payoff which is *no more* than the payoff of the agent playing $R2$. It is precisely for this difference in payoffs for the same given pure strategy Nash equilibrium for cases (N4) and (N6) that calls for designing different punishment strategies to implement the cyclically fair path.

Consider first (N4), that is $\max\{u_3, \frac{u_1}{3}\} < u_2 < \frac{u_1}{2}$. Consider the strategy profile $\sigma^a = (\sigma_1^a, \sigma_2^a, \sigma_3^a)$ that specifies the following.

(i) Without loss of generality at $t = 1$, each agent $i \in \{1, 2, 3\}$ plays $Ri$.
(ii) If agent $i$ plays $R1$ in period $t - 1$, then $i$ plays $R3$ in period $t$.
(iii) If agent $i$ plays $Rk \neq R1$ in period $t - 1$, then $i$ plays $R(k - 1)$ in period $t$.
(iv) If any agent $i$ violates either of 1, 2 or 3 in some period $t$ then in all future periods $t + 1, t + 2, \ldots$, all the non-deviating agents $(N \setminus \{i\})$ plays $R1$.

Conditions (i)–(iii) in the strategy profile $\sigma^a$ ensures that agents follow the cyclically fair norm. Condition (iv) is the punishment requirement that specifies that, if an agent deviates, then the non-deviating agents punish the deviating agent by playing $R1$, for all future periods.

**Proposition 13.3** *If* $\max\{u_3, \frac{u_1}{3}\} < u_2 < \frac{u_1}{2}$, *then there exists* $\bar{\delta} \in (0, 1)$ *such that for all* $\delta \in (\bar{\delta}, 1)$, $\sigma^a = (\sigma_1^a, \sigma_2^a, \sigma_3^a)$ *is a sub-game perfect equilibrium of* $G_\delta^\infty(u_1, u_2, u_3)$.

*Proof* Fix agent 2's strategy at $\sigma_2^a$ and agent 3's strategy at follows $\sigma_3^a$. We first show that, given this specification, $\sigma_1^a$ is the best response for agent 1 provided agent 1 is sufficiently patient. Consider agent 1 at any history and at any time $t$. Given the utility restriction $\max\{u_3, \frac{u_1}{3}\} < u_2 < \frac{u_1}{2}$, agent 1 has an incentive to deviate only if at time $t$, agent 1 is supposed to play either $R3$ or $R2$ (otherwise agent 1 has no profitable deviation in the stage game at $t$ when the strategy prescribes $R1$). Also, the best deviation for agent 1 is to play $R1$ and get a payoff of $\frac{u_1}{2}$. If agent 1 deviates, then, following the strategy profile $\sigma^a$, agents 2 and 3 play $R1$ for all future periods. At each of such future periods, the best response of agent 1 is to play $R2$. Therefore, the present discounted payoff of agent 1 from deviation is

$$D_1(\delta) = \frac{u_1}{2} + \left(\delta u_2 + \delta^2 u_2 + \cdots\right) = \frac{u_1(1-\delta) + 2\delta u_2}{2(1-\delta)}. \tag{13.1}$$

By not deviating in a period $t$ where agent 1 had to play $R2$ (under conditions (i)–(iii)), agent 1's present discounted value of payoff from period $t$ onwards is

$$E_2(\delta) = \left(u_2 + \delta u_1 + \delta^2 u_3\right) + \left(\delta^3 u_2 + \delta^4 u_1 + \delta^5 u_3\right) + \cdots = \frac{u_2 + \delta u_1 + \delta^2 u_3}{(1-\delta^3)}. \tag{13.2}$$

Similarly, by not deviating in a period $t'$ where agent 1 had to play $R3$, agent 1's present discounted value of payoff from period $t'$ onwards is

$$E_3(\delta) = \left(u_3 + \delta u_2 + \delta^2 u_1\right) + \left(\delta^3 u_3 + \delta^4 u_2 + \delta^5 u_1\right) + \cdots = \frac{u_3 + \delta u_2 + \delta^2 u_1}{(1-\delta^3)}. \tag{13.3}$$

A sufficient condition for $\sigma_1^a$ to be a best response for agent 1 (given the strategies $\sigma_2^a$ and $\sigma_3^a$ of agents 2 and 3 respectively) is that $\min\{E_2(\delta), E_3(\delta)\} > D_1(\delta)$. Note that $\min\{E_2(\delta), E_3(\delta)\} = E_3(\delta)$ since $E_2(\delta) - E_3(\delta) = \frac{(u_2-u_3)(1-\delta^2)+(u_1-u_2)\delta(1-\delta)}{(1-\delta^3)} > 0$. Therefore, for any $\delta \in (0,1)$ such that $E_3(\delta) - D_1(\delta) > 0$, $\sigma_1^a$ is the best response for agent 1. Observe that $E_3(\delta) - D_1(\delta) = \frac{F(\delta)}{2(1-\delta^3)}$ where $F(\delta) = 2(u_3 + \delta u_2 + \delta^2 u_1) - (u_1(1-\delta) + 2\delta u_2)(1 + \delta + \delta^2)$. Note that $F(\delta)$ is continuous in $\delta$ and, given (N4), $F(0) = 2u_3 - u_1 < 0$ and $F(1) = 4(\frac{u_1}{2} + \frac{u_3}{2} - u_2) > 0$. Hence, there exists a $\bar{\delta} \in (0,1)$ such that for all $\delta \in (\bar{\delta}, 1)$, $F(\delta) > 0$ and $\sigma_1^a$ is the best response for agent 1. Using similar arguments it is easy to show that $\sigma_2^a$ is the best response for agent 2 against $\sigma_1^a$ and $\sigma_3^a$ and $\sigma_3^a$ is the best response for agent 3 against $\sigma_1^a$ and $\sigma_2^a$. Hence $\sigma^a$ is a Nash equilibrium of $G_\delta^\infty(u_1, u_2, u_3)$ for all $\delta \in (\bar{\delta}, 1)$. Finally, since the punishment play induced by $\sigma^a$ is either of the three pure strategy Nash equilibria $(R1, R2, R1)$, $(R2, R1, R1)$ and $(R1, R1, R2)$; it is credible. Hence, for all $\delta \in (\bar{\delta}, 1)$, the strategy profile $\sigma^a = (\sigma_1^a, \sigma_2^a, \sigma_3^a)$ is also a sub-game perfect equilibrium. □

To implement the cyclically fair norm for the KPR problem for the case (N6), that is for the stage game $G(u_1, u_2, u_3)$ satisfying $u_3 < \frac{u_1}{2} \le u_2 \le u_1$, we consider the strategy profile $\sigma^b = (\sigma_1^b, \sigma_2^b, \sigma_3^b)$ that specifies the following conditions.

(i) Without loss of generality at $t = 1$, each agent $i \in \{1, 2, 3\}$ plays $Ri$.
(ii) If agent $i$ plays $R1$ in period $t-1$, then $i$ plays $R3$ in period $t$.
(iii) If agent $i$ plays $Rk \neq R1$ in period $t-1$, then $i$ plays $R(k-1)$ in period $t$.
(iv) If any agent $i$ violates either of 1, 2 or 3 in some period $t$, then we have the following:

a. If the deviation is initiated by agent 1, then, for all future periods, agent 2 plays $R2$ and agent 3 plays $R1$.
b. If the deviation is initiated by agent 2, then, for all future periods, agent 1 plays $R1$ and agent 3 plays $R2$.
c. If the deviation is initiated by agent 3, then, for all future periods, agent 1 plays $R2$ and agent 2 plays $R1$.

The first three conditions of strategy profile $\sigma^b$ are identical to that of the strategy profile $\sigma^a$ since these three conditions are meant to induce cooperative behavior across agents in order to implement the cyclically fair norm. However, the punishment scheme (that is, condition (iv)) under the strategy profile $\sigma^b$ is different and more subtle compared the punishment scheme under $\sigma^a$. Under $\sigma^b$, the two non-deviating agents punish the deviating agent by going to two different restaurant by playing $R1$ and $R2$. The best response to this behavior, at any stage game (irrespective of the identity of the deviating agent) is to go to $R1$. Thus, the punishment scheme generates any one of the three pure strategy Nash equilibria—$(R1, R2, R1)$, $(R2, R1, R1)$ and $(R1, R1, R2)$; where the deviating agent gets stage game payoff of $\frac{u_1}{2}$ for all future periods after the deviation period.

**Proposition 13.4** *If* $u_3 < \frac{u_1}{2} \leq u_2 \leq u_1$, *then there exists* $\delta^* \in (0, 1)$ *such that for all* $\delta \in (\delta^*, 1)$, *the strategy profile* $\sigma^b = (\sigma_1^b, \sigma_2^b, \sigma_3^b)$ *is a sub-game perfect equilibrium of* $G_\delta^\infty(u_1, u_2, u_3)$.

*Proof* We first show that if agent 2 plays $\sigma_2^b$ and agent 3 plays $\sigma_3^b$ then playing $\sigma_1^b$ is the best response for agent 1. Observe that the most profitable deviation at any time $t$ available to 1 is to play $R1$ in that period $t$ where the prescribed strategy under $\sigma^b$ for agent 1 is $R3$. If agent 1 decides to deviate then, as per $\sigma^b$, for all future periods, agent 2 plays $R2$ and agent 3 plays $R1$. Given this punishment strategy followed by agents 2 and 3, the best response of agent 1, in all future periods, is to play $R1$. Therefore, the resultant punishment play at each period in future is $(R1, R2, R1)$ with each stage payoff of $\frac{u_1}{2}$ to 1. So, the present discounted value of the payoff sequence that results after deviation for agent 1 is $D_2(\delta) = \frac{u_1}{2} + \delta\frac{u_1}{2} + \delta^2\frac{u_1}{2} + \cdots = \frac{u_1}{2(1-\delta)}$. By not deviating in a period $t$ (where agent 1 had to play $R3$) and following $\sigma_1^b$, agent 1 gets a present discounted value payoff that equals $E_3(\delta) = (u_3 + \delta u_2 + \delta^2 u_1) + \cdots = \frac{u_3+\delta u_2+\delta^2 u_1}{1-\delta^3}$. If for any $\delta \in (0, 1)$, $E_3(\delta) - D_2(\delta) > 0$, then $\sigma_1^b$ is the best response for agent 1. Observe that $E_3(\delta) - D_2(\delta) = \frac{G(\delta)}{2(1-\delta^3)}$ where $G(\delta) = 2(u_3 + \delta u_2 + \delta^2 u_1) - u_1(1 + \delta + \delta^2)$. Note that $G(\delta)$ is continuous in $\delta$ and, given (N6), $G(0) = 2u_3 - u_1 < 0$ and $G(1) = 2(u_2 + u_3 - \frac{u_1}{2}) > 0$. Hence, there exists a $\delta^* \in (0, 1)$ such that for all $\delta \in (\delta^*, 1)$, $G(\delta) > 0$ and $\sigma_1^b$ is the best response for agent 1. Using similar arguments it is easy to show that $\sigma_2^b$ is the best response for agent 2 and $\sigma_3^b$ is the best response for agent 3. Hence $\sigma^b$ is a Nash equilibrium of $G_\delta^\infty(u_1, u_2, u_3)$ for all $\delta \in (\delta^*, 1)$. Finally, since the punishment play induced by $\sigma^b$ is a pure strategy Nash equilibrium, it is credible. Hence, for all $\delta \in (\delta^*, 1)$, the strategy profile $\sigma^b = (\sigma_1^b, \sigma_2^b, \sigma_3^b)$ is also a sub-game perfect equilibrium. □

## 13.6 Summary

Using simple and basic techniques from infinitely repeated games with discounting we have established how with small number of players one can sustain the cyclically fair norm as an equilibrium in a KPR problem with general preference structures. In particular, we have highlighted how to design pure strategies, that at times requires careful designing of the punishment scheme for perpetrators, to sustain the cyclically fair norm that requires cyclical stage game sacrifices on part of the agents. We have established the following results.

(i) If for the KPR problem the set of pure strategy Nash equilibria of the stage game includes the set of all Pareto efficient strategies then there is no need to design punishment schemes to implement the cyclically fair norm as a subgame perfect equilibrium of the KPR problem.
(ii) Sufficiently high patience level of the agents and the design of appropriate punishment strategies become mandatory when, for the KPR problem with either two agents or three agents, the set of pure strategy Nash equilibria of the stage game does not include the set of all Pareto efficient strategies.
(iii) The punishment scheme that works for the two agent case is one where the deviating agent is punished by shifting to the inefficient Nash equilibrium of the stage game for all future periods after the deviation. This kind of punishment is enough to deter a rational agent with sufficiently high patience level from unilateral deviation.
(iv) For the three agent KPR problem one needs to design different types of punishment schemes as, depending on the restrictions on the (common) preferences, we have different sets of pure strategy Nash equilibria of the stage game. The restrictions on preferences that are of interest are the following — (a) $\frac{u_1}{2} > u_2 > u_3$ and (b) $u_2 \geq \frac{u_1}{2} > u_3$. For both these cases the set of pure strategy Nash equilibria of the stage game are identical and yet one needs to design different pure strategies to implement the cyclically fair norm. For both (a) and (b), the pure strategy Nash equilibrium of the stage game requires two agents going to the first restaurant and one agent going to the second restaurant. However, for case (a), the expected payoff associated with going to the first restaurant is more than the payoff obtained from going to the second restaurant, but, for case (b), the expected payoff associated with going to the first restaurant is weakly less than the payoff obtained from going to the second restaurant. Therefore, while designing the punishment scheme for the perpetrators one needs to incorporate this payoff difference across (a) and (b) and hence we require two different strategies to implement the same cyclically fair norm.

We believe that for the KPR problems with more than three agents and general preference structure, the designing of punishment schemes to implement the cyclically fair norm is an important issue that needs to be addressed in greater detail.

## References

1. Becker JG, Damianov DS (2006) On the existence of symmetric mixed strategy equilibria. Econ Lett 90:84–87
2. Chakrabarti AS, Chakrabarti BK, Chatterjee A, Mitra M (2009) The kolkata paise restaurant problem and resource utilization. Physica A 388(12):2420
3. Chakrabarti BK, Chakraborti A, Chatterjee A (eds) (2006) Econophysics and sociophysics. Wiley-VCH, Berlin
4. Friedman J (1971) A non-cooperative equilibrium for supergames. Rev Econ Stud 38:1–12
5. Ghosh A, Chakrabarti AS, Chakrabarti BK (2010) Kolkata paise restaurant problem in some uniform learning strategy. In: Basu B, Chakrabarti BK, Ghosh A, Chakravarty SR, Gangopadhyay K (eds) Econophysics and economics of games, social choices and quantitative techniques. Springer, Milan
6. Ghosh A, Chakrabarti BK (2009) Kolkata paise restaurant (KPR) problem. http://demonstrations.wolfram.com/KolkataPaiseRestaurantKPRProblem
7. Ghosh A, Chatterjee A, Mitra M, Chakrabarti BK (2010) Statistics of the kolkata paise restaurant problem. New J Phys 12:075033
8. Gibbons R (1992) Game theory for applied economists. Princeton University Press, Princeton
9. Mailath GJ, Samuelson L (2006) Repeated games and reputations: long-run relationships. Oxford University Press, London

# Chapter 14
# An Introduction to Multi-player, Multi-choice Quantum Games: Quantum Minority Games & Kolkata Restaurant Problems

**Puya Sharif and Hoshang Heydari**

**Abstract** We give a self contained introduction to a few quantum game protocols, starting with the quantum version of the two-player two-choice game of Prisoners dilemma, followed by an $n$-player generalization trough the quantum minority games, and finishing with a contribution towards an $n$-player $m$-choice generalization with a quantum version of a three-player Kolkata restaurant problem. We have omitted some technical details accompanying these protocols, and instead laid the focus on presenting some general aspects of the field as a whole. This review contains an introduction to the formalism of quantum information theory, as well as to important game theoretical concepts, and is aimed to work as a review suiting economists and game theorists with limited knowledge of quantum physics as well as to physicists with limited knowledge of game theory.

## 14.1 Introduction

Quantum game theory is the natural intersection between three fields. Quantum mechanics, information theory and game theory. At the center of this intersection stands one of the most brilliant minds of the 20th century, John von Neumann. As one of the early pioneers of quantum theory, he made major contributions to the mathematical foundation of the field, many of them later becoming core concepts in the merger between quantum theory and information theory, giving birth to quantum computing and quantum information theory [1], today being two of the most active fields of research in both theoretic and experimental physics. Among economists may he be mostly known as the father of modern game theory [2–4], the study of rational interactions in strategic situations. A field well rooted in the influential book *Theory of Games and Economic Behavior* (1944), by Von Neumann and Oscar Morgenstern. The book offered great advances in the analysis of strategic games

P. Sharif (✉) · H. Heydari
Department of Physics, Stockholm University, 10691 Stockholm, Sweden
e-mail: ps@puyasharif.net

H. Heydari
e-mail: hoshang@fysik.su.se

F. Abergel et al. (eds.), *Econophysics of Systemic Risk and Network Dynamics*, New Economic Windows,
DOI 10.1007/978-88-470-2553-0_14, © Springer-Verlag Italia 2013

and in the axiomatization of measurable utility theory, and drew the attention of economists and other social scientists to these subjects. For the last decade or so there has been an active interdisciplinary approach aiming to extend game theoretical analysis into the framework of quantum information theory, through the study of quantum games [5–10]; offering a variety of protocols where use of quantum peculiarities like entanglement in quantum superpositions, and interference effects due to quantum operations has shown to lead to advantages compared to strategies in a classical framework. The first papers appeared in 1999. Meyer showed with a model of a penny-flip game that a player making a *quantum move* always comes out as a winner against a player making a *classical* move regardless of the classical players choice [11]. The same year Eisert et al. published a quantum protocol in which they overcame the dilemma in Prisoners dilemma [12]. In 2003 Benjamin and Hayden generalized Eisert's protocol to handle multi-player quantum games and introduced the quantum minority game together with a solution for the four player case which outperformed the classical randomization strategy [13]. These results were later generalized to the $n$-players by Chen et al. in 2004 [14]. Multi-player minority games has since then been extensively investigated by Flitney et al. [15–17]. An extension to multi-choice games, as the Kolkata restaurant problem was offered by the authors of this review, in 2011 [18].

### *14.1.1 Games as Information Processing*

Information theory is largely formulated independent of the physical systems that contains and processes the information. We say that the theory is substrate independent. If you read this text on a computer screen, those bits of information now represented by pixels on your screen has traveled through the web encoded in electronic pulses through copper wires, as burst of photons trough fiber-optic cables and for all its worth maybe on a piece of paper attached to the leg of a highly motivated raven. What matters from an information theoretical perspective is the existence of a differentiation between some states of affairs. The general convention has been to keep things simple and the smallest piece of information is as we all know a *bit* $b \in \{0, 1\}$, corresponding to a binary choice: *true* or *false*, *on* or *off*, or simply *zero* or *one*. Any chunk of information can then be encoded in strings of bits: $\mathbf{b} = b_{n-1}b_{n-2}\cdots b_0 \in \{0, 1\}^n$. We can further define functions on strings of bits, $f : \{0, 1\}^n \rightarrow \{0, 1\}^k$ and call these functions computations or actions of information processing.

In a similar sense games are in their most general form independent of a physical realization. We can build up a formal structure for some strategic situation and model cooperative and competitive behavior within some constrained domain without regards to who or what these game playing agents are or what their actions actually is. No matter if we consider people, animals, cells, multinational companies or nations, simplified models of their interactions and the accompanied consequences can be formulated in a general form, within the framework of game theory.

Lets connect these two concepts with an example. We can create a one to one correspondence with between the conceptual framework of game theory and the formal structure of information processing. Let there be $n$ agents faced with a binary choice of joining one of two teams. Each choice is represented by a binary bit $b_i \in \{0, 1\}$. The final outcome of these individual choices is then given by an $n$-bit output string $\mathbf{b} \in \{0, 1\}^n$. We have $2^n$ possible outcomes, and for each agent we have some preference relation over these outcomes $\mathbf{b}_j$. For instance, agent 1 may prefer to have agent 3 in her team over agent 4, and may prefer any configuration where agent 5 is on the other team over any where they are on the same and so on. For each agent $i$, we'll have a preference relation of the following form, fully determining their objectives in the given situation:

$$\mathbf{b}_{x_1} \succeq \mathbf{b}_{x_2} \succeq \cdots \succeq \mathbf{b}_{x_m}, \quad m = 2^n, \tag{14.1}$$

where $\mathbf{b}_{x_i} \succeq \mathbf{b}_{x_j}$ means that the agent in question prefers $\mathbf{b}_{x_i}$ to $\mathbf{b}_{x_j}$, or is at least indifferent between the choices. To formalize things further we assign a numerical value to each outcome $\mathbf{b}_{x_j}$ for *each* agent, calling it the *payoff* $\$_i(\mathbf{b}_{x_j})$ to agent $i$ due to outcome $\mathbf{b}_{x_j}$. This allows us to move from the preference relations in (14.1) to a sequence of inequalities. $\mathbf{b}_{x_i} \succeq \mathbf{b}_{x_j} \iff \$(\mathbf{b}_{x_i}) \geq \$(\mathbf{b}_{x_j})$. The aforementioned binary choice situation can now be formulated in terms of functions $\$_i(\mathbf{b}_{x_j})$ of the output strings $\mathbf{b}_{x_j}$, where each entry $b_i$ in the strings corresponds to the choice of an agent $i$.

So far has the discussion only regarded the output string without mentioning any input. We could without loss of generality define an input as string where all the entries are initialized as 0's, and the individual choices being encoded by letting each participant either leave their bit unchanged or performing a NOT-operation, where NOT(0) = 1.

More complicated situations with multiple choices could be modeled by letting each player control more than one bit or letting them manipulate strings of information bearing units with more states than two; of which we will see an example of later.

### *14.1.2 Quantization of Information*

Before moving on to the quantum formalism of operators and quantum states, there is one intermediate step worth mentioning, the *probabilistic* bit, which has a certain probability $p$ of being in one state and a probability of $1 - p$ of being in the other. If we represent the two states '0' and '1' of the ordinary bit by the two-dimensional vectors $(1, 0)^T$ and $(0, 1)^T$, then a probabilistic bit is given by a linear combination of those basis vectors, with real positive coefficients $p_0$ and $p_1$, where $p_0 + p_1 = 1$. In this formulation, randomization between two different choices in a strategic situation would translate to manipulating an appropriate probabilistic bit.

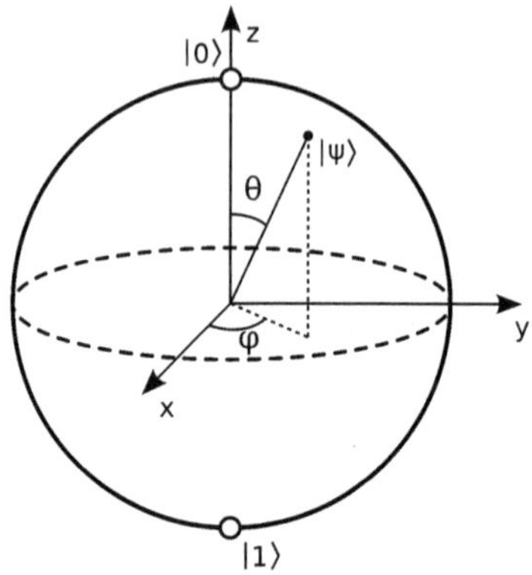


**Fig. 14.1** The Bloch sphere. A geometric representation of the state space of a single qubit

**The Quantum Bit** Taking things a step further, we introduce the quantum bit or the *qubit*, which is a representation of a two level quantum state, such as the spin state of an electron or the polarization of a photon. A qubit lives in a two dimensional complex space spanned by two basis states denoted $|0\rangle$ and $|1\rangle$, corresponding to the two states of the classical bit.

$$|0\rangle = \begin{pmatrix} 1 \\ 0 \end{pmatrix}, \qquad |1\rangle = \begin{pmatrix} 0 \\ 1 \end{pmatrix}. \tag{14.2}$$

Unlike the classical bit, the qubit can be in any superposition of $|0\rangle$ and $|1\rangle$:

$$|\psi\rangle = a_0|0\rangle + a_1|1\rangle, \tag{14.3}$$

where $a_0$ and $a_1$ are complex numbers obeying $|a_0|^2 + |a_1|^2 = 1$. $|a_i^2|$ is simply the probability to find the system in the state $|i\rangle, i \in \{0, 1\}$. Note the difference between this and the case of the probabilistic bit! We are now dealing with complex coefficients, which means that if we superpose two qubits, then some coefficients might be eliminated. This interference is one of many effects without counterpart in the classical case. The state of an arbitrary qubit can be written in the *computational basis* as:

$$|\psi\rangle = \begin{pmatrix} a_0 \\ a_1 \end{pmatrix}. \tag{14.4}$$

The state of a general qubit can be parameterized as:

$$|\psi\rangle = \cos\left(\frac{\vartheta}{2}\right)|0\rangle + e^{i\varphi}\sin\left(\frac{\vartheta}{2}\right)|1\rangle, \tag{14.5}$$

where we have factored out and omitted a global phase due to the physical equivalence between the states $e^{i\phi}|\psi\rangle$ and $|\psi\rangle$. This so called *state vector* describes a point on a spherical surface with $|0\rangle$ and $|1\rangle$ at its poles, called the Bloch-sphere, parameterized by two real numbers $\theta$ and $\varphi$, depicted in Fig. 14.1. A simple comparison between the state space of the classical, probabilistic and quantum bit is shown in Fig. 14.2.

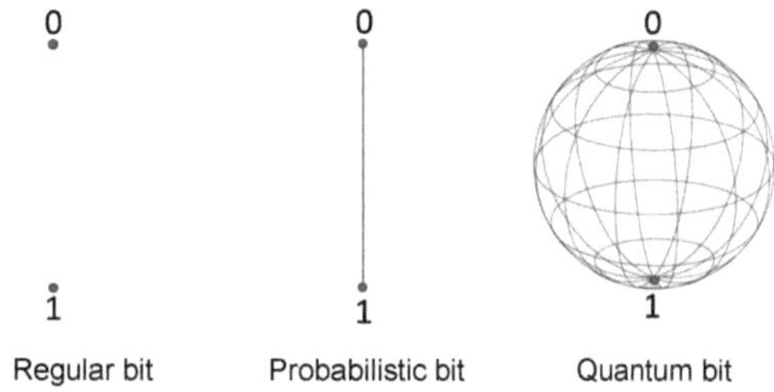


**Fig. 14.2** The classical bit has only two distinct states, the probabilistic bit can be in any normalized convex combination of those states, whereas the quantum bit has a much richer state space

### 14.1.2.1 Hilbert Spaces and Composite Systems

The state vector of a quantum system is defined in a complex vector space called *Hilbert space* $\mathscr{H}$. Quantum states are represented in common Dirac notation as "ket's", written as the right part $|\psi\rangle$ of a bracket ("bra-ket"). Algebraically a "ket" is column vector in our state space. This leaves us to define the set of "bra's" $\langle\phi|$ on the dual space of $\mathscr{H}$, $\mathscr{H}^\star$. The dual Hilbert space $\mathscr{H}^\star$ is defined as the set of linear maps $\mathscr{H} \to \mathbf{C}$, given by

$$\langle\phi| : |\psi\rangle \mapsto \langle\phi|\psi\rangle \in \mathbb{C}, \tag{14.6}$$

where $\langle\phi|\psi\rangle$ is the inner product of the vectors $|\psi\rangle$, $|\phi\rangle \in \mathscr{H}$. We can now write down a more formal definition of a Hilbert space: It is a complex inner product space with the following properties:

(i) $\langle\phi|\psi\rangle = \langle\psi|\phi\rangle^\dagger$, where $\langle\psi|\phi\rangle^\dagger$ is the complex conjugate of $\langle\psi|\phi\rangle$.
(ii) The inner product$\langle\phi|\psi\rangle$ is linear in the first argument: $\langle a\phi_1 + b\phi_2|\psi\rangle = a^\dagger\langle\phi_1|\psi\rangle + b^\dagger\langle\phi_2|\psi\rangle$.
(iii) $\langle\psi|\psi\rangle \geq 0$.

The space of an $n$-qubit system is spanned by a basis of $2^n$ orthogonal vectors $|e_i\rangle$; one for each possible combination of the basis-states of the individual qubits, obeying the orthogonality condition:

$$\langle e_i|e_j\rangle = \delta_{ij}, \tag{14.7}$$

where $\delta_{ij} = 1$ for $i = j$ and $\delta_{ij} = 0$ for $i \neq j$. We say that the Hilbert space of a composite system is the tensor products of the Hilbert spaces of its parts. So the space of an $n$-qubit system is simply the tensor product of the spaces of the $n$ qubits.

$$\mathscr{H}_{\mathscr{Q}} = \mathscr{H}_{\mathscr{Q}_n} \otimes \mathscr{H}_{\mathscr{Q}_n - 1} \otimes \mathscr{H}_{\mathscr{Q}_{n-2}} \otimes \cdots \otimes \mathscr{H}_{\mathscr{Q}_1}, \tag{14.8}$$

where $\mathscr{Q}_i$ the quantum system $i$ is a vector in $\mathbf{C}^2$. A general $n$-qubit system can therefore be written

$$|\psi\rangle = \sum_{x_n,\ldots,x_1=0}^{1} a_{x_n\cdots x_1}|x_n\cdots x_1\rangle, \tag{14.9}$$

where

$$|x_n \cdots x_1\rangle = |x_n\rangle \otimes |x_{n-1}\rangle \otimes \cdots \otimes |x_1\rangle \in \mathscr{H}_{\mathscr{Q}} \tag{14.10}$$

with $x_i \in \{0, 1\}$ and complex coefficients $a_{x_i}$. For a two qubit system, $|x_2\rangle \otimes |x_1\rangle = |x_2\rangle|x_1\rangle = |x_2 x_1\rangle$, we have

$$|\psi\rangle = \sum_{x_2, x_1=0}^{1} a_{x_2 x_1} |x_2 x_1\rangle = a_{00}|00\rangle + a_{01}|01\rangle + a_{10}|10\rangle + a_{11}|11\rangle. \tag{14.11}$$

This state space is therefore spanned by four basis vectors:

$$|00\rangle, \quad |01\rangle, \quad |10\rangle, \quad |11\rangle, \tag{14.12}$$

which are represented by the following 4-dimensional column vectors respectively:

$$\begin{pmatrix} 1 \\ 0 \\ 0 \\ 0 \end{pmatrix}, \begin{pmatrix} 0 \\ 1 \\ 0 \\ 0 \end{pmatrix}, \begin{pmatrix} 0 \\ 0 \\ 1 \\ 0 \end{pmatrix}, \begin{pmatrix} 0 \\ 0 \\ 0 \\ 1 \end{pmatrix}. \tag{14.13}$$

#### 14.1.2.2 Operators

A linear operator on a vector space $\mathscr{H}$ is a linear transformation $\mathrm{T} : \mathscr{H} \to \mathscr{H}$, that maps vectors in $\mathscr{H}$ to vectors in the same space $\mathscr{H}$. Quantum states are normalized, and we wish to keep the normalization; we are therefore interested in transformations that can be regarded as rotations in $\mathscr{H}$. Such transformations are given by *unitary operators* $U$. An operator $U$ is called unitary if $U^{-1} = U^{\dagger}$. They preserve inner products between vectors, and thereby their norm. A *projection operator* $P$ is Hermitian i.e. $P = P^{\dagger}$ and satisfies $P^2 = P$. We can create a projector $P$, by taking the outer product of a vector with itself:

$$P = |\phi\rangle\langle\phi|. \tag{14.14}$$

$P$ is a matrix with every element $P_{ij}$ being the product of the elements $i, j$ of the vectors in the outer product. This operator projects any vector $|\gamma\rangle$ onto the 1-dimensional subspace of $\mathscr{H}$, spanned by $|\phi\rangle$:

$$P|\gamma\rangle = |\phi\rangle\langle\phi||\gamma\rangle = \langle\phi|\gamma\rangle|\phi\rangle. \tag{14.15}$$

It simply gives the portion of $|\gamma\rangle$ along $|\phi\rangle$.

We will often deal with unitary operators $U \in \mathrm{SU}(2)$, i.e operators from the *special unitary group* of dimension 2. The group consists of $2 \times 2$ unitary matrices with determinant 1. These matrices will be operating on single qubits (often in systems of 2 or more qubits). The generators of the group are the *Pauli spin matrices* $\sigma_x$, $\sigma_y$, $\sigma_z$, shown together with the identity matrix $I$:

$$I = \begin{pmatrix} 1 & 0 \\ 0 & 1 \end{pmatrix}, \quad \sigma_x = \begin{pmatrix} 0 & 1 \\ 1 & 0 \end{pmatrix}, \quad \sigma_y = \begin{pmatrix} 0 & -i \\ i & 0 \end{pmatrix}, \quad \sigma_z = \begin{pmatrix} 1 & 0 \\ 0 & -1 \end{pmatrix}. \tag{14.16}$$

Note that $\sigma_x$ is identical to a classical (bit-flip) 'NOT'-operation. General $2 \times 2$ unitary operators can be parameterized with three parameters $\theta, \alpha, \beta$, as follows:

$$U(\theta,\alpha,\beta) = \begin{pmatrix} e^{i\alpha}\cos(\theta/2) & ie^{i\beta}\sin(\theta/2) \\ ie^{-i\beta}\sin(\theta/2) & e^{-i\alpha}\cos(\theta/2) \end{pmatrix}. \tag{14.17}$$

An operation is said to be local if it only affects a part of a composite (multi-qubit) system. Connecting this to the concept of the bit-strings in the previous section; a local operation translates to just controlling one such bit. This is a crucial point in the case of modeling the effect of individual actions, since each agent in a strategic situation is naturally constrained to decisions regarding their own choices. The action of a set of local operations on a composite system is given by the tensor product of the local operators. For a general $n$-qubit $|\psi\rangle$ as given in (14.9) and (14.10) we get:

$$U_n \otimes U_{n-1} \otimes \cdots \otimes U_1|\psi\rangle = \sum_{x_n,\ldots,x_1=0}^{1} a_{x_n\ldots x_1} U_n|x_n\rangle \otimes U_{n-1}|x_{n-1}\rangle \otimes \cdots \otimes U_1|x_1\rangle. \tag{14.18}$$

#### 14.1.2.3 Mixed States and the Density Operator

We have so far only discussed *pure states*, but sometimes we encounter quantum states without a definite state vector $|\psi\rangle$, these are called *mixed states* and consists of a states that has certain probabilities of being in some number of different pure states. So for example a state that is in $|\psi_1\rangle = a_0^1|0\rangle + a_1^1|1\rangle$ with probability $p_1$ and in $|\psi_2\rangle = a_0^2|0\rangle + a_1^2|1\rangle$ with probability $p_2$ is mixed. We handle mixed states by defining a density operator $\rho$, which is a hermitian matrix with unit trace:

$$\rho = \sum_i p_i|\psi_i\rangle\langle\psi_i|, \tag{14.19}$$

where $\sum_i p_i = 1$. A pure state in this representation is simply a state for which all probabilities, except one is zero. If we apply a unitary operator $U$ on a pure state, we end up with $U|\psi\rangle$ which has the density operator $U\rho U^\dagger = U|\psi\rangle\langle\psi|U^\dagger$. Regardless if we are dealing with pure or mixed states, we take the expectation value of upon measurement ending up in a $|\phi\rangle$ by calculating $\mathrm{Tr}(|\phi\rangle\langle\phi|\rho)$, where $|\phi\rangle\langle\phi|$ is a so called projector. For calculating the expectation values of a state to be in *any* of a number of states $|\phi_i\rangle$, we construct a projection operator $P = \sum_i |\phi_i\rangle\langle\phi_i|$ and take the trace over $P$ multiplied by $\rho$.

#### 14.1.2.4 Entanglement

Entanglement is the resource our game-playing agents will make use of in the quantum game protocols to achieve better than classical performance. Non-classical correlations are thus introduced, by which the players can synchronize their behavior without any additional communication. An entangled state is basically a quantum system that *cannot* be written as a tensor product of its subsystems, we'll thus define two classes of quantum states. Examples below refers to two-qubit states.

Product states:

$$|\Psi_{\mathscr{Q}}\rangle = |\Psi_{\mathscr{Q}_2}\rangle \otimes |\Psi_{\mathscr{Q}_1}\rangle, \quad \text{or using density matrix} \quad \rho_{\mathscr{Q}} = \rho_{\mathscr{Q}_2} \otimes \rho_{\mathscr{Q}_1}, \tag{14.20}$$

and entangled states

$$|\Psi_{\mathscr{Q}}\rangle \neq |\Psi_{\mathscr{Q}_2}\rangle \otimes |\Psi_{\mathscr{Q}_1}\rangle, \quad \text{or using density matrix} \quad \rho_{\mathscr{Q}} \neq \rho_{\mathscr{Q}_2} \otimes \rho_{\mathscr{Q}_1}. \tag{14.21}$$

For a mixed state, the density matrix is defined as mentioned by $\rho_{\mathscr{Q}} = \sum_{i=1}^{N} p_i |\psi_i\rangle\langle\psi_i|$ and it is said to be separable, which we will denote by $\rho_{\mathscr{Q}}^{sep}$, if it can be written as

$$\rho_{\mathscr{Q}}^{sep} = \sum_i p_i \left(\rho^i_{\mathscr{Q}_2} \otimes \rho^i_{\mathscr{Q}_1}\right), \qquad \sum_i p_i = 1. \tag{14.22}$$

A set of very important two-qubit entangled states are the Bell states

$$|\Phi^{\pm}_{\mathscr{Q}}\rangle = \frac{1}{\sqrt{2}}\big(|00\rangle \pm |11\rangle\big), \qquad |\Psi^{\pm}_{\mathscr{Q}}\rangle = \frac{1}{\sqrt{2}}\big(|01\rangle \pm |10\rangle\big). \tag{14.23}$$

The GHZ-type-states

$$|\mathrm{GHZ}_n\rangle = \frac{1}{\sqrt{2}}\big(|00\cdots 0\rangle + e^{i\phi}|11\cdots 1\rangle\big) \tag{14.24}$$

could be seen as an $n$-qubit generalization of $|\Phi^{\pm}_{\mathscr{Q}}\rangle$-states.

### *14.1.3 Classical Games*

It is instructive to review the theory of classical games and some major solution concepts before moving on to examples of quantum games. We'll start by defining classical pure and mixed strategy games, and then move on to introducing some relevant solution concepts and finish off with a definition of quantum games.

A game is a formal model over the interactions between a number of agents (*agents, players, participants*, and *decision makers* may be used interchangeably) under some specified sets of choices (*choices, strategies, actions* and *moves*, may be used interchangeably). Each combination of choices made, or strategies chosen by the different players leads to an outcome with some certain level of desirability for each of them. The level of desirability is measured by assigning a real number, a so called *payoff* \$ for each game outcome for each player. Assuming rational players, each will choose actions that maximizes their expected payoff $E(\$)$, i.e. in an deterministic as well as in an probabilistic setting acting in a way that, based on the known information about the situation, maximizes the expectation value of their payoff. The structure of the game is fully specified by the relations between the different combinations of strategies and the payoffs received by the players. A key point is the interdependence of the payoffs with the strategies chosen by the other players. A situation where the payoff of one player is independent of the strategies of the others would be of little interest from a game theoretical point of view. It is natural to extend the notion of payoffs to *payoff functions* whose arguments are the chosen strategies of all players and ranges are the real valued outputs that assigns a level of desirability for each player to each outcome.

**Pure Strategy Classical Game** We have a set of $n$ players $\{1, 2, \ldots, n\}$, $n$ strategy sets $S_i$, one for each player $i$, with $s_i^j \in S_i$, where $s_i^j$ is the $j$th strategy of player $i$. The strategy space $S = S_1 \times S_2 \times \cdots \times S_n$ contains all $n$-tuples pure strategies, one from each set. The elements $\sigma \in S$ are called strategy profiles, some of which will earn them the status of being a *solution* with regards to some solution concept.

We define a game by its payoff-functions $\$_i$, where each is a mapping from the strategy space $S$ to a real number, the payoff or utility of player $i$. We have:

$$\$_i : S_1 \times S_2 \times \cdots \times S_n \rightarrow \mathbf{R}. \tag{14.25}$$

**Mixed Strategy Classical Game** Let $\Delta(S_i)$ be the set of convex linear combinations of the elements $s_i^j \in S_i$. A mixed strategy $s_i^{mix} \in \Delta(S_i)$ is then given by:

$$\sum_{s_i^j \in S_i} p_i^j s_i^j \quad \text{with} \sum_j p_i^j = 1, \tag{14.26}$$

where $p_i^j$ is the probability player $i$ assigns to the choice $s_i^j$. The space of mixed strategies $\Delta(S) = \Delta(S_1) \times \Delta(S_2) \times \cdots \times \Delta(S_n)$ contains all possible mixed strategy profiles $\sigma_{mix}$. We now have:

$$\$_i : \Delta(S_1) \times \Delta(S_2) \times \cdots \times \Delta(S_n) \rightarrow \mathbf{R}. \tag{14.27}$$

Note that the pure strategy games are fully confined within the definition of mixed strategy games and can be accessed by assigning all strategies except one, the probability $p^j = 0$. This class of games could be formalized in a framework using probabilistic information units, such as the probabilistic bit.

### 14.1.4 Solution Concepts

We will introduce two of many game theoretical solution concepts. A solution concept is a strategy profile $\sigma^* \in S$, that has some particular properties of strategic interest. It could be a strategy profile that one would expect a group of rational self-maximizing agents to arrive at in their attempt to maximize their minimum expected payoff. Strategy profiles of this form i.e. those that leads to a combination of choices where each choice is the best possible response to any possible choice made by other players tend to lead to an equilibrium, and are good predictors of game outcomes in strategic situations. To see how such equilibria can occur we'll need to develop the concept of *dominant strategies*.

**Definition 14.1** (Strategic dominance) A strategy $s^{dom} \in S_i$ is said to be dominant for player $i$, if for any strategy profile $\sigma_{-i} \in S/S_i$, and any other strategy $s^j \neq s^{dom} \in S_i$:

$$\$_i\left(s^{dom}, \sigma_{-i}\right) \geq \$_i\left(s^j, \sigma_{-i}\right) \quad \text{for all } i = 1, 2, \ldots, n. \tag{14.28}$$

Lets look at a simple example. Say that we have two players, Alice with legal strategies $s^1_{Alice}$, $s^2_{Alice} \in S_{Alice}$ and Bob with $s^1_{Bob}$, $s^2_{Bob} \in S_{Bob}$. Now, if the payoff Alice receives when playing $s^1_{Alice}$ against any of Bob's two strategies is higher than (or at least as high as) what she receives by playing $s^2_{Alice}$, then $s^1_{Alice}$ is her dominant strategy. Her payoff can of course vary depending on Bob's move but regardless what Bob does, her dominant strategy is the *best response*. Now there is no guarantee that such dominant strategy exists in a pure strategy game, and often must the strategy space be expanded to accommodate for mixed strategies for them to exist.

If both Alice and Bob has a dominant strategy, then this strategy profile becomes a *Nash Equilibrium*, i.e. a combination of strategies for which none of them can gain by unilaterally deviating from. The Nash equilibrium profile acts as an attractor in the strategy space and forces the players into it, even though it is not always an optimal solution. Combinations can exist that can lead to better outcomes for both (all) players.

**Definition 14.2** (Nash equilibrium) Let $\sigma^{NE}_{-i} \in S/S_i$ be a strategy profile containing the dominant strategies of every player except player $i$, and let $s^{NE}_i \in S_i$ be the dominant strategy of player $i$. Then for all $s^j_i \neq s^{NE}_i \in S_i$:

$$\$_i\left(s^{NE}_i, \sigma^{NE}_{-i}\right) \geq \$_i\left(s^j_i, \sigma^{NE}_{-i}\right) \quad \text{for all } i = 1, 2, \ldots, n. \tag{14.29}$$

If we have a situation where an agent can increase its payoff without decreasing any others, then this would per definition mean that nobody would mind if that agent would do so. Each such increase in payoff is called a *Pareto improvement*. When no such improvement can be done, then the strategy profile is said to be *Pareto optimal*.

**Definition 14.3** (Pareto efficiency) A Pareto efficient or Pareto optimal strategy profile is one where none of the participating agents can increase their payoff without decreasing the payoff of someone else.

## 14.2 Quantum Games

In the quantum game protocols (*protocol* and *scheme* may be used interchangeably) presented in this paper, the $m_i$ different choices available to a player $i$ will be encoded in the basis states of an $m_i$-level quantum system, where the $m_i$ denotes the dimensionality of the Hilbert space $\mathscr{H}_{\mathscr{Q}_i}$ associated with that subsystem. Each of the $n$ player holds one subsystem leading to a total system with a state vector $a$ in an $\prod_{i=1}^{n} \dim(\mathscr{H}_{\mathscr{Q}_i})$—dimensional space. The definition of a quantum game must therefore include a Hilbert space of a multipartite multilevel system $\mathscr{H}_{\mathscr{Q}} = \mathscr{H}_{\mathscr{Q}_n} \otimes \mathscr{H}_{\mathscr{Q}_{n-1}} \otimes \cdots \otimes \mathscr{H}_{\mathscr{Q}_1}$.

The different subsystems must in general be allowed to have a have a common origin to accommodate entanglement in the shared initial state $\rho_{in} \in \mathscr{H}_{\mathscr{Q}}$. This is often modeled by including a referee that prepares an initial state and distributes the subsystems among the players. Whether or not this step invokes on the non-communication criteria certain games have, is under debate. We justify it by the fact that no communication is done under the crucial step of choosing a strategy. The strategies are applied by local quantum operations on the quantum state held by each player. No player has any access to any part of the system except its own subsystem, and no information can be sent between the players with aid of the shared quantum resource. Classical strategies becomes quantum strategies by expanding the strategy sets:

$$s_i \in S_i \Rightarrow U_i \in \mathrm{S}(m_i), \tag{14.30}$$

where the set of allowed quantum operations $\mathrm{S}(m_i)$ is some subset of the special unitary group $\mathrm{SU}(m_i)$. We will later see that the nature of the game can be determined by restrictions on $\mathrm{S}(m_i)$. It is an important point to be able to show that the classical version of a game is recoverable just by restricting the set of allowed operators. At least if we want it to be a *proper quantization* [9], i.e. an extension of the classical game into the quantum realm, and not a whole new game without a classical counterpart.

We define a quantum game in two steps:

$$\begin{aligned} U_n \otimes U_{n-1} \otimes \cdots \otimes U_1 : \mathscr{H}_{\mathscr{Q}_n} \otimes \mathscr{H}_{\mathscr{Q}_{n-1}} \otimes \cdots \otimes \mathscr{H}_{\mathscr{Q}_1} \\ \to \mathscr{H}_{\mathscr{Q}_n} \otimes \mathscr{H}_{\mathscr{Q}_{n-1}} \otimes \cdots \otimes \mathscr{H}_{\mathscr{Q}_1}, \end{aligned} \tag{14.31}$$

$$\$_i : \mathscr{H}_{\mathscr{Q}_n} \otimes \mathscr{H}_{\mathscr{Q}_{n-1}} \otimes \cdots \otimes \mathscr{H}_{\mathscr{Q}_1} \to \mathbf{R}, \tag{14.32}$$

where the first step is a transformation of the state of the complete system by local operations, and the second is a mapping from the Hilbert space of the quantum state to a real number, the expected payoff of player $i$.

### 14.2.1 The Quantum Game Protocol

- The game begins with an entangled initial state $|\psi_{in}\rangle$. Each subsystem has a dimensionality $m$ that equal to the number of pure strategies in each players strategy set. In the protocols covered in this paper, all players will face the same number of choices. The number of subsystems equals the number of players. One can assume that $|\psi_{in}\rangle$ has been prepared at some location by a referee that then has distributed the subsystems among the players [12, 13].
- The players then chooses an unitary operator $U$ from a subset of SU($m$), and applies it to their subsystem. The initial state $\rho_{in}$ transforms to a final state $\rho_{fin}$, given by:

$$\rho_{fin} = U \otimes U \otimes \cdots \otimes U \rho_{in} U^\dagger \otimes U^\dagger \otimes \cdots \otimes U^\dagger. \tag{14.33}$$

  In the absence of communication, and due to the symmetry of these games, all players are expected to do the same operation.
- The players then measures their own subsystem, collapsing their quantum states to units of classical information. For the case of a two-choice protocol, each player ends up with a classical bit $b_i$, and the complete system has thus collapsed into a classical string $\mathbf{b}$, corresponding to a pure strategy profile $\sigma \in S$. For the quantum game to have an advantage over a classical game, the collective action of the players must have decreased the probability of the final state $\rho_{fin}$ to collapse into such basis states (classical information strings/strategy profiles) that are undesired, i.e. leading to lower or zero payoff \$.
- To calculate the expected payoffs $E(\$)$, we define for each player $i$ a payoff-operator $P_i$, which contains the sum of orthogonal projectors associated with the states for which player $i$ receives a payoff \$. We have:

$$P_i = \sum_j \$_i^j |\chi_i^j\rangle\langle\chi_i^j|, \tag{14.34}$$

  where the states $|\chi_i^j\rangle$ are those sates that leads to a payoff for player $i$, and $\$_i^j$ the associated payoffs. The expected payoff $E(\$_i)$ of player $i$ is calculated by taking the trace of the product of the final state $\rho_{fin}$ and the payoff-operator $P_i$:

$$E_i(\$) = \mathrm{Tr}(P_i \rho_{fin}). \tag{14.35}$$

### 14.2.2 Prisoners Dilemma

The prisoners dilemma is one of the most studied game theoretical problems. It was introduced in 1950 by Merrill Flood and Melvin Dresher, and has been widely used ever since to model a variety of situations, including oligopoly pricing, auction bidding, salesman effort, political bargaining and arms races. In is in its standard form, a symmetric simultaneous game of complete information. Two players, Alice and Bob (A and B) are faced with a choice to *cooperate* or to *defect*, without any

**Table 14.1** The normal-form representation of prisoners dilemma

| | | Bob | |
|---|---|---|---|
| | | Cooperate | Defect |
| Alice | Cooperate | (3, 3) | (0, 5) |
| | Defect | (5, 0) | (1, 1) |

information about the action taken by the other. The payoffs they receive due to any combination of choices is shown in Table 14.1, where the first entry in each parenthesis shows the payoff $\$_A$ of Alice and the second entry the payoff $\$_B$ of Bob. Given that Bob chooses to cooperate, Alice receives $\$_A = 3$ if she chooses to do the same, and she receives $\$_A = 5$ if she chooses to defect. If Bob instead defects, then Alice receives $\$_A = 0$ by cooperating and $\$_A = 1$ by choosing to defect. No matter what Bob does, Alice will always gain by choosing to defect, equipping her with a strictly dominant strategy! Due to the symmetry of the game, the same is true for Bob, forcing them into a Nash equilibrium strategy profile of (defect, defect), which pays out $\$_{AB} = 1$ to each. This outcome is clearly far from efficient, since there is a Pareto optimal strategy profile (cooperate, cooperate) that would have given them $\$_{AB} = 3$, and hence the dilemma.

Quantum prisoners dilemma was introduced by J. Eisert, M. Wilkens, and M. Lewenstein in 1999 [11]. Here Alice and Bob are equipped with a quantum resource, a maximally entangled Bell-type-state, and each of them are in possession of a subsystem. The Hilbert space of the game is given by: $\mathscr{H} = \mathscr{H}_B \otimes \mathscr{H}_A$, with $\mathscr{H}_A = \mathscr{H}_B = \mathbf{C}^2$. We'll identify the following relations, mapping classical outcomes with basis states of the Hilbert space: (cooperate, cooperate) $\rightarrow |00\rangle$, (cooperate, defect) $\rightarrow |01\rangle$, (defect, cooperate) $\rightarrow |10\rangle$ and (defect, cooperate) $\rightarrow |11\rangle$. The entangled initial state is created by acting with an entangling operator $J = \frac{1}{\sqrt{2}} I^{\otimes 2} + i\sigma_x^{\otimes 2}$ on a product state initialized as (cooperate, cooperate):

$$J|00\rangle = \frac{1}{\sqrt{2}}\big(|00\rangle + i|11\rangle\big). \tag{14.36}$$

Note that the entangling operator performs a global operation, i.e. an operation performed on both subsystems simultaneously. One can consider it to be performed by a referee, loyal to both parties. The game proceeds by Alice and Bob performing their local strategies $U_A$ and $U_B$, and the state is turned into its final form: $|\psi_{fin}\rangle = (U_B \otimes U_A)J|00\rangle$. Before measurement is performed, an disentangling operator $J^\dagger$ is applied. The inclusion of $J$ and $J^\dagger$ into the protocol assures that the classical game is embedded into the quantum version, whereby the classical prisoners dilemma can be accessed by restricting the set of allowed operators to $U_A$, $U_B \in \{I, \sigma_x\}$. It is a simple task to show that any combination of the identity operator $I$ and the bit-flip operator $\sigma_x$ commutes with $J$, and together with the fact that $JJ^\dagger = I$, one concludes that this restriction turns the protocol into classical (one-bit) operations on a bit string '00'. The complete protocol is shown as a circuit diagram in Fig. 14.3.

It is now left to define a set of operators $U$, representing allowed *quantum* strategies, and the payoff operators $P_A$ and $P_B$. Eisert et al. considered a two parameter subset of SU(2) as the strategy space:

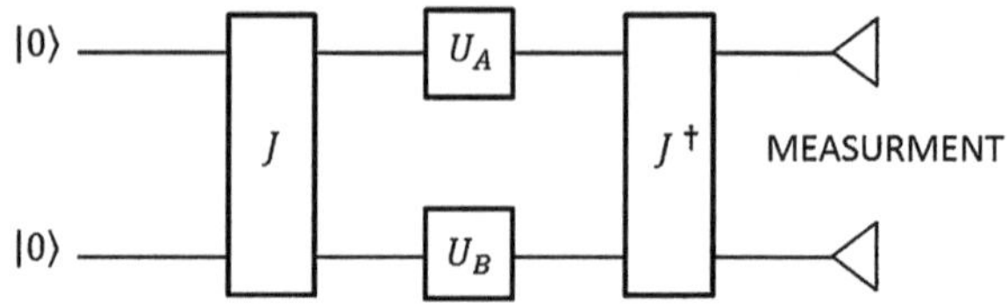


**Fig. 14.3** Circuit diagram of the quantum prisoners dilemma protocol

$$U(\theta,\alpha)=\begin{pmatrix} e^{i\alpha}\cos(\theta/2) & \sin(\theta/2) \\ -\sin(\theta/2) & e^{-i\alpha}\cos(\theta/2)\end{pmatrix}. \tag{14.37}$$

The classical strategies are represented by $U(0,0)=I$ and $U(0,\pi)=\sigma_x$. We construct Alice's payoff operator $P_A$ as defined in (14.34) with values from the payoff matrix:

$$P_A = 3|00\rangle\langle 00| + 5|01\rangle\langle 01| + 1|11\rangle\langle 11|. \tag{14.38}$$

Her expected payoff is calculated by taking the trace of the final state and the payoff operator: $E(\$_A)=\mathrm{Tr}(P_A\rho_{fin})$, where $\rho_{fin}=|\psi_{fin}\rangle\langle\psi_{fin}|$. It can be shown that when the set of strategies are expanded to allow any $U(\theta,\alpha)$, the old Nash equilibrium (defect, defect) $\rightarrow U(0,\pi)\otimes U(0,\pi)$ ceases to exist! Instead a new Nash equilibrium emerges at

$$U_A = U_B = U(0,\pi/2)=\begin{pmatrix} i & 0 \\ 0 & -i\end{pmatrix}. \tag{14.39}$$

This strategy leads to an expected payoff $E(\$_A)=E(\$_A)=3$. Thereby they both receive an expected payoff that equals the Pareto optimal solution in the classical pure strategy version, with the addition that this solution is also a Nash equilibrium. Dilemma resolved. It should be added that if the strategy sets are further expanded to include all SU(2) operations, this solution vanishes, and there is no Nash equilibrium strategy profile in pure quantum strategies, whereby one has to include mixed quantum operations to find an equilibrium [19].

### *14.2.3 Minority Games*

We extend the previous protocol to ones with multiple agents, by introducing the minority game. The game consists of $n$ of non-communicating players that must independently make up their mind between two choices. We could regard these players as investors on a market deciding between two equally attractive securities, as commuters choosing between two equally fast routes to a suburb, or any collection of agents facing situations where they wish to make the minority choice. The core objective of the players are thus to avoid the crowd. We encode the two choices as $|0\rangle$ and $|1\rangle$ in the computational basis like before. The players receive payoff a $\$=1$ if they happen to be in the smaller group. So if the number of players choosing $|0\rangle$ is less than the number of players choosing $|1\rangle$, the first group receives payoff whereas the second group is left with nothing. Would the players happen to be evenly distributed between the two choices, then they'll all go empty handed.

The Nash equilibrium solution is to randomize between $|0\rangle$ and $|1\rangle$ using a fair coin. The *one shoot* version we are considering will necessarily have a mixed strategy solution, since any deterministic strategy would lead all players to the same choice and thus a maximally undesired outcome. The expected payoff $E(\$)$ for a player is simply the number outcomes with that player in the minority group divided by the number of different possible outcomes. For a four player game, there are two minority outcomes for each player, out of sixteen possible. This gives a expected payoff of $1/8$.

A quantum version of a four player minority game was presented by Benjamin and Hayden in 2000 [13], offering a solution that significantly outperformed the classical version of the game. The advantage comes from the possibility of eliminating (or reducing the probability of) such final outcomes where the players are evenly distributed among the two choices. The collective application of local unitary operators on the subsystems of an entangled state can thus transform this initial state in such a way that a better-than-classical result is achieved. This transformation does not have a classical analogue, and the performance is due to interference effects from the local phases added to the qubits by the players local operations. We are not including the action of an entangling operator $J$ in this section, we simply assume the initial state to be entangled at the start of the protocol, and it can again be assumed that the state has been prepared by an unbiased referee and distributed among the players. Considering the four-player case, we begin the protocol with an GHZ-type state similar to the one used in the previous two-player game, but now consisting of *four* entangled qubits.

$$|\psi_{in}\rangle = \frac{1}{\sqrt{2}}\big(|0000\rangle + |1111\rangle\big). \tag{14.40}$$

The Hilbert space of the game is sixteen dimensional, accounting for all possible game outcomes. $\mathscr{H}_{\mathscr{Q}} = \mathscr{H}_{\mathscr{Q}_4} \otimes \mathscr{H}_{\mathscr{Q}_3} \otimes \mathscr{H}_{\mathscr{Q}_2} \otimes \mathscr{H}_{\mathscr{Q}_1}$, with $\mathscr{H}_{\mathscr{Q}_i} = \mathbf{C}^2$. Each player $i = 1, 2, 3, 4$ is permitted to manipulate its subsystem with the full machinery of local quantum operations: $U_i \in SU(2)$ given in (14.17). The payoff operator $P_i$ projects the final state onto the desired states of player $i$, and is given by

$$P_i = \sum_{j=1}^{k} |\xi_i^j\rangle\langle\xi_i^j|. \tag{14.41}$$

The sum is over all the $k$ different states $|\xi_i^j\rangle$, for which player $i$ is in the minority. Its worth to note that the sums are always over a even number $k$, and that they run over the states of the following form:

$$P_i = \sum_{j=1}^{k} |\xi_i^j\rangle\langle\xi_i^j| = \sum_{j=1}^{k/2} |\vartheta_i^j\rangle\langle\vartheta_i^j| + \sum_{j=1}^{k/2} |\overline{\vartheta_i^j}\rangle\langle\overline{\vartheta_i^j}|, \tag{14.42}$$

where $|\overline{\vartheta_i^j}\rangle$ is the bit-flipped version of $|\vartheta_i^j\rangle$, i.e 0's and 1's are interchanged. The payoff operator $P_1$ for player 1 in the four player case is given by:

$$P_1 = |0001\rangle\langle 0001| + |1110\rangle\langle 1110|. \tag{14.43}$$

By playing $U(\theta, \alpha, \beta) = U(\frac{\pi}{2}, -\frac{\pi}{8}, \frac{\pi}{8})$, the four players can completely eliminate the risk of upon measurement ending up with an outcome where none of them receives a payoff. This quantum strategy leads to an expected payoff $E(\$) = \frac{1}{4}$ that is twice as good as in the classical case $E(\$) = \frac{1}{8}$. The strategy profile is a Nash equilibrium as well as Pareto optimal. Quantum minority games has been extensively studied for cases of arbitrary $n$, and it can be shown that the quantum versions gives rise to better than classical payoffs for any game with an even number of players [14].

### *14.2.4 Kolkata Restaurant Problem*

The Kolkata restaurant problem is an extension of the minority game [20–24], where the $n$ players now has $m$ choices. As the story goes, the choice is between $m$ restaurants. The players receive a payoff if their choice is not too crowded, i.e the number of agents that chose the same restaurant is under some limit. We will discuss the case for which this limit is one. Just like in the minority game previously discussed, the Kolkata restaurant problem offers a way for modeling heard behavior and market dynamics, where visiting a restaurant translates to buying a security, in which case an agent wishes to be the only bidder. In our simplified model there are just three agents, Alice, Bob and Charlie. They have three possible choices: security 0, security 1 and security 2. They receive a payoff $\$ = 1$ if their choice is unique, i.e that nobody else has made the same choice, otherwise they receive $\$ = 0$. The game is so called *one shoot*, which means that it is non-iterative, and the agents have no information from previous rounds to base their decisions on. Under the constraint that they cannot communicate, there is nothing left to do other than randomizing between the choices just like in the minority games in the previous section. Given the symmetric nature of the problem, any deterministic strategy would lead all three agents to the same strategy, which in turn would mean that all three would leave empty handed. There are 27 different strategy profiles possible, i.e combinations of choices. 12 of which gives a payoff of $\$ = 1$ to each one of them. Randomization gives therefore agent $i$ an expected payoff of $E(\$) = \frac{4}{9}$.

In the quantum version we let Alice, Bob and Charlie share a quantum resource [18]. Each has a part of a multipartite quantum state. They play their strategy by manipulating their own part of the combined system, before measuring their subsystems and choosing accordingly. Whereas classically the players would be allowed randomizing over a discrete set of choices, in the quantum version each subsystem is allowed to be transformed with arbitrary local quantum operations, just like before. In the absence of entanglement, quantum games of this type usually yield the same payoffs as their classical counterparts, whereas the combination of unitary operators (or a subset therein) and entanglement, will be shown to outperform the classical randomization strategy.

When moving from quantum game protocols with two choices into ones with three, we'll need some additional structure. Instead of qubits will we be dealing with qutrits, which are their three level versions. The local operations on qutrits are now represented by a more complicated group of matrices, the SU(3) group. Everything else will essentially be similar to that of the quantum minority game.

A qutrit is a 3-level quantum system on 3-dimensional Hilbert space $\mathscr{H}_{\mathscr{Q}} = \mathbf{C}^3$, written in the computational basis as:

$$|\psi\rangle = a_0|0\rangle + a_1|1\rangle + a_2|2\rangle \in \mathbf{C}^3, \tag{14.44}$$

with $a_0, a_1, a_2 \in \mathbf{C}$ and $|a_0|^2 + |a_1|^2 + |a_2|^2 = 1$. A general $n$-qutrit system $|\Psi\rangle$ is a vector on $3^n$-dimensional Hilbert space, and is written as a linear combination of $3^n$ orthonormal basis vectors.

$$|\Psi\rangle = \sum_{x_n,\ldots,x_1=0}^{2} a_{x_n\ldots x_1}|x_n\cdots x_1\rangle, \tag{14.45}$$

where

$$|x_n\cdots x_1\rangle = |x_n\rangle \otimes |x_{n-1}\rangle \otimes \cdots \otimes |x_1\rangle \in \mathscr{H}_{\mathscr{Q}} = \overbrace{\mathbf{C}^3 \otimes \cdots \otimes \mathbf{C}^3}^{n\text{-times}}, \tag{14.46}$$

with $x_i \in \{0, 1, 2\}$ and complex coefficients $a_{x_i}$, obeying $\sum |a_{x_n\cdots x_1}|^2 = 1$.

Single qutrits are transformed with unitary operators $U \in \mathrm{SU}(3)$, i.e operators from the special unitary group of dimension 3, acting on $\mathscr{H}_{\mathscr{Q}}$ as $U : \mathscr{H}_{\mathscr{Q}} \to \mathscr{H}_{\mathscr{Q}}$. In a multi-qutrit system, operations on single qutrits are said to be local. They affect the state-space of the corresponding qutrit only. The SU(3) matrix is parameterized by defining three general, mutually orthogonal complex unit vectors $\bar{x}$, $\bar{y}$, $\bar{z}$, such that $\bar{x} \cdot \bar{y} = 0$ and $\bar{x}^* \times \bar{y} = \bar{z}$. We construct a SU(3) matrix by placing $\bar{x}$, $\bar{y}^*$ and $\bar{z}$ as its columns [25]. Now a general complex unit vector is given by:

$$\bar{x} = \begin{pmatrix} \sin\theta\cos\phi e^{i\alpha_1} \\ \sin\theta\sin\phi e^{i\alpha_2} \\ \cos\theta e^{i\alpha_3} \end{pmatrix}, \tag{14.47}$$

and one complex unit vector orthogonal to $\bar{x}$ is given by:

$$\bar{y} = \begin{pmatrix} \cos\chi\cos\theta\cos\phi e^{i(\beta_1-\alpha_1)} + \sin\chi\sin\phi e^{i(\beta_2-\alpha_1)} \\ \cos\chi\cos\theta\sin\phi e^{i(\beta_1-\alpha_2)} - \sin\chi\cos\phi e^{i(\beta_2-\alpha_2)} \\ -\cos\chi\sin\theta e^{i(\beta_1-\alpha_3)} \end{pmatrix}, \tag{14.48}$$

where $0 \le \phi, \theta, \chi, \le \pi/2$ and $0 \le \alpha_1, \alpha_2, \alpha_3, \beta_1, \beta_2 \le 2\pi$. We have a general SU(3) matrix $U$, given by:

$$U = \begin{pmatrix} x_1 & y_1^* & x_2^* y_3 - y_3^* x_2 \\ x_2 & y_2^* & x_3^* y_1 - y_1^* x_3 \\ x_3 & y_3^* & x_1^* y_2 - y_2^* x_1 \end{pmatrix}, \tag{14.49}$$

and it is controlled by eight real parameters $\phi, \theta, \chi, \alpha_1, \alpha_2, \alpha_3, \beta_1, \beta_2$.

The initial state, a maximally entangled GHZ-type state

$$|\psi_{in}\rangle = \frac{1}{\sqrt{3}}\big(|000\rangle + |111\rangle + |222\rangle\big) \in \mathscr{H}_{\mathscr{Q}} = \mathbb{C}^3 \otimes \mathbb{C}^3 \otimes \mathbb{C}^3, \tag{14.50}$$

is symmetric and unbiased in regards to permutation of player position and has the property of letting us embed the classical version of the game, accessible trough restrictions on the strategy sets. To show this, we define a set of operators corresponding to classical pure strategies that gives raise to deterministic payoffs when applied to $|\psi_{in}\rangle$. The cyclic group of order three, $C_3$, generated by the matrix:

$$s = \begin{pmatrix} 0 & 0 & 1 \\ 1 & 0 & 0 \\ 0 & 1 & 0 \end{pmatrix} \tag{14.51}$$

where $s^3 = s^0 = I$ and $s^2 = s^T$, has the properties we are after. The set of classical strategies $S = \{s^0, s^1, s^2\}$ with $s^i \otimes s^j \otimes s^k|000\rangle = |ijk\rangle$ acts on the initial state $|\psi_{in}\rangle$ as:

$$\begin{aligned} s^i \otimes s^j \otimes s^k \frac{1}{\sqrt{3}}\big(|000\rangle + |111\rangle + |222\rangle\big) \\ = \frac{1}{\sqrt{3}}\big(|0+i0+j0+k\rangle + |1+i1+j1+k\rangle + |2+i2+j2+k\rangle\big). \end{aligned} \tag{14.52}$$

Note that the superscripts denotes powers of the generator and that the addition is modulo 3. In the case under study, where there is no preference profile over the different choices, any combination of the operators in $S = \{s^0, s^1, s^2\}$ leads to the same payoffs when applied to $|\psi_{in}\rangle$ as to $|000\rangle$. We form a density matrix $\rho_{in}$ out of the initial state $|\psi_{in}\rangle$ and add noise that can be controlled by the parameter $f$ [17]. We get:

$$\rho_{in} = f|\psi_{in}\rangle\langle\psi_{in}| + \frac{1-f}{27} I_{27}, \tag{14.53}$$

where $I_{27}$ is the $27 \times 27$ identity matrix. Alice, Bob and Charlie now applies a unitary operator $U$ that maximizes their chances of receiving a payoff $\$ = 1$, and thereby the initial state $\rho_{in}$ is transformed into the final state $\rho_{fin}$.

$$\rho_{fin} = U \otimes U \otimes U \rho_{in} U^\dagger \otimes U^\dagger \otimes U^\dagger. \tag{14.54}$$

We define for each player $i$ a payoff-operator $P_i$, which contains the sum of orthogonal projectors associated with the states for which player $i$ receives a payoff $\$ = 1$. For Alice this would correspond to

$$P_A = \left( \sum_{x_3,x_2,x_1=0}^{2} |x_3x_2x_1\rangle\langle x_3x_2x_1|, x_3 \neq x_2, x_3 \neq x_1, x_2 \neq x_1 \right) + \left( \sum_{x_3,x_2,x_1=0}^{2} |x_3x_2x_1\rangle\langle x_3x_2x_1|, x_3 = x_2 \neq x_1 \right). \tag{14.55}$$

The expected payoff $E_i(\$)$ of player $i$ is as usual calculated by taking the trace of the product of the final state $\rho_{fin}$ and the payoff-operator $P_i$:

$$E(\$_i) = \mathrm{Tr}(P_i \rho_{fin}). \tag{14.56}$$

It can be shown that if Alice, Bob and Charlie acts with a general SU(3), there exist a $U^{opt}(\phi, \theta \cdot \chi \cdot \alpha_1, \alpha_2, \alpha_3, \beta_1, \beta_2) \in \mathrm{SU}(3)$, given by $U^{opt}(\frac{\pi}{4}, \cos^{-1}(\frac{1}{\sqrt{3}}), \frac{\pi}{4}, \frac{5\pi}{18}, \frac{5\pi}{18}, \frac{5\pi}{18}, \frac{\pi}{3}, \frac{11\pi}{6})$, that outperforms classical randomization. The strategy profile $U^{opt} \otimes U^{opt} \otimes U^{opt}$ leads to a payoff of $E(\$) = \frac{6}{9}$, assuming $(f = 1)$, compared to the classical $E^c(\$) = \frac{4}{9}$. Letting the payoff function depend on the fidelity parameter $f$, we get a payoff function $E(\$(f)) = \frac{2}{9}(f + 2)$ where we can clearly see that the expected payoff reaches the classical value as $f \to 0$.

## 14.3 Outlook

As the field of quantum information theory matures and information processing moves into the quantum realm, will it be increasingly important to study the broad spectrum of effects of this transition. Game theory is the study of strategic decision making under limited information. How decision making should or will change as situations are played out in a world where this information is *quantum* information, will be some of many conceptual challenges to address if classical communication and computing, is due to be replaced by systems governed by the peculiar and counter-intuitive laws of quantum mechanics.

**Acknowledgements** The work was supported by the Swedish Research Council (VR).

## References

1. Nielsen M, Chuang I (2000) Quantum computation and quantum information. Cambridge University Press, Cambridge
2. Hargreaves Heap S, Varoufakis Y (2004) Game theory – a critical introduction. Routledge, London

3. Fudenberg D, Tirole J (1991) Game theory. MIT Press, Cambridge
4. Osbourne MJ, Rubinstein A (1994) A course in game theory. MIT Press, Cambridge
5. Flitney AP (2008) Review of quantum games. In: Haugen IN, Nilsen AS (eds) Game theory: strategies, equilibria, and theorems, p 140
6. Piotrowski EW, Sladkowski J (2003) An invitation to quantum game theory. Int J Theor Phys 42:1089
7. Khan FS, Phoenix SJD (2011) Nash equilibrium in quantum superpositions. In: Proceedings of SPIE, vol 8057 80570K-1
8. Landsburg SE (2011) Quantum game theory. Wiley encyclopedia of operations research and management science
9. Bleiler SA (2008) A formalism for quantum games and an application. Preprint http://arxiv.org/abs/0808.1389
10. Marinatto L, Weber T (2000) A quantum approach to static games of complete information. Phys Lett A 272:291–303
11. Meyer D (1999) Quantum strategies. Phys Rev Lett 82:1052–1055
12. Eisert J, Wilkens M, Lewenstein M (1999) Quantum games and quantum strategies. Phys Rev Lett 83:3077–3080
13. Benjamin S, Hayden P (2001) Multiplayer quantum games. Phys Rev A 64:030301
14. Chen Q, Wang Y (2004) N-player quantum minority game. Phys Lett A 327(98):102
15. Flitney A, Hollenberg LCL (2007) Multiplayer quantum minority game with decoherence. Quantum Inf Comput 7:111–126
16. Flitney A, Greentree A (2007) Coalitions in the quantum minority game: classical cheats and quantum bullies. Phys Lett A 362:132137
17. Schmid C, Flitney AP, Wieczorek W, Kiesel N, Weinfurter H, Hollenberg LCL (2010) Experimental implementation of a four-player quantum game. New J Phys 12:063031
18. Sharif P, Heydari H (2011) Quantum solution to a three player Kolkata restaurant problem using entangled qutrits. Preprint http://arxiv.org/abs/1111.1962
19. Benjamin S, Hayden P (2001) Comment on quantum games and quantum strategies. Preprint http://arxiv.org/abs/quant-ph/0003036
20. Chakrabarti BK (2007) Kolkata restaurant problem as a generalised El Farol Bar problem. In: Chatterjee A, Chakrabarti BK (eds) Econophysics of markets & business networks. New economic windows series. Springer, Milan, pp 239–246
21. Chakrabarti AS, Chakrabarti BK, Chatterjee A, Mitra M (2009) The Kolkata paise restaurant problem and resource utilization. Physica A 388:2420–2426
22. Ghosh A, Chakrabarti AS, Chakrabarti BK (2010) Kolkata paise restaurant problem in some uniform learning strategy limits. In: Basu B, Chakrabarti BK, Chakravarty SR, Gangopadhyay K (eds) Econophysics & economics of games, social choices & quantitative techniques, New economic windows. Springer, Milan, pp 3–9
23. Ghosh A, Chatterjee A, Mitra M, Chakrabarti BK (2010) New J Phys 12:075033
24. Arthur WB (1994) Inductive reasoning and bounded rationality: El Farol problem. Am Econ Assoc Papers & Proc 84:406
25. Mathur M, Sen D (2001) Coherent states for SU(3). J Math Phys 42:4181–4196

# Part III
# Miscellaneous Reports

# Chapter 15
# Cluster Analysis and Gaussian Mixture Estimation of Correlated Time-Series by Means of Multi-dimensional Scaling

**Takero Ibuki, Sei Suzuki, and Jun-ichi Inoue**

**Abstract** We investigate cross-correlations between typical Japanese stocks collected through Yahoo!Japan website (http://finance.yahoo.co.jp/). By making use of multi-dimensional scaling (MDS) for the cross-correlation matrices, we draw two-dimensional scattered plots in which each point corresponds to each stock. To make a clustering for these data plots, we utilize the mixture of Gaussians to fit the data set to several Gaussian densities. By minimizing the so-called Akaike Information Criterion (AIC) with respect to parameters in the mixture, we attempt to specify the best possible mixture of Gaussians. It might be naturally assumed that all the two-dimensional data points of stocks shrink into a single small region when some economic crisis takes place. The justification of this assumption is numerically checked for the empirical Japanese stock data, for instance, those around 11 March 2011.

## 15.1 Introduction

We sometimes encounter the problem to find a non-trivial structure in correlated time series observed from multi channels. For instance, in neuroscience, we should extract the meaningful structure from brain waves and specify what kind of the structure corresponds to human behavior by using various techniques of pattern recognition. This type of knowledge is required for achieving the so-called Brain Machine Interface (BMI) (which is sometimes refereed to as the Brain Computer

T. Ibuki · J. Inoue
Graduate School of Information Science and Technology, Hokkaido University, N14-W9, Kita-Ku, Sapporo 060-0814, Japan

T. Ibuki
e-mail: ibuki@complex.ist.hokudai.ac.jp

J. Inoue (✉)
e-mail: jinoue@cb4.so-net.ne.jp

S. Suzuki
Department of Physics and Mathematics, Aoyama-Gakuin University, 5-10-1 Fuchinobe, Sagamihara 252-5258, Japan
e-mail: sei@phys.aoyama.ac.jp

F. Abergel et al. (eds.), *Econophysics of Systemic Risk and Network Dynamics*, New Economic Windows,
DOI 10.1007/978-88-470-2553-0_15, 

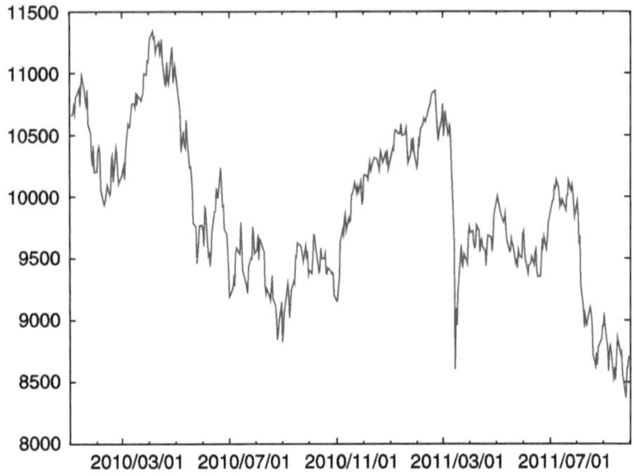


**Fig. 15.1** The Nikkei stock average around 11th March 2011. It is easily found that the curve suddenly drops after the crisis

Interface (BCI)), which is one of the remarkable applications of neuroscience to engineering. Such time series obtained from multi-channel measurements have been widely provided in both natural and social sciences. Hence, it is now important for us to carry out empirical data analysis extensively to solve various modern and serious problems around us.

As everybody knows, in our country on 11th March 2011, a massive earthquake and huge TSUNAMI hit the northeast coast and, as a result, Fukushima nuclear power plant was seriously damaged. People who was living in that area has taken refuge from the nuclear radiation. These unpredictable disasters caused by both nature and/or human errors have made our country in a lot of difficulties that we have never encountered before.

Of course, the disaster affected seriously on Japanese economy. In fact, after the earthquake, Japanese NIKKEI stock market quickly responded to the crisis and quite a lot of traders sold their stocks of companies whose branches or plants are located in that disaster stricken area. As a result, the Nikkei stock average suddenly drops after the crisis (see Fig. 15.1). As is shown in Fig. 15.1, 'crushes' actually took place due to the reaction of the stock market to the earthquake. Obviously, the curve of the Nikkei stock average shown in Fig. 15.1 is a 'macroscopic' aspect (behavior) of the market. Namely, microscopic information about the behavior of each stock is averaged out. Actually, the curve was obtained as an average of several (more than 1000) stocks. It is important for us to make an attempt to bring out more 'microscopic' useful information, which is never obtained from the averaged macroscopic quantities such as stock average, about the market.

As a candidate of such 'microscopic information', we can use the (linear) correlation coefficient based on the two-body interactions between stocks [1, 2]. To figure out the mechanism of financial crisis, it might be helpful for us to visualize such correlations in stocks and compare the dynamical behavior of the correlation before and after crisis.

In this paper, we attempt to visualize the correlation of each stock in two-dimension. In [2], the authors constructed 'minimum spanning tree' to visualize such correlations in stocks, however, here we use a different approach. Namely, we specify each location of $N$ stocks from a given set of the $N(N-1)/2$ distances by making use of the so-called multi-dimensional scaling (MDS) [3]. We next make a clustering of these scattered data points by fitting the data with a mixture of Gaussians under the Akaike Information Criterion (AIC). In order to maximize the likelihood function (minimizing the AIC), we use both simulated annealing (SA) and the so-called EM (Expectation and Maximization) algorithm. It might be naturally assumed that all the two-dimensional scattered points shrink into a single small region when economic crisis takes place. The justification of this assumption is numerically checked for the empirical Japanese stock data, for instance, those around 11 March 2011.

This paper is organized as follows. In Sect. 15.2, we explain our measurement of stock correlation, that is linear correlation coefficient. We also mention how we convert the coefficient to the distance between stocks. The set of the distances is an input data for the multi-dimensional scaling (MDS). The detail procedure of the MDS is explained in Sect. 15.3. Then, we show the resulting scattered plots obtained for the Japanese stock data around 11th March 2011. In Sect. 15.4, we attempt to make a clustering of data by fitting the scattered plots by a mixture of Gaussians. To estimate the parameters appearing in the mixture, we utilize both simulated annealing and EM algorithm. We also introduce the Akaike Information Criterion (AIC) to determine the number of components in the mixture so as to fit the data as plausible as possible. The result for the empirical data is shown in Sect. 15.5. The last section is summary.

## 15.2 Linear Correlation Coefficient

In this paper, we utilize the linear correlation coefficient (Pearson product-moment correlation coefficient) to measure the strength of correlation between stocks [1, 2]. The coefficient is calculated as follows.

Let us define $p_i(t)(\geq 0)$ as a price of stock $i$ at time $t$. Then, we evaluate the return of the price $p_i(t)$ in terms of logarithmic measurement as

$$r_i^{(t)} \equiv \log p_i(t) - \log p_i(t-1). \tag{15.1}$$

For this logarithmic return, we calculate the moving average over the time window with width $M$ as

$$\overline{r_i^{(t)}} \equiv \frac{1}{M} \sum_{l=t-M+1}^{t} r_i^{(l)} \tag{15.2}$$

for stock $i$, and also evaluate the two-body correlation between stocks $i, j$ by the following definition

$$\overline{r_i^{(t)} r_j^{(t)}} \equiv \frac{1}{M} \sum_{l=t-M+1}^{t} r_i^{(l)} r_j^{(l)}. \tag{15.3}$$

Then, the linear correlation coefficient is given by

$$\rho_{ij}^{(t)} = \frac{\overline{r_i^{(t)} r_j^{(t)}} - (\overline{r_i^{(t)}})(\overline{r_j^{(t)}})}{\sqrt{[\overline{(r_i^{(t)})^2} - (\overline{r_i^{(t)}})^2][\overline{(r_j^{(t)})^2} - (\overline{r_j^{(t)}})^2]}}. \tag{15.4}$$

We should keep in mind that the above coefficient (15.4) satisfies

$$-1 \le \rho_{ij}^{(t)} \le 1 \tag{15.5}$$

and apparently it cannot be treated as a 'distance'. Hence, we transform the coefficient $\rho_{ij}^{(t)}$ into the distance $d_{ij}^{(t)}$ between the stocks $i, j$ as

$$d_{ij}^{(t)} = \sqrt{\frac{(1 - \rho_{ij}^{(t)})}{2}}. \tag{15.6}$$

Then, the distance $d_{ij}^{(t)}$ actually satisfies

$$0 \le d_{ij}^{(t)} \le 1 \tag{15.7}$$

and

- $d_{ij}^{(t)} \ge 0$ (non-negative),
- $d_{ij}^{(t)} = 0$ if $i = j$ (identity of indiscernibles),
- $d_{ij}^{(t)} = d_{ji}^{(t)}$ (symmetry),
- $d_{ij}^{(t)} + d_{jk}^{(t)} \ge d_{ik}^{(t)}$ (triangle inequality).

In next section, we use a set of $d_{ij}^{(t)}$ as an input data to the multi-dimensional scaling.

## 15.3 Multi-dimensional Scaling

In this section, we shall explain the procedure of making a scattered plot by multi-dimensional scaling (MDS). Let us first specify the location of stock $i$ by means of $P$-dimensional vector $\boldsymbol{X}_i \equiv (x_{i1}, x_{i2}, \ldots, x_{iP})$, $i = 1, \ldots, N$. Naturally, the distance between stocks $i, j$ is given by

$$d_{ij} = \sqrt{\sum_{m=1}^{P} (x_{im} - x_{jm})^2}. \tag{15.8}$$

We also define the inner product of vectors of stocks $i$ and $j$ as

$$\boldsymbol{X}_i \cdot \boldsymbol{X}_j = z_{ij} = \frac{1}{2}\left(\frac{1}{N}\sum_{k=1}^{N} d_{kj}^2 + \frac{1}{N}\sum_{k=1}^{N} d_{ik}^2 - \frac{1}{N^2}\sum_{k=1}^{N}\sum_{l=1}^{N} d_{kl}^2 - d_{ij}^2\right) \quad (15.9)$$

where we should notice that we chose the origin of axis as the 'center of mass' for $N$ stocks points, that is, $\sum_{i=1}^{N} \boldsymbol{X}_i/N$.

Then, to find the location $\boldsymbol{X}_i$; $i = 1, \ldots, N$ which generates a set of distances $\{d_{ij}\}$ properly, we should minimize the following energy function:

$$E = \sum_i \sum_j \left(z_{ij} - \sum_{m=1}^{P} x_{im} x_{jm}\right)^2 \quad (15.10)$$

with respect to $\boldsymbol{X}_i$; $i = 1, \ldots, N$. Thus, our problem to find the best possible locations for stocks is rewritten in terms of an optimization problem to find the ground state of the energy function $E$.

We easily notice that the solution to minimize $E$ is given in terms of a symmetric $N \times N$ matrix $\boldsymbol{Z} \equiv \{z_{ij} \mid i, j = 1, \ldots, N\}$ and a $P \times N$ matrix $\boldsymbol{X} = (\boldsymbol{X}_1, \ldots, \boldsymbol{X}_N)^T$ as

$$\boldsymbol{Z} = \boldsymbol{X}\boldsymbol{X}^T. \quad (15.11)$$

It should be noted that the $T$ appearing in the shoulder of vector such as $(\cdots)^T$ stands for the transposing. As the matrix $Z$ is real and symmetric, we can write

$$\boldsymbol{Z} = \boldsymbol{Y}\boldsymbol{\Lambda}\boldsymbol{Y}^T \quad (15.12)$$

where we defined $\boldsymbol{Y} = (\boldsymbol{y}_1, \ldots, \boldsymbol{y}_N)^T$ for eigenvector $\boldsymbol{y}_k$ of $\boldsymbol{Z}$ satisfying

$$\boldsymbol{Z}\boldsymbol{y}_k = \lambda_k \boldsymbol{y}_k, \quad k = 1, \ldots, N. \quad (15.13)$$

$\boldsymbol{\Lambda}$ is defined as

$$\boldsymbol{\Lambda} = \mathrm{diag}(\lambda_1, \ldots, \lambda_N). \quad (15.14)$$

Hence, we immediately obtain the solution

$$\boldsymbol{X} = \boldsymbol{Y}\boldsymbol{\Lambda}^{1/2}. \quad (15.15)$$

As we need two-dimensional plot ($P = 2$), we should pick up the two eigenvectors for the largest and the second largest eigenvalues, and construct $\boldsymbol{X}$ in terms of (15.15).

### 15.3.1 Empirical Data Analysis

We apply the above procedure to the financial data provided on the web site [4]. Especially, we pick up 200 stocks including the so-called TOPIX (TOkyo stock

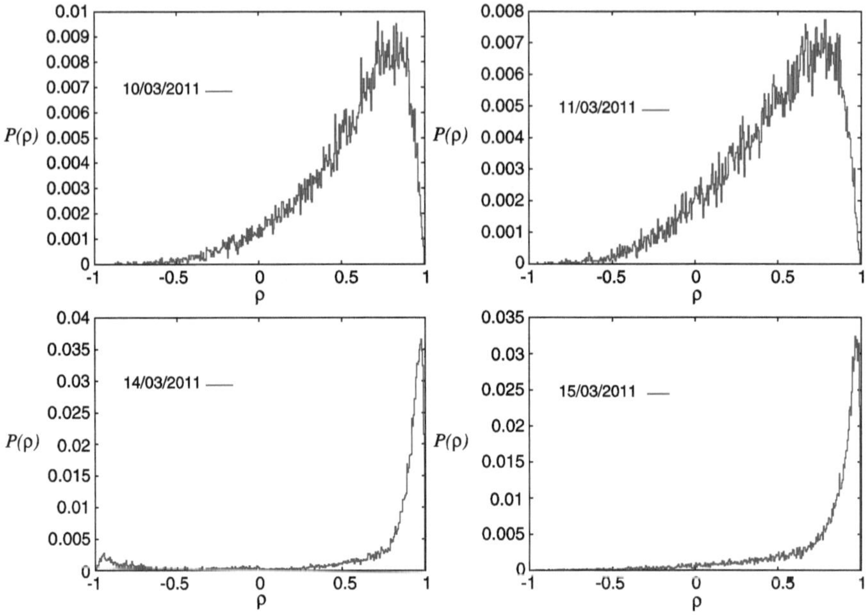


**Fig. 15.2** The distribution $P(\rho)$ of the linear correlation coefficient from the *upper left* to the *lower right*, the results on 10th, 11th, 14th and 15th March 2011 are plotted

Price IndeX) Core30, which consists of typical 30 stock indices being picked up from the view point of 'current price' or 'liquidity' from the Nikkei stock market. It should be kept in mind that the data is not given as 'tick-by-tick', the minimal time interval of the data is one day (the closing price is given in the data set).

We first plot the distribution $P(\rho)$ of the linear coefficient $\{\rho_{ij}^{(t)}\}, i, j = 1, \ldots, N$ for specific four days around the crisis in Fig.15.2. From these panels, we find that after the crisis on 11th March 2011, the correlation is strongly enhanced. More precisely, on 14th March, a single bulk of the distribution on 11th splits into two bulks, namely, positively and negatively correlated parts. However, the next day 15th March, the negative part vanishes and a single strongly correlated bulk survives.

The distribution does not necessarily mention something about microscopic information about the behavior of stocks. Thus, for this set of Japanese stocks, we next draw the scattered plots by making use of the MDS in Fig. 15.3. From these panels, we clearly find that after the crisis, the scattered plots actually shrink into a small region centered at the origin (the center of mass) as we expected before. As time goes on, the plots change their structure into very curious one. Apparently, the shape is completely different from the usual structure such as a plot on 10th March 2011 (Fig. 15.3a).

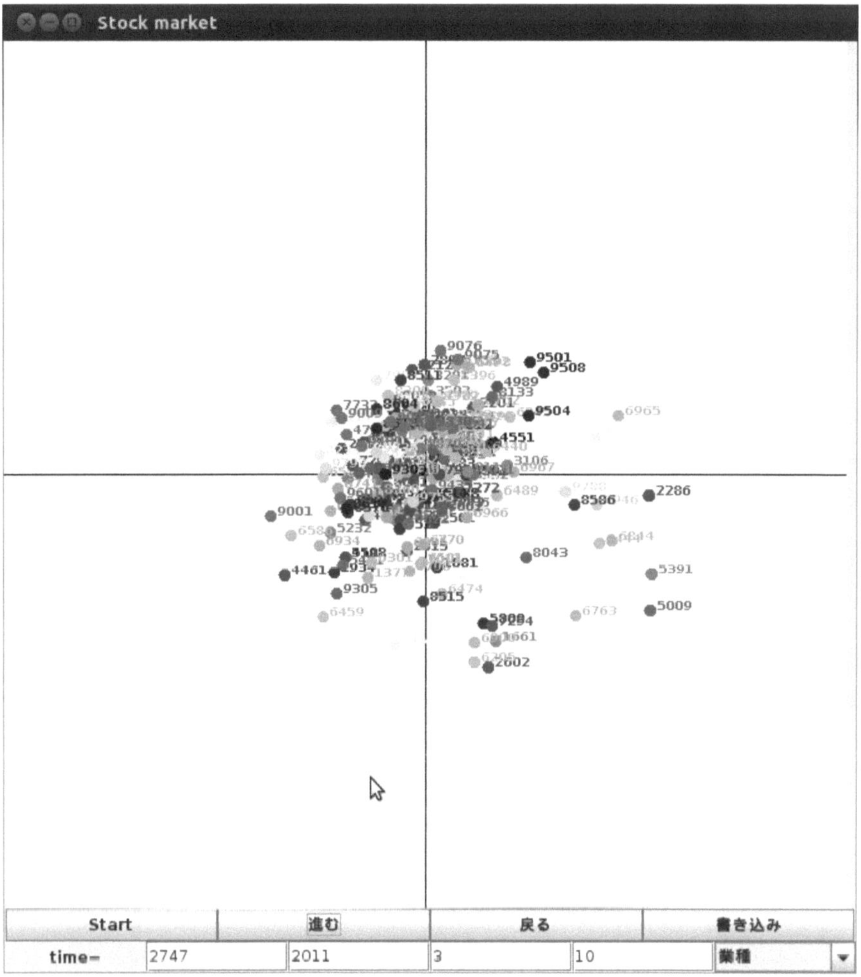


(a)

**Fig. 15.3** The result of the MDS. We pick up 200 stocks including the so-called TOPIX Core30 and the Nikkei stock average as empirical data set. From (**a**) to (**d**), the result on 10th, 11th, 14th and 15th March 2011 are shown. Different colors indicate different types of business. The numbers accompanying the dots show company IDs. We set the width of time window to evaluate the correlation coefficient as $M = 7$ (days)

## 15.4 Clustering by Mixture of Gaussians

In this section, we make a clustering from scattered plots which was obtained by the MDS in the previous section. We first approximate the plots by a mixture of two-dimensional Gaussians. Let us consider that the number of Gaussians is $K$ and each component is given by a multivariate normal Gaussian $\mathcal{N}(\boldsymbol{x} \mid \boldsymbol{\mu}^{(k)}, \boldsymbol{\Sigma}^{(k)})$, $k =$

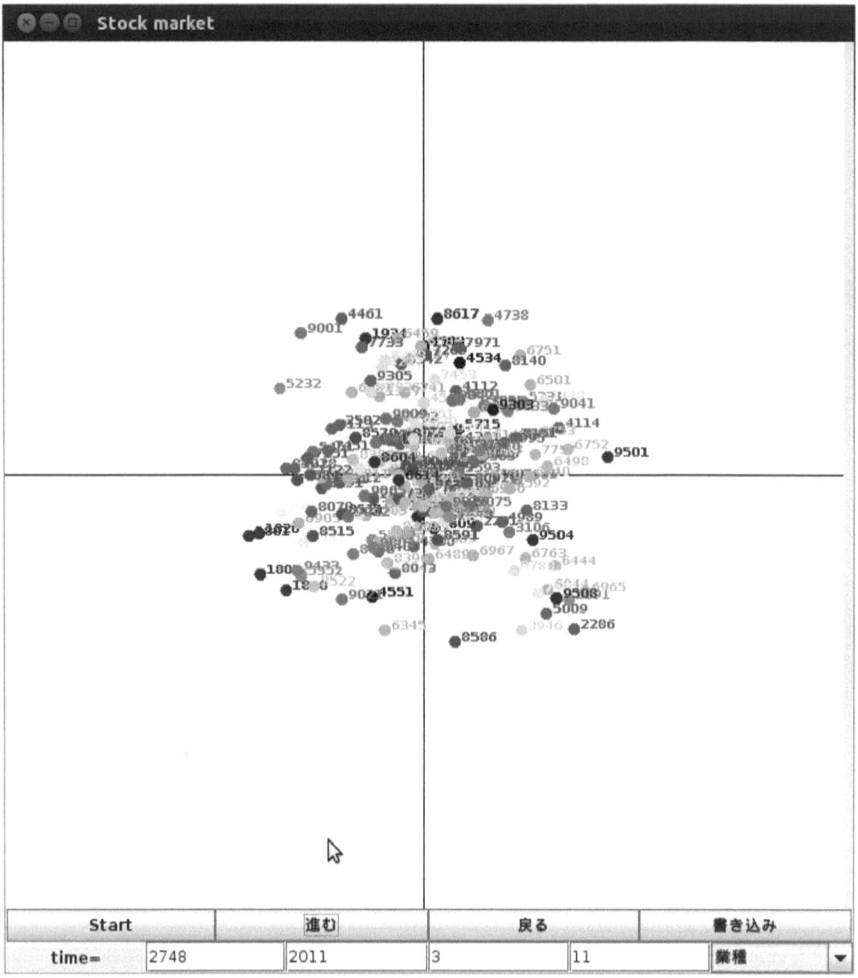


(b)

**Fig. 15.3** (Continued)

$1, \ldots, K$, where $\boldsymbol{\mu}^{(k)}$ stands for the mean and $\boldsymbol{\Sigma}^{(k)}$ denotes a matrix of variance-covariance. The vector $\boldsymbol{x}$ corresponds to the $N$ points in two-dimension, namely, $\boldsymbol{x}_i = (x_i, y_i)^T; i = 1, \ldots, N$. Hence, a mixture of these Gaussians is written by

$$p(\boldsymbol{x}) = \sum_{k=1}^{K} \pi_k \mathcal{N}\big(\boldsymbol{x} \mid \boldsymbol{\mu}^{(k)}, \boldsymbol{\Sigma}^{(k)}\big) \tag{15.16}$$

where $\pi_k$ is an weight of the $k$th Gaussian and it should satisfy $\sum_{k=1}^{L} \pi_k = 1$ for normalization.

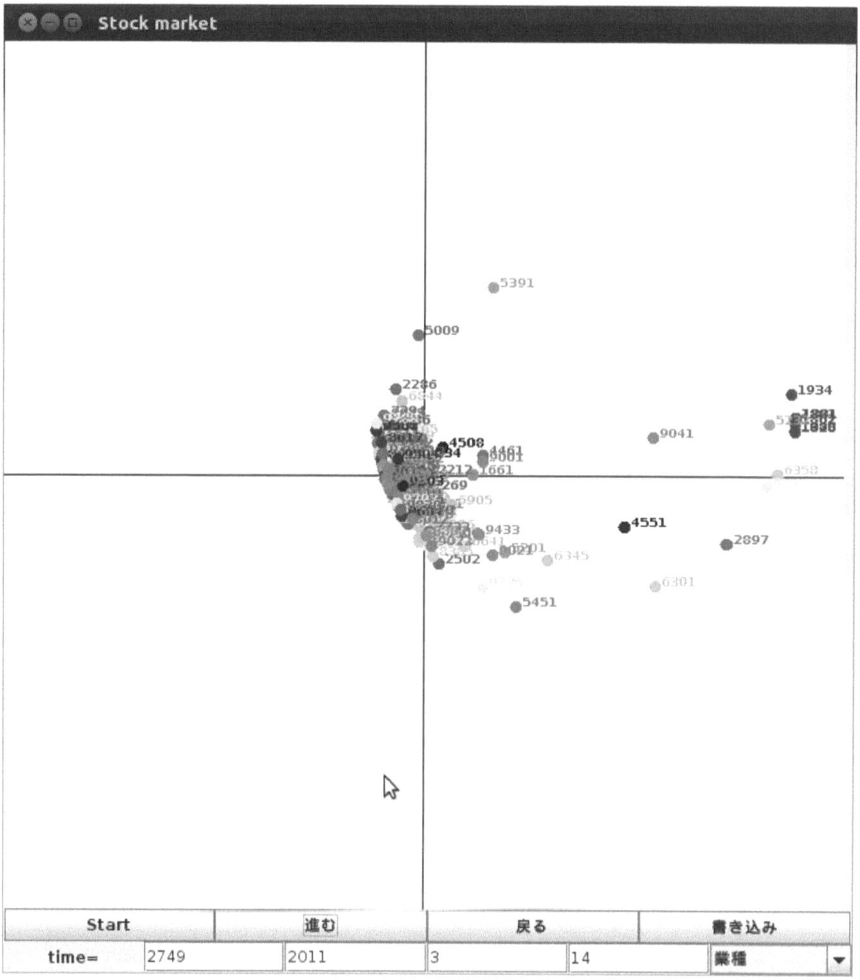


(c)

**Fig. 15.3** (Continued)

Then, for a given set of $N$ data points $\boldsymbol{x}_1, \boldsymbol{x}_2, \ldots, \boldsymbol{x}_N$, the likelihood function $L(\boldsymbol{\pi}, \{\boldsymbol{\mu}\}, \{\boldsymbol{\Sigma}\}) \equiv \log \prod_{i=1}^{N} p(\boldsymbol{x}_i)$ is written as

$$L\big(\boldsymbol{\pi}, \{\boldsymbol{\mu}\}, \{\boldsymbol{\Sigma}\}\big)$$

$$= \sum_{i=1}^{N} \log \sum_{k=1}^{K} \frac{\pi_k}{2\pi |\boldsymbol{\Sigma}^{(k)}|^{1/2}} \exp\left[-\frac{1}{2}\big(\boldsymbol{x}_i - \boldsymbol{\mu}^{(k)}\big)^T \big(\boldsymbol{\Sigma}^{(k)}\big)^{-1} \big(\boldsymbol{x}_i - \boldsymbol{\mu}^{(k)}\big)\right] \quad (15.17)$$

where we defined

$$\boldsymbol{\pi} \equiv (\pi_1, \pi_2, \ldots, \pi_K), \quad (15.18)$$

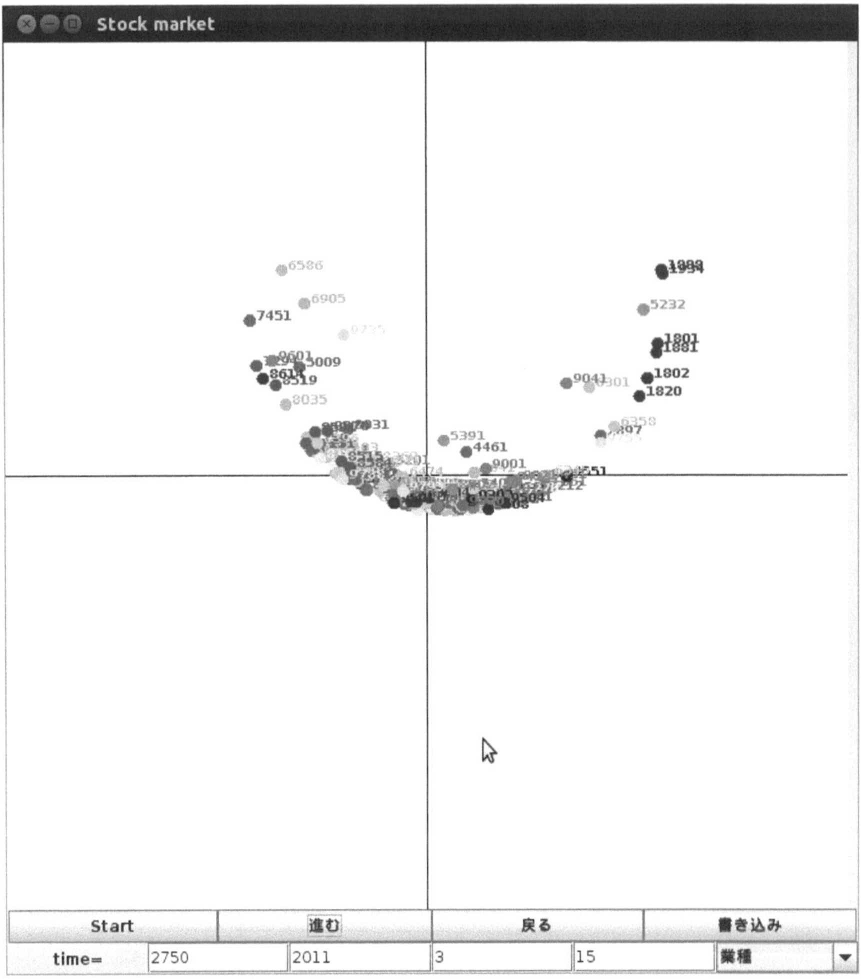


(d)

**Fig. 15.3** (Continued)

$$\{\boldsymbol{\mu}\} \equiv \left(\boldsymbol{\mu}^{(1)}, \boldsymbol{\mu}^{(2)}, \cdots, \boldsymbol{\mu}^{(K)}\right), \tag{15.19}$$

$$\{\boldsymbol{\Sigma}\} \equiv \left(\boldsymbol{\Sigma}^{(1)}, \boldsymbol{\Sigma}^{(2)}, \cdots, \boldsymbol{\Sigma}^{(K)}\right) \tag{15.20}$$

for simplicity.

Therefore, we might use the gradient learning algorithm with respect to the above parameters by regarding the likelihood $L$ as a cost function, namely,

$$\frac{d\boldsymbol{\pi}}{dt} = \frac{\partial L}{\partial \boldsymbol{\pi}}, \qquad \frac{d\boldsymbol{\mu}^{(k)}}{dt} = \frac{\partial L}{\partial \boldsymbol{\mu}^{(k)}}, \qquad \frac{d\boldsymbol{\Sigma}^{(k)}}{dt} = \frac{\partial L}{\partial \boldsymbol{\Sigma}^{(k)}}, \quad k = 1, \ldots, K. \tag{15.21}$$

However, the number of parameters amounts to $6K-1$, and even for $K=2$ mixtures, we should have 11 variables. Obviously, it is very difficult for us to deal with the differential equations to obtain the solution numerically within a certain precision. Hence, in the next subsections, we use simulated annealing (SA) and the so-called EM algorithm to maximize the likelihood function.

### 15.4.1 Simulated Annealing

We attempt to minimize the likelihood $L$ by simulated annealing. According to the conventional Gibbs sampler, we update each parameter in such a way as $\pi_k \to \pi_k + \delta\pi_k$, where $\delta\pi_k$ etc. are given randomly from some suitable range. We also use the linear scheduling of temperature as $T \to T - \Delta T$.

For the resulting maximum of the likelihood $L(\boldsymbol{\pi}^*, \{\boldsymbol{\mu}\}^*, \{\boldsymbol{\Sigma}\}^*)$, the $K$ ellipses which approximate the scattered plot are given by

$$\begin{aligned}&\left(\sigma_y^{(k)*}\right)^2\left(x-\mu_x^{(k)*}\right)^2 - 2\sigma_{xy}^{(k)*}\left(x-\mu_x^{(k)*}\right)\left(y-\mu_y^{(k)*}\right) + \left(\sigma_x^{(k*)}\right)^2\left(y-\mu_y^{(k)*}\right)^2 \\ &\quad = 2\left(\left(\sigma_x^{(k)*}\sigma_y^{(k)*}\right)^2 - \left(\sigma_{xy}^{(k)*}\right)^2\right) \qquad (15.22)\end{aligned}$$

for $k=1,\dots,K$.

### 15.4.2 EM Algorithm

We can maximize the likelihood function indirectly by making use of the EM algorithm [5]. In the algorithm, the so-called 'E-step' (Expectation step) evaluates the following Q function:

$$Q\left(\{\boldsymbol{\mu}\},\{\boldsymbol{\Sigma}\} \mid \{\boldsymbol{\mu}_t\},\{\boldsymbol{\Sigma}_t\}\right) = \sum_{\boldsymbol{x}} \frac{\sum_k \mathcal{N}(\boldsymbol{x} \mid \boldsymbol{\mu}_t^{(k)}, \boldsymbol{\Sigma}_t^{(k)}) \log \mathcal{N}(\boldsymbol{x} \mid \boldsymbol{\mu}^{(k)}, \boldsymbol{\Sigma}^{(k)})}{\sum_k \mathcal{N}(\boldsymbol{x} \mid \boldsymbol{\mu}_t^{(k)}, \boldsymbol{\Sigma}_t^{(k)})}. \qquad (15.23)$$

Then, we update each parameter by the following 'M-Step' (Maximization step):

$$\begin{cases} \boldsymbol{\mu}_{t+1}^{(k)} = \arg\max\limits_{\boldsymbol{\mu}^{(k)}} Q\left(\{\boldsymbol{\mu}\},\{\boldsymbol{\Sigma}\} \mid \{\boldsymbol{\mu}_t\},\{\boldsymbol{\Sigma}_t\}\right), & k=1,\dots,K, \\ \boldsymbol{\Sigma}_{t+1}^{(k)} = \arg\max\limits_{\boldsymbol{\Sigma}^{(k)}} Q\left(\{\boldsymbol{\mu}\},\{\boldsymbol{\Sigma}\} \mid \{\boldsymbol{\mu}_t\},\{\boldsymbol{\Sigma}_t\}\right), & k=1,\dots,K. \end{cases} \qquad (15.24)$$

Repeating the above 'E-step' and 'M-step' until the parameters converge to the steady state, namely, $\boldsymbol{\mu}^{(k)*}$, $\boldsymbol{\Sigma}^{(k)*}$, the cluster $C(\boldsymbol{x})$ to which the vector $\boldsymbol{x}$ belong is determined by

$$C(\boldsymbol{x}) = \arg\max_k g_k(\boldsymbol{x}) \qquad (15.25)$$

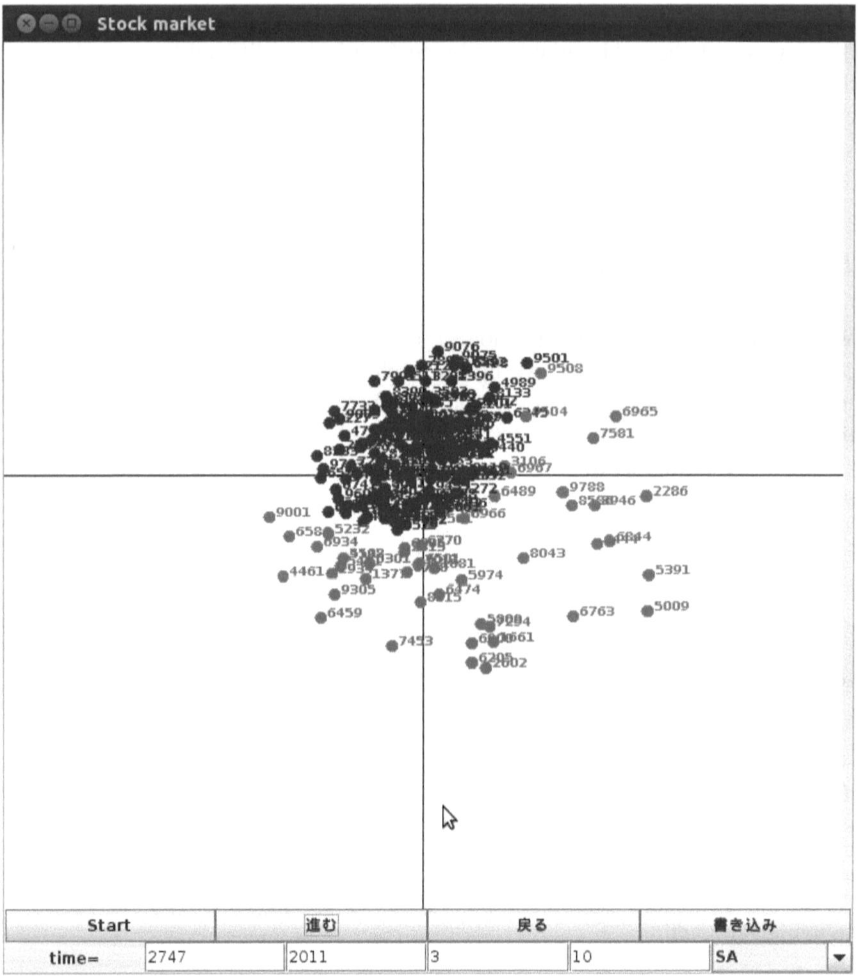


(a)

**Fig. 15.4** The resulting clusters obtained by SA. From (**a**) to (**d**), the results on 10th, 11th, 14th and 15th March 2011 are shown. The scattered plots we make a clustering are the same as shown in Fig. 15.3

where we defined

$$g_k(\boldsymbol{x}) \equiv \frac{\mathcal{N}(\boldsymbol{x} \mid \boldsymbol{\mu}^{(k)*}, \boldsymbol{\Sigma}^{(k)*})}{\sum_k \mathcal{N}(\boldsymbol{x} \mid \boldsymbol{\mu}^{(k)*}, \boldsymbol{\Sigma}^{(k)*})}. \tag{15.26}$$

### *15.4.3 The Number of Parameters and AIC*

In the previous subsections, we fixed the number of parameters $K$ during the SA and the EM algorithm. However, we should also determine the number from the

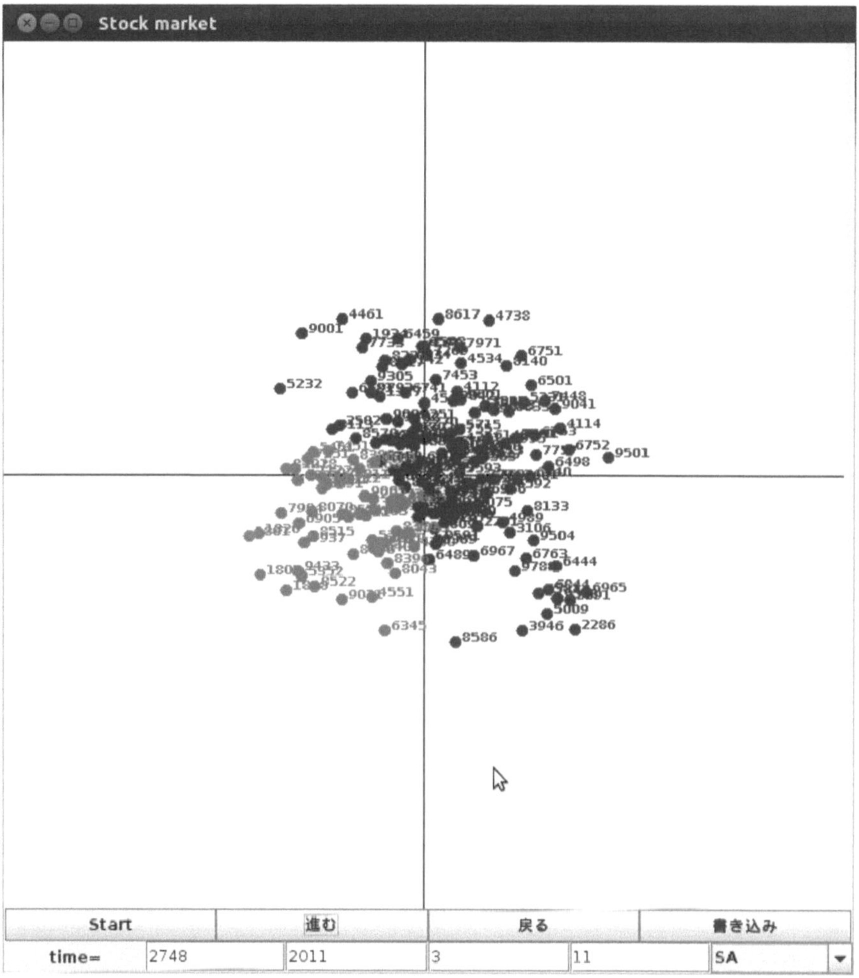


(b)

**Fig. 15.4** (Continued)

empirical data. For this purpose, we utilize the Akaike Information Criterion (AIC). The AIC is defined as a function of $K$ by

$$\begin{aligned} \mathrm{AIC} &\equiv (-2)\big\{L\big(\boldsymbol{\pi}^*, \{\boldsymbol{\mu}\}^*, \{\boldsymbol{\Sigma}\}^*\big) - (\text{\# of parameters in the probabilistic model})\big\} \\ &= -2L\big(\boldsymbol{\pi}^*, \{\boldsymbol{\mu}\}^*, \{\boldsymbol{\Sigma}\}^*\big) + 12K - 2. \end{aligned} \tag{15.27}$$

Hence, we select the reasonable number $K$ after evaluating the AIC for $K = 1, 2, \ldots, K_{\max}$ and selecting the best possible $K$ that minimizes the AIC.

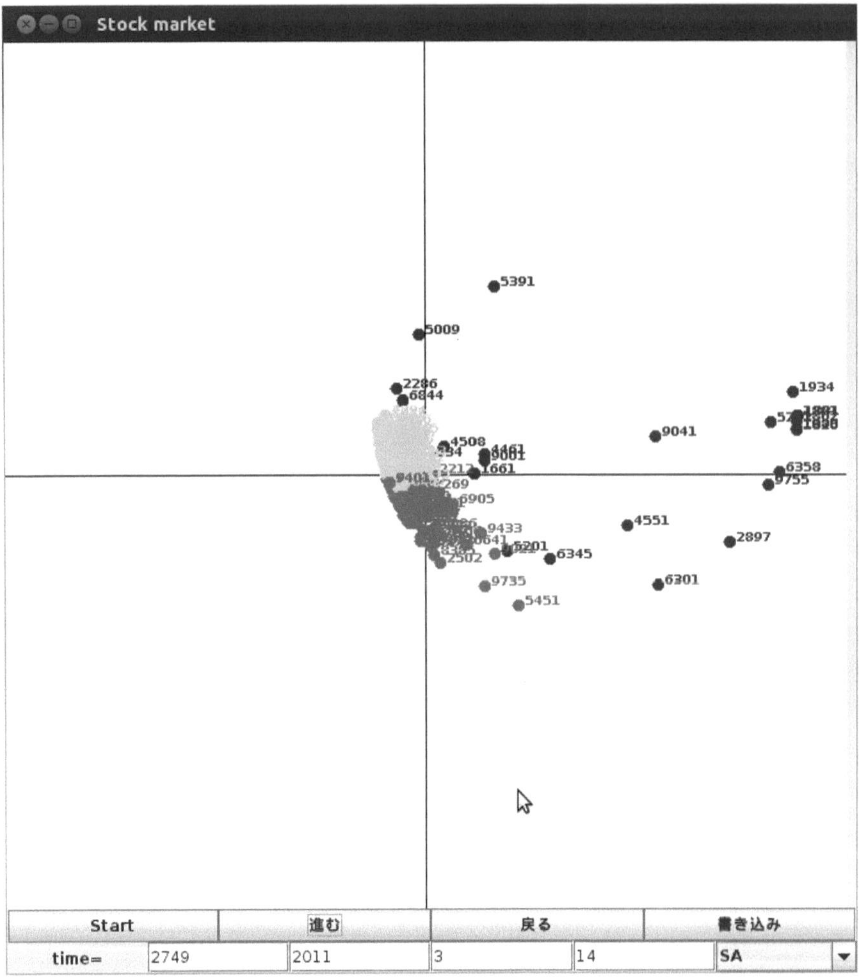


(c)

**Fig. 15.4** (Continued)

## 15.5 Empirical Data Analysis

By using the above theoretical framework, we carry out numerical analysis for the empirical data which was shown in Fig. 15.3.

### *15.5.1 Simulated Annealing*

We apply simulated annealing (SA) to maximize the likelihood function and to estimate the parameters appearing in the mixture of Gaussians. In our numerical analy-

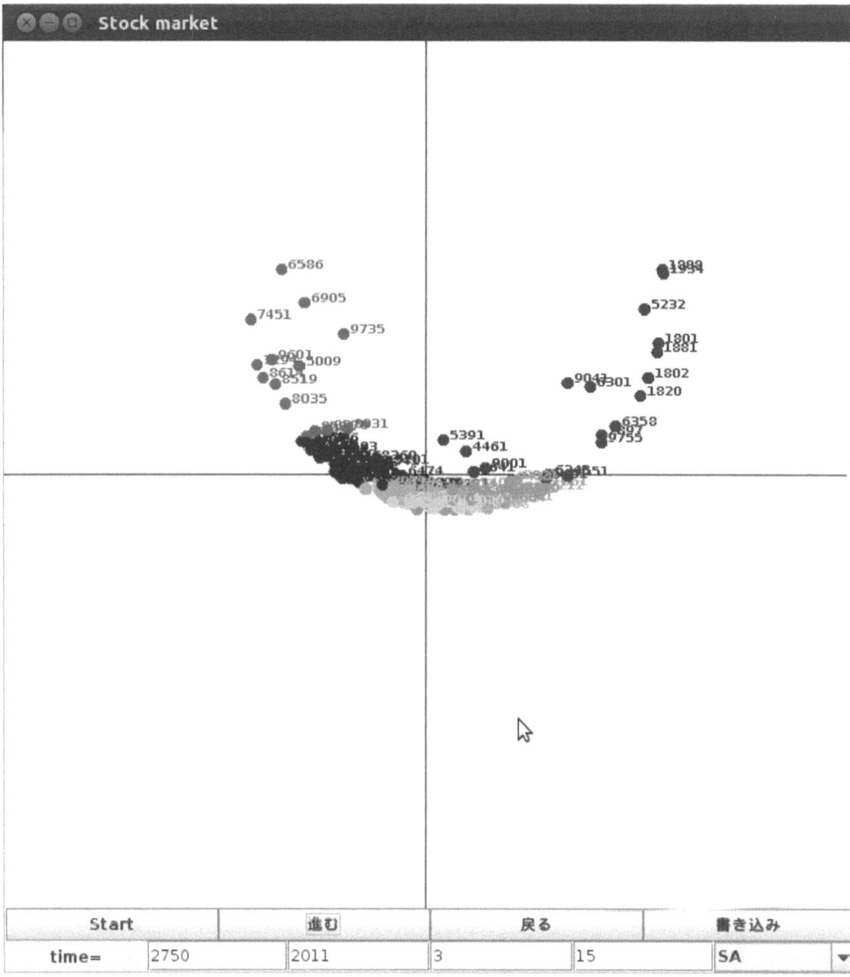


(d)

**Fig. 15.4** (Continued)

sis, we set the parameters to control the SA as $T_0 = 20$ (initial value of temperature), $\Delta T = 0.001$, $\delta\pi_k = \frac{0.01}{K} \times$ a uniform random number in the interval [0, 1]. The increments of the other parameters are chosen from a uniform distribution of the range [0, 1]. We show the result in Fig. 15.4.

From this figure, we find that before crisis, those widely distributed scattered plots are well-fitted by two large clusters. Just after the crisis, we also find that two clusters are needed to explain the scattered plots and after the crisis, say, 14th and 15th March 2011, the number of clusters changes as $K = 3$ and $K = 6$, respectively. We should notice that the density of each dominant cluster does not decreases (rather increases) although the number $K$ increases.

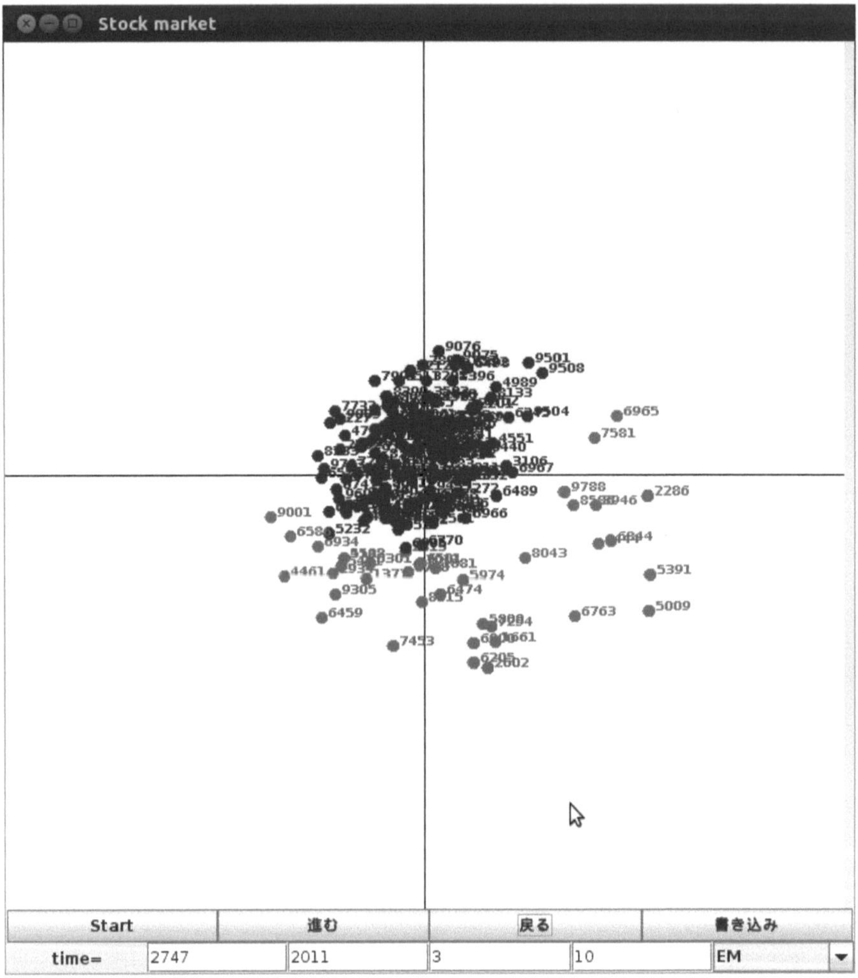


(a)

**Fig. 15.5** The resulting clusters obtained by EM. From (**a**) to (**d**), the results on 10th, 11th, 14th and 15th March 2011 are shown. The scattered plots we make a clustering are the same as shown in Fig. 15.3

## *15.5.2 EM Algorithm*

We next show the result of clustering achieved by the EM algorithm in Fig. 15.5. The number of clusters is slightly different from that of the SA. Actually, we find $K = 2, 4, 5$ and $K = 6$ for 10th, 11th, 14th and 15th March 2011, respectively. However, after crisis on 15th March 2011, a large amount of members in each cluster is overlapping with the result by SA.

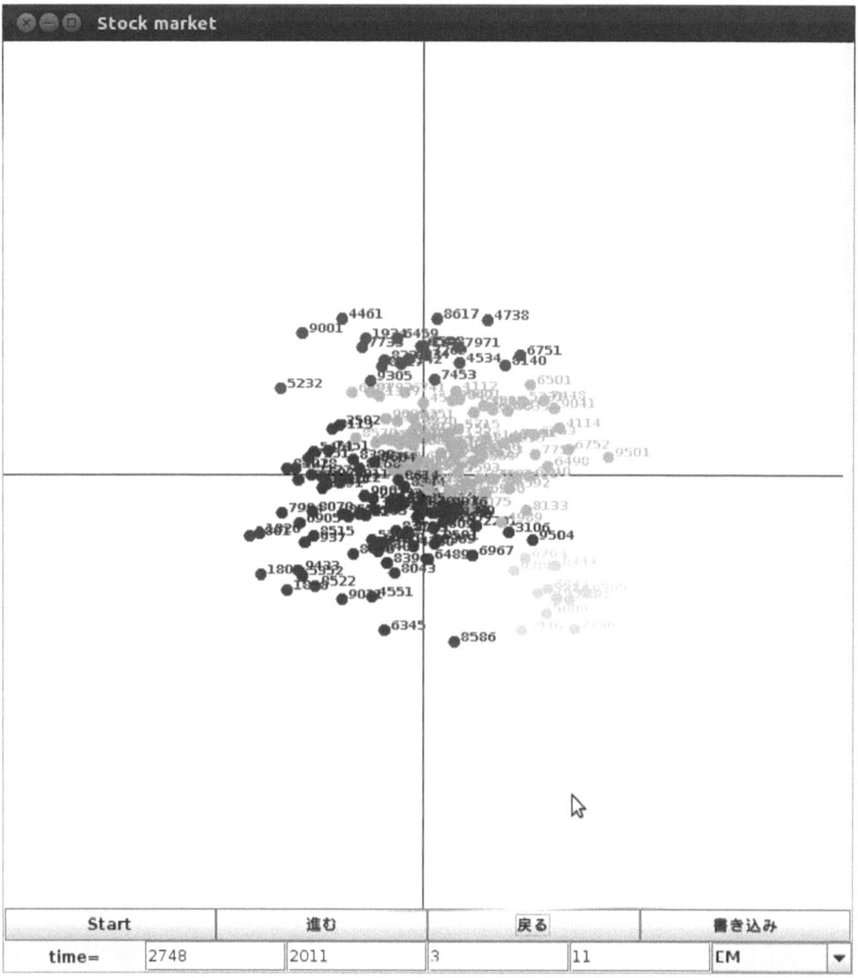


(b)

**Fig. 15.5** (Continued)

## *15.5.3 Time Dependence of AIC*

As we saw in the previous subsection, we found that the number of clusters itself is not an indicator to specify the time of crushes. Here we examine the AIC to characterize the crises because the AIC could be evaluated at each time step. We show the result in Fig. 15.6. From this figure, we clearly find that for both procedure to maximize the likelihood, the AIC takes a sharp and deep minimum at the crisis. This result implies that we can utilize the AIC instead of the stock average to specify the time of the crisis.

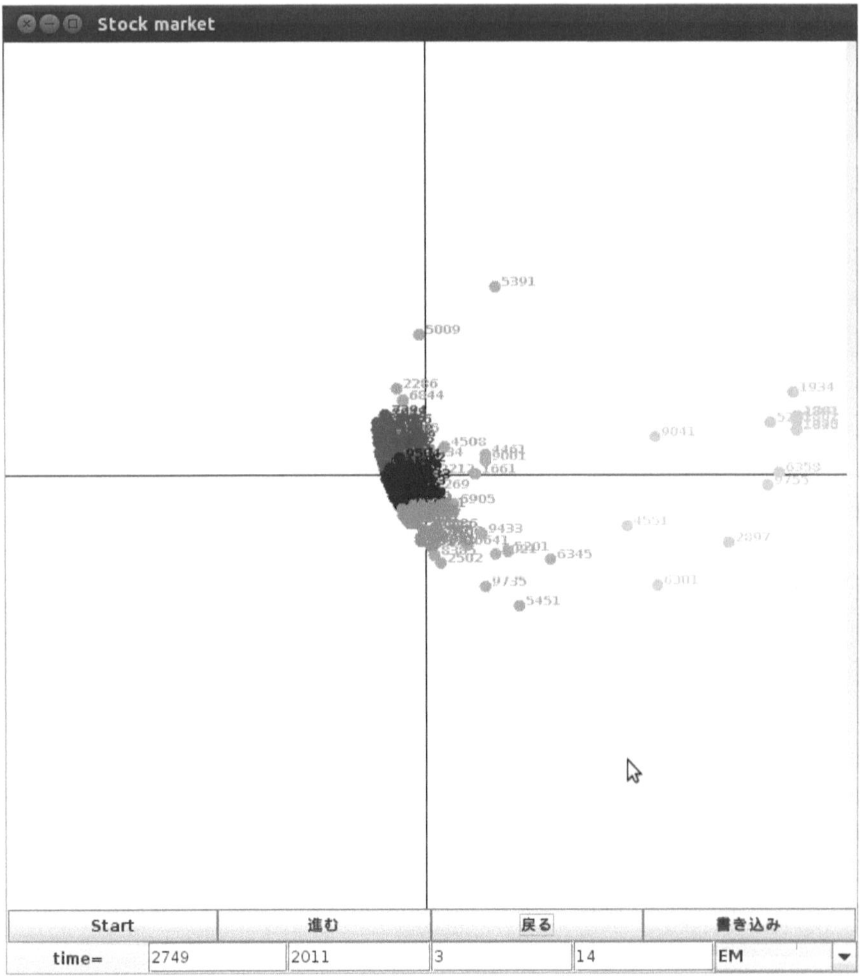


(c)

**Fig. 15.5** (Continued)

To conform the conjecture, we plot the AIC from 14th January 2001 to 18th October 2011 in Fig. 15.7. In this figure, the AIC takes its minimum at socially serious affair around which the stock sharply drops. For instance, we recognize that the AIC takes its keen minimum around 16th January 2008 and 10th October 2008. These points correspond to the crush caused by 'subprime lending problem' and the crisis due to 'Lehman shock', respectively. Thus, we can specify the signal of crisis from the behavior of AIC.

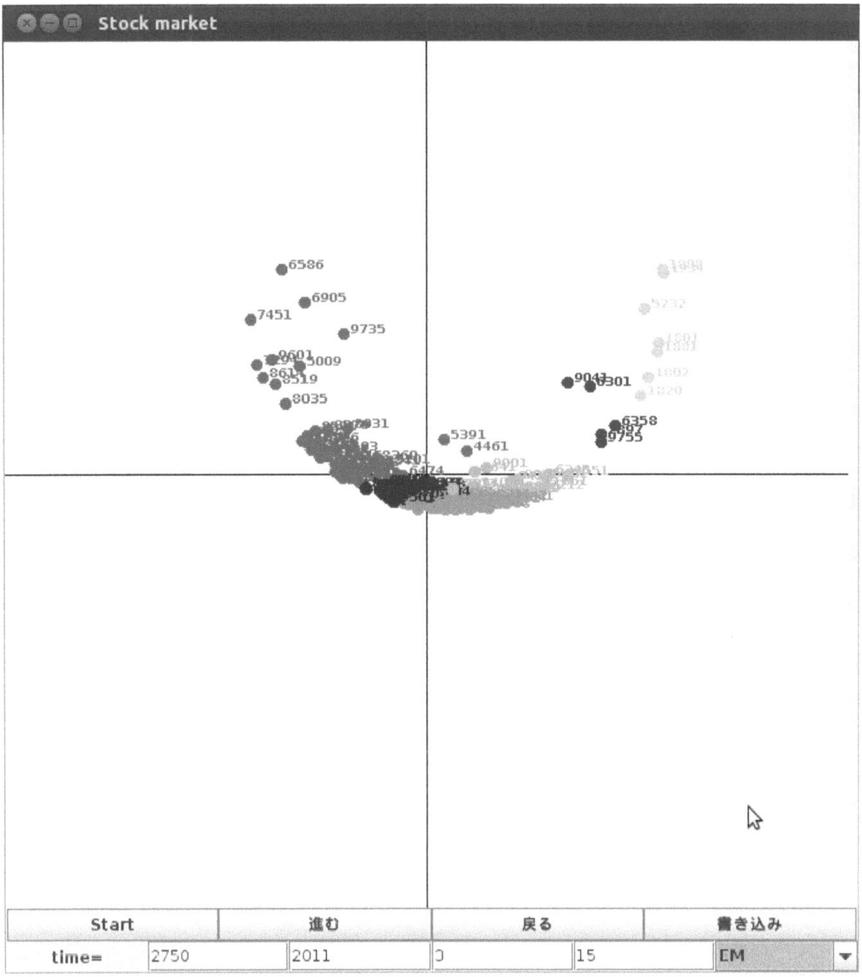


(d)

**Fig. 15.5** (Continued)

## 15.6 Summary

In this paper, we investigated cross-correlations between typical Japanese stocks collected through Yahoo!Japan website (http://finance.yahoo.co.jp/). By making use of MDS for the cross-correlation matrices, we drew two-dimensional scattered plots in which each point corresponds to each stock. To make a clustering for these data plots, we utilized the mixture of Gaussians to fit the data set to several Gaussian densities. By minimizing the AIC with respect to parameters in the mixture via simulated annealing and EM algorithm, we specified the best possible mixture of Gaussians. We actually found that all the scattered plots shrink into a single small

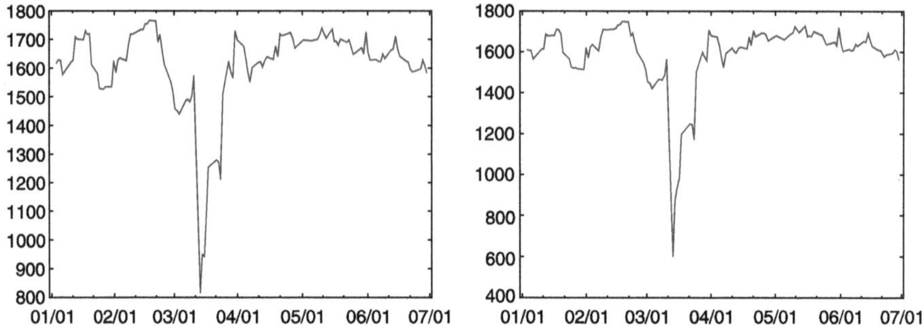


**Fig. 15.6** Time evolution of the AIC around 11th March 2011. The *left panel* is result for the SA, whereas the *right panel* shows the result by the EM algorithm. We clearly find that for both procedure to maximize the likelihood, the AIC takes a sharp and deep minimum at the crisis

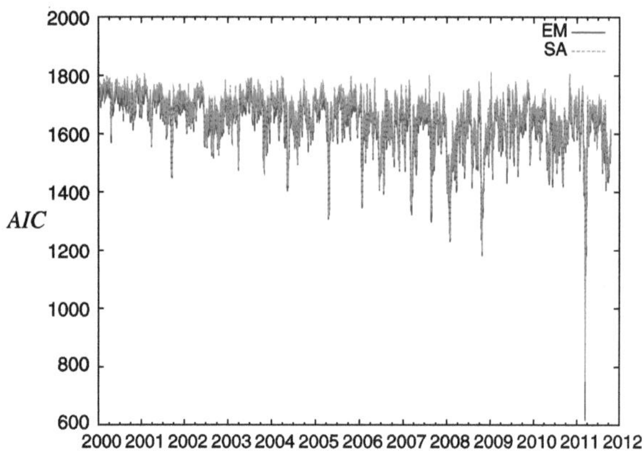


**Fig. 15.7** The behavior of AIC from 14th January 2001 to 18th October 2011

region when some economic crisis takes place. The shape of clusters after the crisis seems to be quite non-trivial. By applying the clustering procedure, we determined the number of clusters before and after the crisis, however, the number itself can not be used for characterizing the crisis. However, we also found the AIC might be utilized for the indicator to specify the crush.

Relationship between our approach and random matrix theory [6, 7] is important and it should be addressed as our future study. Although we here applied the procedure to financial data set, however, it could be applicable to the other problems including some marketing data.

**Acknowledgements** We thank Frederic Abergel, Anirban Chakraborti, Asim K. Ghosh and Bikas K. Chakrabarti, and all the other organizers of *Econophysics-Kolkata VI*. One of the authors (JI) thanks Enrico Scalas, Giacomo Livan for valuable discussion. This work was financially supported by Grant-in-Aid for Scientific Research (C) of Japan Society for the Promotion of Science, No. 22500195.

## References

1. Mantegna R (1999) Eur J Phys B 11:193
2. Onnela J-O, Chakrabarti A, Kaski K, Kertesz J, Kanto A (2003) Phys Rev E 68:056110
3. Borg I, Groenen P (2005) Modern multidimensional scaling: theory and applications. Springer, New York
4. http://finance.yahoo.co.jp/
5. Bishop CM (2006) Pattern recognition and machine learning. Springer, New York
6. Bouchaud J-P, Potters M (2000) Theory of financial risk and derivative pricing. Cambridge University Press, Cambridge
7. Livan G, Alfarano S, Scalas E (2011) Phys Rev E 84:016113

# Chapter 16
# Analyzing Crisis in Global Financial Indices

**Sunil Kumar and Nivedita Deo**

**Abstract** We apply the Random Matrix Theory and complex network techniques to 20 global financial indices and study the correlation and network properties before and during the financial crisis of 2008 respectively. We find that the largest eigenvalue deviate significantly from the upper bound which shows a strong correlation between financial indices. By using a sliding window of 25 days we find that largest eigenvalue represent the collective information about the correlation between global financial indices and its trend indicate the market conditions. It is confirmed that eigenvectors corresponding to second largest eigenvalue gives useful information about the sector formation in the global financial indices. We find that these clusters are formed on the basis of the geographical location. The correlation network is constructed using threshold method for different values of threshold $\theta$ in the range 0 to 0.9, at $\theta = 0.2$ the network is fully connected. At $\theta = 0.6$, the Americas, Europe and Asia/Pacific form different clusters before the crisis but during the crisis Americas and Europe are strongly linked. If we further increase the threshold to 0.9 we find that European countries France, Germany and UK consistently constitute the most tightly linked markets before and during the crisis. We find that the structure of Minimum Spanning Tree before the crisis is more star like whereas during the crisis it changes to be more chain like. Using the multifractal analysis, we find that Hurst exponents of financial indices increases during the period of crisis as compared to the period before the crisis. The empirical results verify the validity of measures, and this has led to a better understanding of complex financial markets.

S. Kumar (✉)
Department of Physics, Ramjas College, University of Delhi, Delhi-110007, India
e-mail: skumar@ramjas.du.ac.in

N. Deo
Department of Physics & Astrophysics, University of Delhi, Delhi-110007, India
e-mail: ndeo@physics.du.ac.in

F. Abergel et al. (eds.), *Econophysics of Systemic Risk and Network Dynamics*, New Economic Windows,
DOI 10.1007/978-88-470-2553-0_16, 

## 16.1 Introduction

The financial markets exhibit very complex dynamics and in recent years have been the focus of some physicists' attempts to apply statistical mechanics to economic problems [15]. The Random Matrix Theory (RMT) tool was developed [16] to deal with the statistics of eigenvalues and eigenvectors of complex many body systems and recently it has been successfully used to investigate the structure of cross-correlation in a large number of financial markets [6, 8, 10, 12, 13, 17, 20, 21, 23, 25]. The largest eigenvalue represents the collective information about the correlation between different stocks and its trend is expected to be dependent on the market conditions, whereas the component of eigenvectors corresponding to remaining large eigenvalues are associated with the formation of different sectors in financial market. Complex network technique in nature have become important method for studying properties of complex systems in the real world and penetrated into statistical physics, social sciences, biological sciences, financial markets [1, 7] and many other fields. The study of complex networks has been initiated by a desire to understand various real systems from the empirical data. Complex network display the spatial topological structure of a system, while the time series is the expression of the temporal dynamics. A network representation is found useful to characterize the system, by associating each element by a node and each interaction by a link. As one of most important advances in statistical physics, complex network theory has become a powerful tool for analyzing financial time series. In this paper, we use threshold and hierarchical method to construct the correlation network of financial indices. The network generated by threshold method [7] are in general disconnected. If the system present a clear cluster organization, threshold methods are typically able to detect them. One of the most common algorithms to detect a possible hierarchical structure hidden in the data is given by the Minimum Spanning Tree (MST) [2, 14] and has been applied successfully [3, 4, 18, 19]. This method selects only the indices with closest interactions among all indices and it generates a visual presentation of the linkage relationship among selected interactions between financial indices [24]. The MST performs better role in identifying the economic sectors from the correlation matrix when it is compared with other more traditional methods, such as spectral methods. In the later procedure one extract the eigenvectors of the correlation matrix and identifies sectors as groups of indices which have a large component (compared to others) in an eigenvector. Despite the fact that this method gives some useful information, the eigenvectors sometimes mix different economic sectors (especially when eigenvalues are close to one another).

## 16.2 Data Analyzed

We investigate the daily closing prices of 20 financial indices around the world traded from the period July 2, 1997 to June 1, 2009. The detail of financial indices and their volatilities are shown in Table 16.1. The data has been obtained from [26].

**Table 16.1** List of financial market indices and their volatilities for the full period

| S.No. | Country | Index | Volatility |
|---|---|---|---|
| 1 | Argentina | MERV | 0.0153 |
| 2 | Brazil | BVSP | 0.0161 |
| 3 | Egypt | CCSI | 0.0055 |
| 4 | India | BSESN | 0.0123 |
| 5 | Indonesia | JKSE | 0.0119 |
| 6 | Malaysia | KLSE | 0.0090 |
| 7 | Mexico | MXX | 0.0115 |
| 8 | South Korea | KS11 | 0.0145 |
| 9 | Taiwan | TWII | 0.0115 |
| 10 | Australia | AORD | 0.0067 |
| 11 | Austria | ATX | 0.0093 |
| 12 | France | FCHI | 0.0108 |
| 13 | Germany | GDAXI | 0.0117 |
| 14 | Hong Kong | HSI | 0.0122 |
| 15 | Israel | TA100 | 0.0090 |
| 16 | Japan | N225 | 0.0111 |
| 17 | Singapore | STI | 0.0101 |
| 18 | Switzerland | SSMI | 0.0093 |
| 19 | United Kingdom | FTSE | 0.0090 |
| 20 | United States | GSPC | 0.0091 |

The daily closing prices and logarithmic returns of 20 financial indices are shown in Fig. 16.1. There are differences in public holidays or weekends among countries so we shifted the data according to the rule that when more than 30 % of markets did not open on a particular day, we remove that day from the data, and when it is below 30 %, we kept existing indices and inserted the last closing price for each of the remaining indices. Also these markets do not operate at the same time zones. It has been studied [8, 17, 20] that correlations of Asian with the USA indices increases when one considers the correlation of the USA indices with the next day indices of the Asian market. We did not considered weekly data to avoid the problem of different operating hours between international market so that we do not miss major changes in markets which tend to occur during a small interval of days. Thus, we consider all indices taken at the same date and filtered the data accordingly.

## 16.3 Random Matrix Theory Approach

Let $S_i(t)$ and $R_i(t)$ denote the daily closing prices and returns of indices $i$ at time $t$ ($i = 1, 2, \ldots, N$; $t = 1, 2, \ldots, L$), respectively. The logarithmic returns $R_i(t)$ can

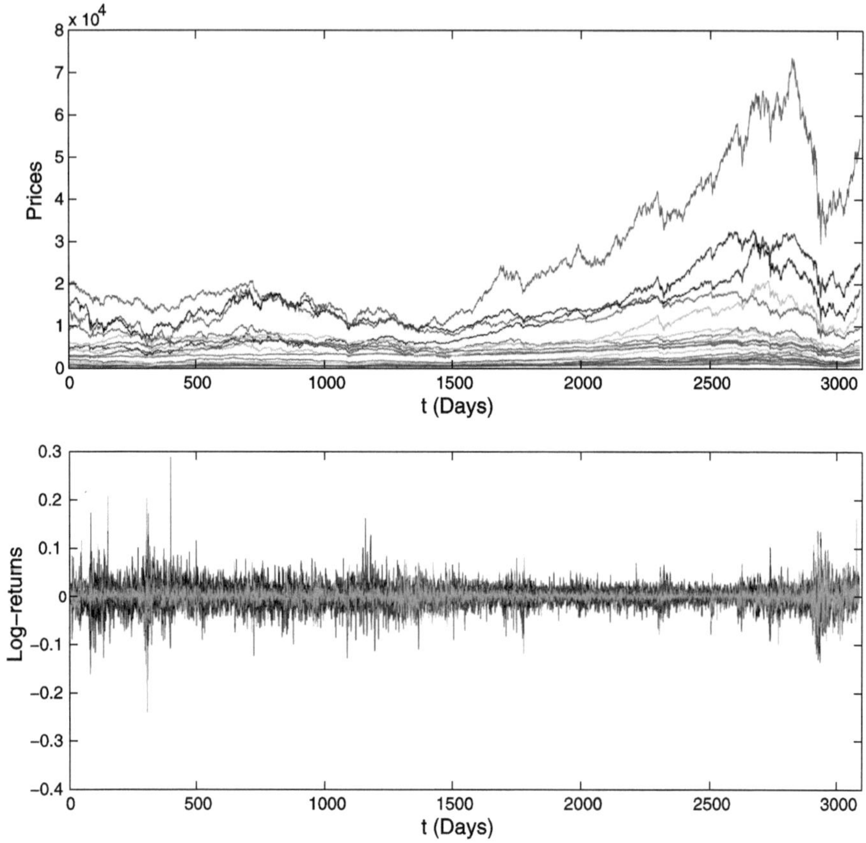


**Fig. 16.1** (**a**) Daily closing prices of financial indices of 20 countries for the period July, 1997 to June, 2009 (**b**) Corresponding log-returns

be defined as,

$$R_i(t) \equiv \ln\big(S_i(t+\Delta t)\big) - \ln\big(S_i(t)\big), \tag{16.1}$$

where $\Delta t = 1$ day is the time lag. The normalized returns for indices $i$ is defined as,

$$r_i(t) \equiv \frac{R_i(t) - \langle R_i \rangle}{\sigma_i}, \tag{16.2}$$

where $\sigma_i \equiv \sqrt{\langle R_i^2 \rangle - \langle R_i \rangle^2}$ is the standard deviation of $R_i$, and $\langle \cdots \rangle$ denotes a time average over the period studied. We then compute the equal-time cross-correlation matrix $C$ with elements,

$$C_{ij} \equiv \big\langle r_i(t) r_j(t) \big\rangle. \tag{16.3}$$

The elements of $C_{ij}$ are limited to the domain $-1 \leq C_{ij} \leq 1$, where $C_{ij} = 1$ defines perfect positive correlations, $C_{ij} = -1$ corresponds to perfect negative correlations,

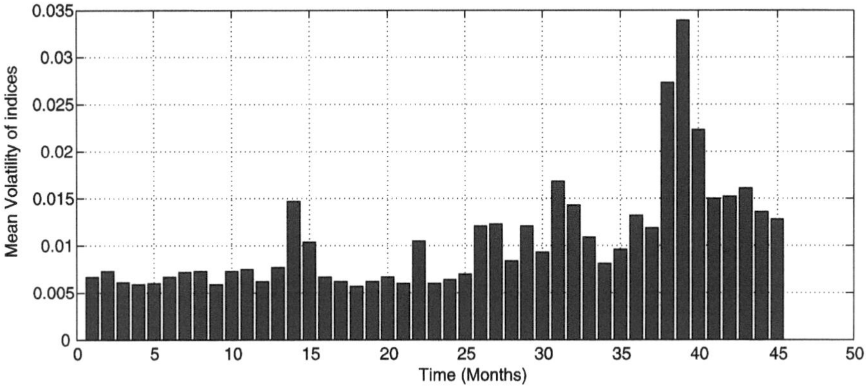


**Fig. 16.2** Mean volatility of 20 financial indices

and $C_{ij} = 0$ corresponds to no correlation. If $N$ time series of length $T$ are mutually uncorrelated, the resulting cross-correlation matrix is termed as a Wishart matrix. Statistical properties of such random matrices are known. In the limit of $N \to \infty$, $L \to \infty$, such that $Q \equiv L/N \geq 1$, the probability distribution $P_{rm}(\lambda)$ of the eigenvalue $\lambda$ is given by,

$$P_{rm}(\lambda) = \frac{Q}{2\pi} \frac{\sqrt{(\lambda_{max}^{rand} - \lambda)(\lambda - \lambda_{min}^{rand})}}{\lambda}, \tag{16.4}$$

for $\lambda$ within the bounds $\lambda_{min}^{rand} \leq \lambda_i \leq \lambda_{max}^{rand}$, where $\lambda_{min}^{rand}$ ($\lambda_{max}^{rand}$) are the lower (upper) bound given by,

$$\lambda_{max(min)}^{rand} = \left[1 \pm (1/\sqrt{Q})\right]^2. \tag{16.5}$$

The volatility gives us a measure of the market fluctuations. We quantify the volatility, as the local average of the absolute value of daily returns of indices in an appropriate time window of $T$ days, as an estimate of volatility in that period $v(t) = \frac{\sum_{t=1}^{T-1} |R(t)|}{T-1}$. We compute the mean volatility of all indices (June 7, 2007 to November 30, 2009) by taking $T = 25$ days which is shown in Fig. 16.2. The volatility for two periods June 7, 2006 to November 30, 2007 and December, 2007 to June, 2009 (before and during the crisis) for individual countries is shown in Fig. 16.3, we consider these two periods as the period before and during the financial crisis of 2008 respectively. We then construct the cross-correlation matrix $C_{ij}$ from daily returns of $N = 20$ indices before and during crisis periods. The probability densities of $C_{ij}$, $P(C_{ij})$ for both periods are compared in Fig. 16.4. The largest eigenvalue deviating from RMT prediction reflects that some influence of the full global market is common to all indices and it alone yields "genuine" information hidden in $C$. The range of eigenvalues within the RMT bounds corresponds to noise and do not yield any system specific information. Therefore, we

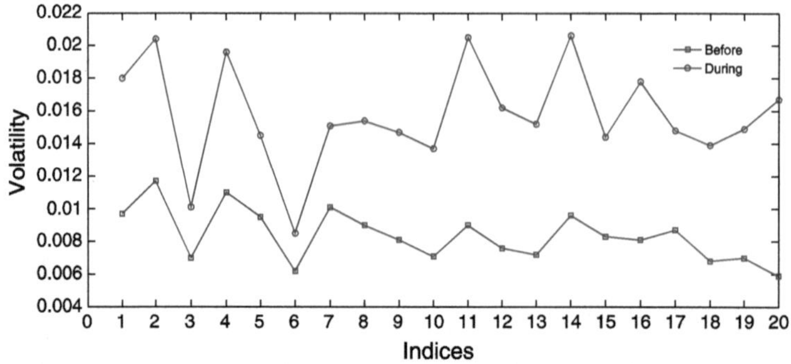


**Fig. 16.3** Volatility of 20 financial indices before and during the crisis

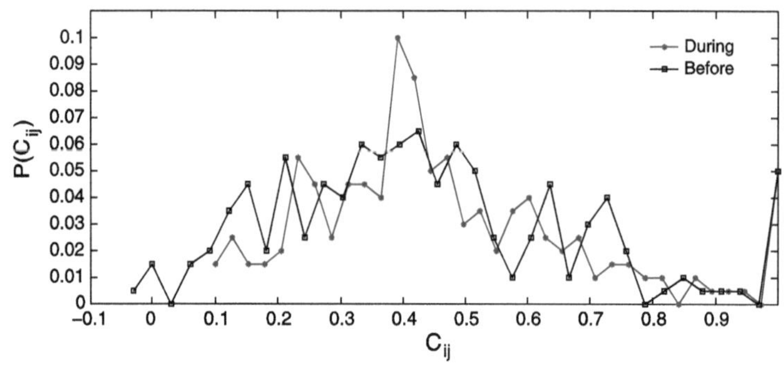


**Fig. 16.4** Plot of the probability density of elements of correlation matrix $C$ calculated using daily returns of 20 indices before and during the crisis. We find the average magnitude of correlation $\langle|C|\rangle = 0.435$ before and $\langle|C|\rangle = 0.463$ during the crisis respectively

compare the properties of $C$ with those of a random correlation matrix in Fig. 16.5 and Fig. 16.6 respectively to extract information about the cross correlations. If there is no correlation between these financial indices, the eigenvalues should be bounded between RMT predictions i.e. $\lambda_{min}^{rand} = 0.597$ and $\lambda_{max}^{rand} = 1.5063$. We find that before the financial crisis period (June 7, 2006 to November 30, 2007), $\lambda_{min}^{real} = 0.0527$ and $\lambda_{max}^{real} = 9.0454$; during financial crisis period (December, 2007 to June, 2009), $\lambda_{min}^{real} = 0.0388$ and $\lambda_{max}^{real} = 9.5282$. Here, we find that largest eigenvalues deviate significantly from the upper bound $\lambda_{max}^{rand}$ which shows a strong correlation between financial indices. We also find an increase in the value of $\langle C_{ij}\rangle = 0.4353$ before the crisis and $\langle C_{ij}\rangle = 0.4634$ during the crisis period. Since the largest eigenvalue represents the collective information about the correlation between different indices therefore we expect its trend to be dependent on the market conditions [12, 21, 23] and can be seen in Fig. 16.7 which is plotted for eigenvectors corresponding to first largest eigenvalue. We find that eigenvectors corresponding to second largest eigenvalue give the information about a sector formation in global financial indices. In Fig. 16.8, we compare eigenvectors corresponding to second largest eigen-

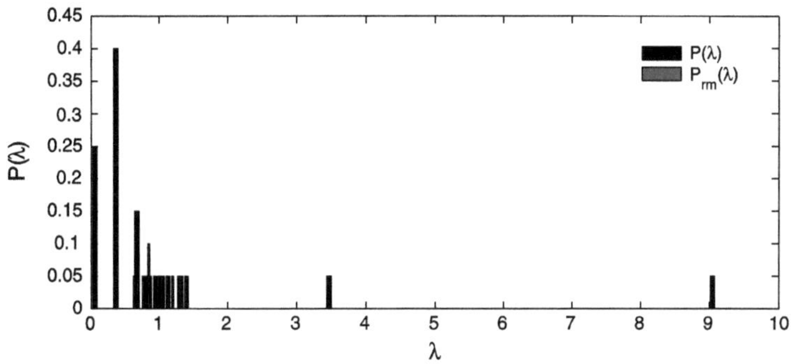


**Fig. 16.5** Comparison of probability density function of 20 financial indices before the crisis. For $N = 20$ indices, $T = 387$ days and $Q = 19.35$, $\lambda_{min}^{rand} = 0.597$ and $\lambda_{max}^{rand} = 1.506$ and $\lambda_{min}^{real} = 0.0527$ and $\lambda_{max}^{real} = 9.045$

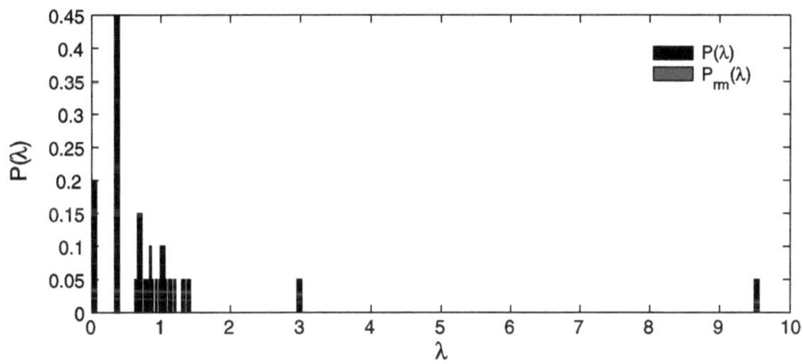


**Fig. 16.6** Comparison of probability density function of 20 financial indices during the crisis. For $N = 20$ indices, $T = 387$ days and $Q = 19.35$, $\lambda_{min}^{rand} = 0.597$ and $\lambda_{max}^{rand} = 1.506$ and $\lambda_{min}^{real} = 0.0388$ and $\lambda_{max}^{real} = 9.528$

value before and during the financial crisis. Countries corresponding to financial indices above eigenvector threshold 0.15 that are contributing most to eigenvectors corresponding to second largest eigenvalues are as follows: Argentina, Brazil, Mexico, France, Germany, Switzerland, UK, US (before the crisis) and Indonesia, Malaysia, South Korea, Taiwan, Australia, Hong Kong, Japan, Singapore (during the crisis). We find that these sectors are forming on the basis of the geographical location. Before crisis indices of Americas (Argentina, Brazil, Mexico, US) and Europe (France, Germany, Switzerland) contribute significantly while during the crisis indices of Asia/Pacific (Indonesia, Malaysia, South Korea, Taiwan, Australia, Hong Kong, Japan, Singapore) contribute significantly to the eigenvectors corresponding to second largest eigenvalue. The classification of major world indices has been considered as [26]. However, eigenvectors corresponding to third largest eigenvalue (Fig. 16.9) does not give so much information as it is near the random matrix bound. We also analyze the eigenvalue dynamics of correlation matrices $C$ constructed by using 3088 daily returns of 20 indices using a sliding window of 25 days. The cor-

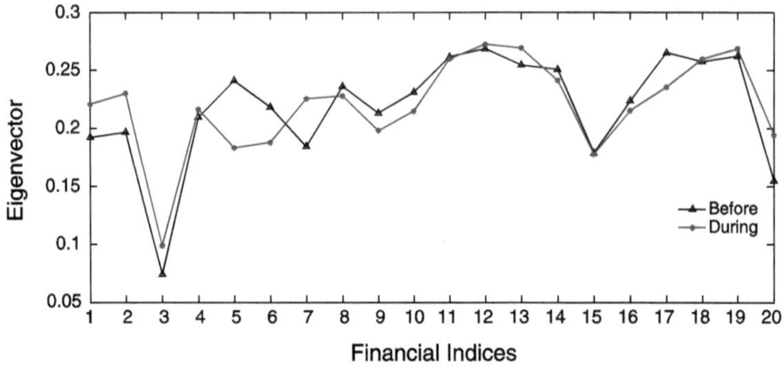


**Fig. 16.7** Comparison of eigenvectors corresponding to first largest eigenvalue before and during the financial crisis of 2008 respectively. No significant difference is observed except the financial indices of Indonesia, Malaysia, and Mexico

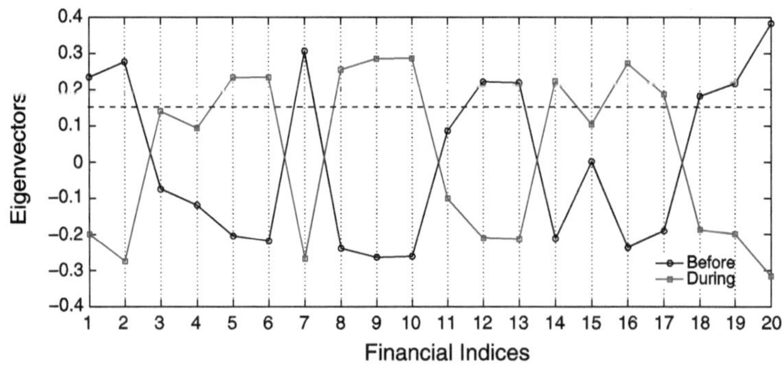


**Fig. 16.8** Comparison of eigenvectors corresponding to second largest eigenvalue. Before crisis indices of Americas (Argentina, Brazil, Mexico, US) and Europe (France, Germany, Switzerland) contribute significantly while during the crisis Asia/Pacific (Indonesia, Malaysia, South Korea, Taiwan, Australia, Hong Kong, Japan, Singapore) contribute significantly. These sectors are formed on the basis of geographical location

relation matrix was constructed from 20 financial indices having the 3088 returns. Fig. 16.10 shows the trend of first, second, and third largest eigenvalue over each of these sliding windows. Here, we find increase in the first and second largest eigenvalues during the financial crisis of 2008 while third largest eigenvalues do not show significant variation. We also analyze the evolution of the structure of the last eigenstate, $U^{20}$ by evaluating the Inverse Participation Ratio (IPR) which allows quantification of the number of components that participate significantly in each eigenvector and tells us more about the level and nature of deviation from RMT. The IPR of the eigenvector $u^k$ is defined by $I^k \equiv \sum_{l=1}^{N}[u_l^k]^4$, where $u_l^k, l = 1, \ldots, N$ are the components of eigenvector $u^k$. Thus IPR allows us to compute the inverse of the number of eigenvector components that contribute significantly to each eigenvector. Fig. 16.11 shows the IPR of 20 financial indices and is closest to 0.05 (=1/20), the value we would expect when all components contribute equally, in the most volatile

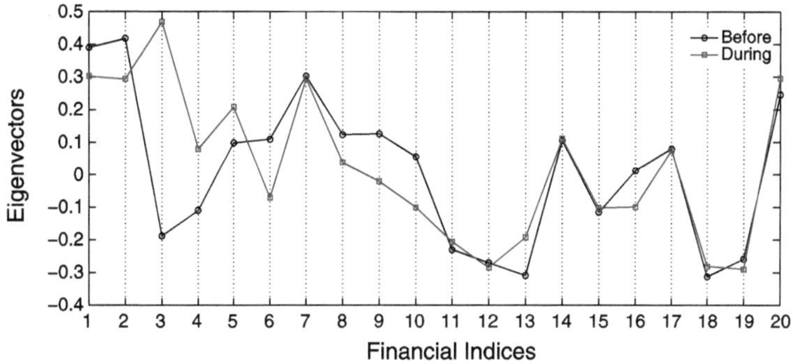


**Fig. 16.9** Comparison of eigenvectors corresponding to third largest eigenvalue before and during the financial crisis of 2008 respectively

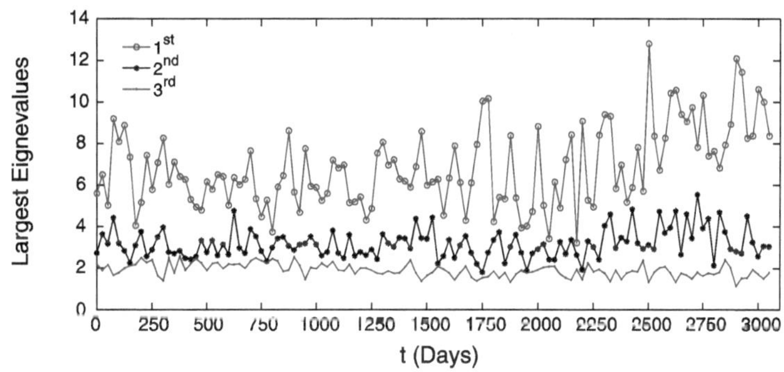


**Fig. 16.10** Largest eigenvalues of the correlation matrices constructed from daily returns of 20 financial indices using a sliding window of 25 days

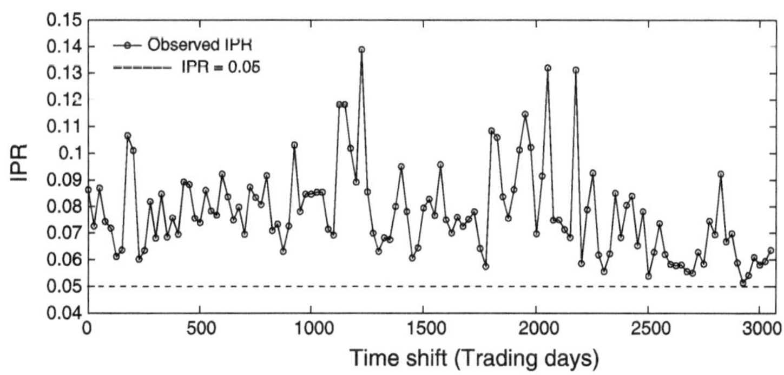


**Fig. 16.11** IPR for the eigenvector $U^{20}$ as a function of time which is obtained from correlation matrix $C$ constructed from daily returns of 20 financial indices for 123 time windows of 25 days each. The *dashed line* marks the value 0.05 of IPR when all components contribute equally

periods of time span. This has similar characteristics to those found for different indices.

## 16.4 Construction and Analysis of the Correlation Network of Financial Indices

### *16.4.1 Threshold Method*

The main idea of constructing the index correlation network is as follows: Let the set of index represent the set of vertices of the network. A certain threshold value $\theta$ is specified such that $-1 \leq \theta \leq 1$. We add an undirected edge connecting the vertices $i$ and $j$ if the correlation coefficient $C_{ij}$ is greater than or equal to $\theta$. Different values of $\theta$ define the networks with the same set of vertices, but different set of edges [7]. We construct networks for different values of threshold $\theta$ in the range 0 to 0.9. The Fruchterman-Reingold layout is used to find the clusters. The Fruchterman-Reingold algorithm [5] is a force-directed layout algorithm. The idea of a force directed layout algorithm is to consider a force between any two nodes. In this algorithm, the nodes are represented by steel rings and the edges are springs between them. The attractive force is analogous to the spring force and the repulsive force is analogous to the electrical force. The basic idea is to minimize the energy of the system by moving the nodes and changing the forces between them. We find that at threshold $\theta = 0.2$ the network is fully connected. In the network at threshold $\theta = 0.6$ (Fig. 16.12) the Americas, Europe and Asia/Pacific forms different clusters before the crisis but during the crisis Americas and Europe forms a combined cluster of strong link between them. If we further increase the threshold $\theta$ up to 0.9 we find that European countries: France, Germany and UK, consistently constitute the most tightly linked markets for both before and during the crisis.

### *16.4.2 Minimum Spanning Tree*

We construct the network of 20 financial indices (before and during 2008 crisis) by using the metric distances [14] $d_{ij} = \sqrt{2(1 - C_{ij})}$ forming an $N \times N$ distance matrix $D$ whose elements varies between 0 and 2. Here $C_{ij}$ is the correlation between indices $i$ and $j$ whose elements varies from $-1$ to 1 thus small values of $d_{ij}$ imply high correlations among indices. The number of possible nodal connections of financial indices is large, $N(N-1)/2$. The MST can reduce this complexity by showing only the $N-1$ most important non-redundant connections in a graphical manner. We use the Prim Algorithm [22] for drawing MST. Prim algorithm is an algorithm in graph theory that finds a minimum spanning tree for a connected graph i.e. it finds a subset of the edges that forms a tree that includes every vertex, where the total weight of all the edges in the tree is minimized. If

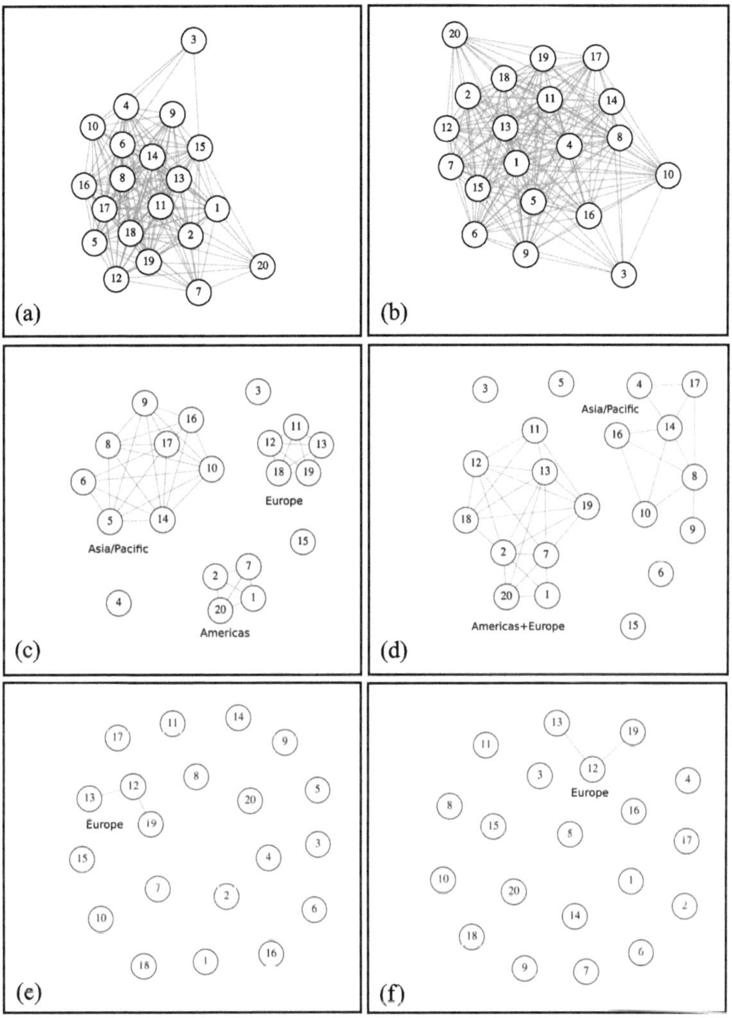


**Fig. 16.12** The financial network of 20 indices at different threshold: (**a**) $\theta = 0.2$ (before) (**b**) $\theta = 0.2$ (during) (**c**) $\theta = 0.6$ (before): Cluster of financial indices of **Americas** (Argentina, Brazil, Mexico, and US), **Europe** (Austria, France, Germany, Switzerland, and UK), and **Asia/Pacific** (Australia, Hong Kong, Indonesia, Malaysia, Japan, Singapore, South Korea, Taiwan). (**d**) $\theta = 0.6$ (during): Clusters of indices of **Asia/Pacific** (Australia, Hong Kong, Japan, India, Singapore, South Korea, Taiwan), **Americas** (Argentina, Brazil, Mexico,US), and **Europe** (Austria, France, Germany, Switzerland, UK). (**e**) $\theta = 0.9$ (before) (**f**) $\theta = 0.9$ (during). At $\theta = 0.9$ indices corresponding to Europe: France, Germany and UK consistently constitute the most tightly linked markets both before and during the crisis

the graph is not connected, then it will only find a MST for one of the connected components. The MST shows the presence of clusters of nodes (indices) which are quite homogeneous and it also displays a structure in subclusters where nodes

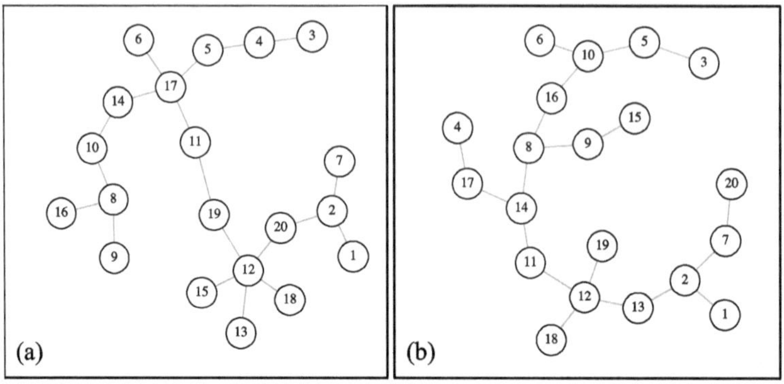


**Fig. 16.13** The MST using Prim algorithm. (**a**) Before the crisis (**b**) During the crisis. There is a strong tendency for financial indices to organize by geographical location

are indices belonging to the same subsector. We find that there is a strong tendency for financial indices to organize by geographical location that can be seen in Fig. 16.13(a) and (b). Before the crisis the structure of MST is more star like whereas during the crises it changes to be more chain like. Using MST, we find that there is a strong tendency for financial indices to organize by geographical location.

## 16.5 Multifractal Analysis of Global Financial Indices

We use the Multifractal Detrended-Fluctuation Analysis (MF-DFA) method [9, 11] to study the multifractal properties of financial indices. The multifractality degree can be quantified by $\Delta h = h(q_{min}) - h(q_{max})$. As the large fluctuations are characterized by smaller scaling exponent $h(q)$ than small fluctuations therefore $h(q)$ for $q < 0$ are larger than those for $q > 0$ and $\Delta h$ is positively defined. We calculate Hurst exponents for financial indices before and during the crisis. Figure 16.14 show increase in the value of Hurst exponents for most of the financial indices during the crisis as compared to period before the crisis. We also find the multifractal degree $(\Delta h)$ of financial indices before and during the crisis and results are compared in Fig. 16.15. We find that there is no significant variation in the multifractal degree except the indices of Egypt, Malaysia, Taiwan, Israel, and Singapore.

## 16.6 Conclusion

We study results obtained before and during the financial crisis of 2008 by using RMT and Network (Threshold and MST) methods. We analyze the eigenvalue dynamics of correlation matrix of 20 financial indices using a sliding window of 25 days. We find that largest eigenvalues deviate significantly from the upper bound

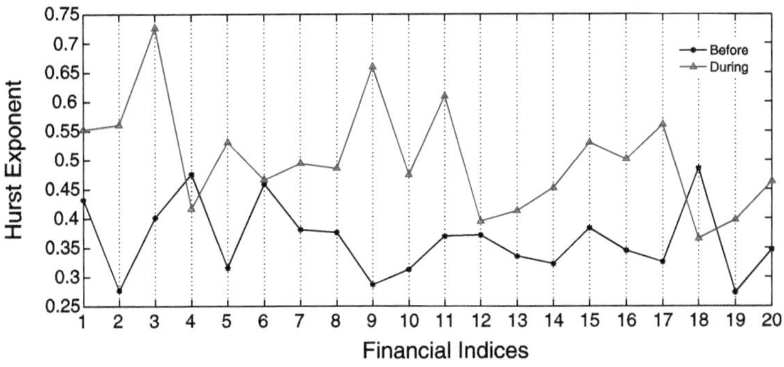


**Fig. 16.14** Hurst exponents for 20 financial indices. Hurst exponents increases for most of the financial indices during the crisis period

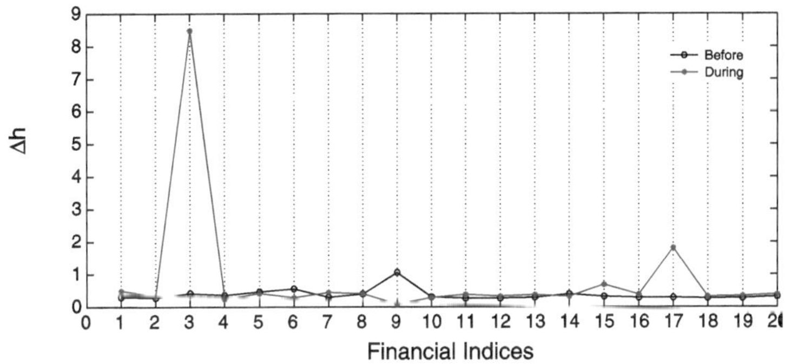


**Fig. 16.15** Multifractal degree ($\Delta h$) before and during the financial crisis for 20 financial indices. A large variation in the value of $\Delta h$ is observed in case of Egypt, Malaysia, Taiwan, Israel and Singapore during the crisis period

$\lambda_{max}^{rand}$ which shows a strong correlation between financial indices. We find that the largest eigenvalue represent the collective information about the correlation between different indices and its trend indicates the market conditions. We also perform the eigenvector analysis corresponding to the first, second and third largest eigenvalue before and during the crisis. It is confirmed that eigenvectors corresponding to second largest eigenvalue gives useful information about the sector formation in the global financial indices. We compare eigenvectors corresponding to second largest eigenvalue before and during the financial crisis. Countries corresponding to financial indices above eigenvector threshold 0.15 that are contributing more are as follows: Argentina, Brazil, Mexico, France, Germany, Switzerland, UK, US (before the crisis) and Indonesia, Malaysia, South Korea, Taiwan, Australia, Hong Kong, Japan, Singapore (during the crisis). We find that these sectors are formed on the basis of the geographical location. However, eigenvectors corresponding to third largest eigenvalue does not give much information as the third largest eigenvalue is near the random matrix bound. We study properties of the correlation networks of financial indices by using the threshold and hierarchical (MST) method respectively. We analyze the effect of financial crisis of 2008 on the correlation network of global financial indices. By constructing networks for different values of threshold $\theta$ in the range 0 to 0.9, we find that at threshold $\theta = 0.2$ the network is fully connected. At threshold $\theta = 0.6$, we find that the Americas, Europe and Asia/Pacific form different clusters before the crisis but during the crisis Americas and Europe are strongly linked. If we further increase the threshold $\theta$ up to 0.9 we find that European countries France, Germany and UK consistently constitute the most tightly linked markets before and during the crisis. We also study the topological properties (mean degree, clustering coefficients, connected components, and clique) of correlation network before and during the crisis. Using MST, we find that there is a strong tendency for financial indices to organize by geographical location. Before the crisis the structure of MST is more star like whereas during the crises it changes to be more chain like. We also study multifractal properties of global financial indices and find that there is an increase in the value of Hurst exponents of financial indices during the crisis period.

**Acknowledgements** We would like to thank Prof. Sanjay Jain for encouragement and discussions. We acknowledge the University Faculty R&D Grant for financial support.

## References

1. Albert R, Barabasi AL (2002) Statistical mechanics of complex networks. Rev Mod Phys 74:47–97
2. Bonanno G, Vandewalle N, Mantegna RN (2000) Taxonomy of stock market indices. Phys Rev E 62:7615
3. Bonanno G, Caldarelli G, Lillo F, Mantegna RN (2003) Topology of correlation-based minimal spanning trees in real and model markets. Phys Rev E 68:046130
4. Coelho R, Gilmore CG, Lucey B, Richmond P, Hutzler S (2007) The evolution of interdependence in world equity markets – evidence from minimum spanning trees. Physica A 376:455–466
5. Fruchterman TMJ, Reingold EM (1991) Graph drawing by force-directed placement. Softw Pract Exp 21(11):1129
6. Gopikrishnan P, Rosenow B, Plerou V, Stanley HE (2001) Quantifying and interpreting collective behavior in financial markets. Phys Rev E 64:035106 (R)
7. Huang WQ, Zhuang XT, Yao S (2009) A network analysis of the Chinese stock market. Physica A 388:2956–2964
8. Junior LS, Franca IDP (2012) Correlation of financial markets in times of crisis. Physica A 391:187–208
9. Kantelhardt JW, Zschiegner SA, Koscielny-Bunde E, Havlin S, Bunde A, Stanley HE (2002) Multifractal detrended fluctuation analysis of nonstationary time series. Physica A 316:87–114
10. Kulkarni V, Deo N (2007) Correlation and volatility in an Indian stock market: a random matrix approach. Eur Phys J B 60:101–109
11. Kumar S, Deo N (2009) Multifractal properties of the Indian financial market. Physica A 388:1593–1602
12. Laloux L, Cizeau P, Bouchaud JP, Potters M (1999) Noise dressing of financial correlation matrices. Phys Rev Lett 83:1467–1470

13. Liu Y, Gopikrishnan P, Cizeau P, Meyer M, Peng CK, Stanley HE (1999) Statistical properties of the volatility of price fluctuations. Phys Rev E 60:1390–1400
14. Mantegna RN (1999) Hierarchical structure in financial markets. Eur Phys J B 11:193–197
15. Mantegna RN, Stanley HE (2000) An introduction to econophysics. Cambridge University Press, Cambridge
16. Mehta ML (1991) Random matrices. Academic Press, Boston
17. Meric I, Kim S, Kim JH, Meric G (2008) Co-movements of U.S., U.K., and Asian stock markets before and after September 11. J Money, Investment Bank 3:47–57. 2001
18. Onnela JP, Chakraborti A, Kaski K, Kertesz J (2003) Dynamic asset trees and black Monday. Physica A 324:247–252
19. Onnela JP, Chakraborti A, Kaski K, Kertesz J, Kanto A (2003) Dynamics of market correlations: taxonomy and portfolio analysis. Phys Rev E 68:056110
20. Pan RK, Sinha S (2007) Collective behavior of stock price movements in an emerging market. Phys Rev E 76:046116
21. Plerou V, Gopikrishnan P, Rosenow B, Amaral LAN, Stanley HE (1999) Universal and nonuniversal properties of cross correlations in financial time series. Phys Rev Lett 83:1471–1474
22. Prim RC (1957) Shortest connection networks and some generalization. Bell Syst Tech J 36:1389–1401
23. Shen J Zheng B (2009) Cross-correlation in financial dynamics. Europhys Lett 86:48005
24. Tumminello M, Aste T, Matteo TD, Mantegna RN (2007) Correlation based networks of equity returns sampled at different time horizons. Eur Phys J B 55:209217
25. Wang D, Podobnik B, Horvatic D, Stanley HE (2011) Quantifying and modeling long-range cross correlations in multiple time series with applications to world stock indices. Phys Rev E 83:046121
26. http://www.finance.yahoo.com

# Chapter 17
# Study of Systemic Risk Involved in Mutual Funds

**Kishore C. Dash and Monika Dash**

**Abstract** Systemic risk, may be defined as the risk that contaminates to the whole system, consisting of many interacting agents that fail one after another. These agents, in an economic context, could be firms, banks, funds, or other financial institutions. Systemic risk is a macroscopic property of a system which emerges due to the nonlinear interaction of agents on a microscopic level. A stock market itself is a system in which there are many sub-systems, like Dowjones, Nifty, Sensex, Nasdaq, Nikkei and other market indices in global perspective. In Indian market, subsystems may be like Sensex, Nifty, BSE200, Bankex, smallcap index, midcap index, S&P CNX 500 and many others. Similarly there are many mutual funds, which have their own portfolio of different stocks, bonds etc. We have attempted to study the systemic risk involved in a fund as a macroscopic object with regard to its microscopic components as different stocks in its portfolio. It is observed that fund managers do manage to reduce the systemic risk just like we take precautions to control the spread of an epidemic.

## 17.1 Introduction

Systemic risk is the macroscopic property of a system which emerges due to the nonlinear interactions of agents on a microscopic level. It may be defined as the risk that contaminates to the whole system, consisting of many interacting agents that fail one after another. A stock market itself is a system in which there are many sub-systems, like Dowjones, Nifty, Sensex, Nasdaq, Nikkei and other market indices in global perspective. In Indian market, subsystems may be like Sensex, Nifty, BSE200, Bankex, smallcap index, midcap index, S&P CNX 500 and many others.

K.C. Dash (✉)
Neelashaila Mahavidyalaya, Rourkela, India
e-mail: kesidash@gmail.com

M. Dash
Indian Institute of Science Education and Research, Pune, India

F. Abergel et al. (eds.), *Econophysics of Systemic Risk and Network Dynamics*, New Economic Windows,
DOI 10.1007/978-88-470-2553-0_17, 

## 17.2 Current Study

There are many mutual funds, which have their own portfolio of different stocks, bonds etc. We have attempted to study the systemic risk involved in a fund as a macroscopic object with regard to its microscopic components as different stocks in its portfolio.

## 17.3 The Model

On the microscopic level, each node $i$ of the network at time $t$ is represented by a dynamic variable $S_i(t) \in \{0, 1\}$ characterizing the failure state. The state is $S_i(t) = 1$, if the node has failed and $S_i(t) = 0$ otherwise. On the macroscopic side, the system state at time $t$ is encoded in the $n$-dimensional state vector $S(t)$, with $n$ being the number of nodes. The macrodynamic variable of interest for systemic risk is the total fraction of failed nodes in the system [1]

$$X(t) = \frac{1}{n}\sum_{i=1}^{n} S_i(t). \tag{17.1}$$

When the value of $X(t)$ is close to one, the system is prone to systemic risk and free of it when the value is close to zero. Here, $X(t)$ is a measure of systemic risk. Agents are represented by nodes. Each of the nodes is characterized by two discrete states 0 and 1, which can be interpreted as, a healthy and a failed state respectively. We assign state '0' while in one phase and state '1' when it completely changes its phase.

## 17.4 Mutual Funds Studied

**HDFC Equity (G)** An open ended growth scheme, DOA—01.01.1995
**HDFC Top 200 (G)** An open ended growth scheme, DOA—11.10.96
**HDFC Taxsaver (G)** An open ended ELSS Fund, DOA—31.03.1996
**HDFC Capital Builder (G)** An open ended growth scheme, DOA—01.02.1994
**Franklin India Prima Plus (G)** An open ended growth scheme, DOA—29.09.1994

### *17.4.1 HDFC Equity Top Ten Holdings*

(From 1st July 2010 to 30th June 2011 for a period of one year) [2]

- SBI: 8.41 % **Lowest on 5.10.11 (1715.3)**, other lows 22.6.11 (2141.05), 20.1.11 (2468.8), 26.7.10 (2408.85)

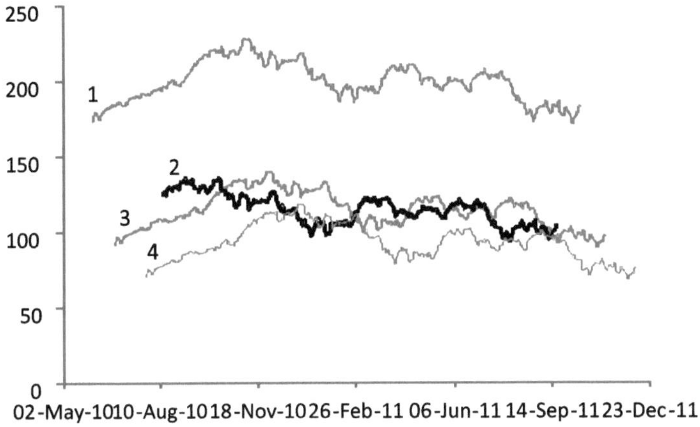


**Fig. 17.1** Daily NAV of HDFC Equity (1), FIPP (2), Top 200 (3) and HDFC Cap Builder (4)

- ICICI Bank: 6.99 %—**Lowest on 8.6.10 (816.9)**, other lows 2.7.10 (840.1), 20.7.10 (889.4), 30.7.10 (904.45), 6.8.10 (951.85)
- TCS: 4.62 %—others—last week of Sept'11
- Infosys: 3.99 %—**Minimum 3rd week of Aug'11**, minimum end of August—11 and 2nd week of Sept'11
- Bank of Baroda: 3.61 %—**Minimum Aug'11 last week**, others—Jan 3rd week, Feb 2nd week and May 3rd week
- CIL: 3.57 %—**Lowest—25.2.11 (291.60)**, 25.11.10 (304.57), 9.12.10 (307.5), 4.10.11 (319.75), 30.3.11 (341.85)
- Bharti Airtel: 3.17 %—**Min 2nd week of Nov'10**, other lows—last week of Jan, 2nd week of March'11
- RIL: 2.90 %—**Lowest (719.5) on 26.08.11**, other lows—between last week of Sept'11 and 1st week of Oct'11
- Tata Motors: 2.76 %—**Minimum—last week of August**, phase changed during 1st week of Feb, 3rd week of June and 3rd and last week of Aug'11
- Cipla: 2.42 %—**Lowest—Sept 1st week '11**

It is observed that all the nodes failed $[S_i(t) = 1]$ between last week of August (2011) and 1st week of October (2011) giving rise to $X(t) = \frac{1}{n}\sum_{i=1}^{n} S_i(t) \cong 1$ and at macroscopic level HDFC Equity does fail during that period as it is clear from the chart (Fig. 17.1)

### *17.4.2 HDFC Top200 Top Ten Holdings*

(From 1st July 2010 to 30th June 2011 for a period of one year) [2]

- SBI: 6.86 %—**Lowest 5.10.11 (1715.3)**, other lows 22.6.11 (2141.05), 20.1.11 (2468.8), 26.7.10 (2408.85) [3]

- ICICI Bank: 6.43 %—**Lowest 8.6.10 (816.9)**, other lows 2.7.10 (840.1), 20.7.10 (889.4), 30.7.10 (904.45), 6.8.10 (951.85)
- Infosys: 5.88 %—**Minimum End of August—11** and 2nd week of Sept'11
- ITC: 4.56 %—**Minimum in last week of Feb'11**
- RIL: 4.13 %—**Lowest (719.5) on 26.08.11**, other lows—between last week of Sept'11 and 1st week of Oct'11
- TCS: 3.95 %—**Min 3rd week of Aug'11**, others—last week of Sept'11
- Bank of Baroda: 3.47 %—**Minimum Aug'11 last week**, others—Jan 3rd week, Feb 2nd week and May 3rd week
- CIL: 3.19 %—**Lowest—25.2.11 (291.60)**, other lows 25.11.10 (304.57), 9.12.10 (307.5), 4.10.11 (319.75), 30.3.11 (341.85)
- Bharti Airtel: 2.91 %—**Min 2nd week of Nov'10**, other lows—last week of Jan, 2nd week of March'11
- Titan Industries: 2.54 %—**Lowest—21.6.10 (2182.41)**, other lows 11.8.10 (2770.05), 31.8.10 (2932.75), 9.2.11 (3086), 21.3.11 (3431.15), 3.5.11 (3771.45), 22.6.11 (4288)

It is observed that all the nodes except ITC fail $[S_i(t) - 1]$ between last week of August (2011) and 1st week of October (2011) giving rise to $X(t) = \frac{1}{n}\sum_{i=1}^{n} S_i(t) \cong 1$ and at macroscopic level HDFC Top 200 fails during that period as it is clear from the chart (Fig. 17.1).

(Titan also fails but the graph has not been updated due to splitting.)

### *17.4.3 HDFC Capital Builder Top Ten Holdings*

(From 1st July 2010 to 30 June 2011 for a period of one year) [2]

- TCS: 6.79 %—**Min 3rd week of Aug'11**, others lows—last week of Sept'11
- Ipca Lab: 5.95 %—**Lowest 4.10.11 (232)**, other lows 8.3.11 (260.5), 16.10.10 (267.15), 30.3.11 (290), 4.5.11 (294)
- SBI: 5.77 %—**Lowest 5.10.11 (1715.3)**, other lows 22.6.11 (2141.05), 20.1.11 (2468.8), 26.7.10 (2408.85)
- BOB: 5.47 %—**Minimum Aug'11 last week**, others—Jan 3rd week, Feb 2nd week and May 3rd week
- RIL: 5.47 %—**Lowest (719.5) on 26.08.11**, other lows—between last week of Sept'11 and 1st week of Oct'11
- Infosys: 5.21 %—**Minimum End of August—11** and Sept'11 2nd week
- Bharti Airtel: 4.95 %—**Minimum 2nd week of Nov'10**, last week of Jan, 2nd week of March'11
- Oil India Ltd: 4.60 %—**Lowest—24.2.11 (1220)**, other lows 21.6.11 (1241.30), 5.9.11 (1288)
- ICICI Bank: 3.92 %—**Lowest 8.6.10 (816.9)**, other lows 2.7.10 (840.1), 20.7.10 (889.4), 30.7.10 (904.45), 6.8.10 (951.85)
- Dr. Reddy's Lab: 3.84 %

It is observed that EIGHT of the nodes fail $[S_i(t) = 1]$ between last week of August (2011) and 1st week of October (2011) giving rise to $X(t) = \frac{1}{n}\sum_{i=1}^{n} S_i(t) \cong 1$ and at macroscopic level HDFC Capital Builder fails during that period as it is clear from the chart (Fig. 17.1). No information about two stocks (IPCA Lab and OIL India, as no charts available).

### *17.4.4 HDFC Tax Saver Top Ten Holdings*

(From 1st July 2010 to 30 June 2011 for a period of one year) [2]

- SBI: 5.38 %—**Lowest 5.10.11 (1715.3)**, other lows 22.6.11 (2141.05), 20.1.11 (2468.8), 26.7.10 (2408.85)
- TCS: 4.94 %—**Min 3rd week of Aug'11**, others—last week of Sept'11
- ITC: 4.88 %—**Minimum in last week of Feb'11**
- Sun Pharma: 4.33 %—**Lowest—17.6.10 (328.98)**, other lows 23.9.10 (378), 24.2.11 (407), 4.10.11 (452)
- Infosys: 4.24 %—**Minimum End of August—11 and 2nd week of Sept'11**
- Bharti Airtel: 3.72 %—**Minimum 2nd week of Nov'10**, last week of Jan, 2nd week of March'11
- ICICI Bank: 3.61 %—**Lowest 8.6.10 (816.9)**, other lows 2.7.10 (840.1), 20.7.10 (889.4), 30.7.10 (904.45), 6.8.10 (951.85)
- Crompton Greaves: 3.33 %—**Lowest in last week of Aug** and first week of Oct'11
- L & T: 2.75 %—**Lowest in last week of Sept'11** and first week of Feb'11. Phase changed during 1st to last week of Jan, 1st week of Feb, last week of Feb, 3rd week of March, mid May.
- Bank of Baroda: 2.74 %—**Minimum Aug'11 last week**, others—3rd week of January, 2nd week of February and 3rd week of May

It is observed that EIGHT of the nodes fail $[S_i(t) = 1]$ between last week of August and 1st week of October of 2011 giving rise to $X(t) = \frac{1}{n}\sum_{i=1}^{n} S_i(t) \cong 1$. (ITC is not failing) and at macroscopic level HDFC Tax Saver fails during that period as it is clear from the chart (Fig. 17.2). No information about SUNPHARMA.

### *17.4.5 Franklin India Prima Plus Top Ten Holdings*

(From Aug' 2010 to 29 July 2011) [3]

- Infosys: 8.24 %—**Minimum End of August—11** and 2nd week of Sept'11
- Bharti airtel: 7.67 %—**Minimum 2nd week of Nov'10**, last week of Jan, 2nd week of March'11

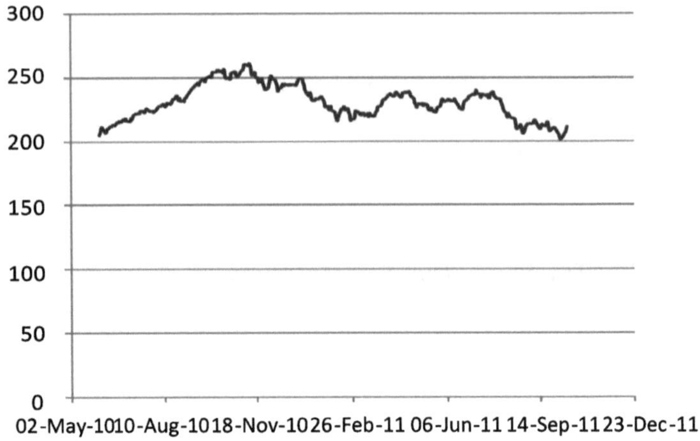


**Fig. 17.2** Daily NAV of HDFC Tax Saver

- ICICI Bank: 6.08 % **Lowest 8.6.10 (816.9)**, other lows 2.7.10 (840.1), 20.7.10 (889.4), 30.7.10 (904.45), 6.8.10 (951.85)
- RIL: 4.81 %—**Lowest (719.5) on 26.08.11**, other lows—between last week of Sept'11 and 1st week of Oct'11 [4]
- GRASIM: 3.76 %—**Lowest 3rd week of June**, other lows—during last week of Aug'11
- IDEA: 3.57 %—**Lowest—end of Feb'11** [4]
- KOTAK MAHINDRA BANK: 3 % **Lowest 1st week of Feb'11**.
- HDFC Bank Ltd: 2.88 %—**Min 1st week of Feb'11**
- DR REDDY'S LAB: 2.27 %—**Lowest—3rd week of July**, other lows—1st week of October
- BOSCH: 2 %—**Lowest 3rd week of March** [4]

It is observed that SIX of the nodes failed $[S_i(t) = 1]$ between last week of August (2011) and 1st week of October (2011) giving rise to $X(t) = \frac{1}{n}\sum_{i=1}^{n} S_i(t) \cong 1$, (IDEA, KOTAK MAHINDRA BANK, BOSCH did not fail while Grasim partially did) and at macroscopic level FRANKLIN INDIA PRIMA PLUS failed during that period as it is clear from the chart (Fig. 17.1).

## 17.5 Conclusion

Thus it is observed that risk is getting transmitted just like an infectious disease from one to another giving rise to systemic risk at the macroscopic level. In case of mutual funds, however it is an indirect way of infection as fund managers choose different stocks at different times, change the percentage of holding and try to minimize the risk, as we have seen in some cases some nodes are not failing. It is also evident from the performance of some funds flashed in Table 17.1.

**Fig. 17.3** Daily closing value. (Courtesy: http://in.finance.yahoo.com)

**Table 17.1** NAV Performance (Courtesy: Intouch mutually, Vol 8, June 2011, HDFC mutual fund [2], Franklin Templeton India monthly factsheet, July 2011 [3])

| Stock | Last 6 months | Last 1 year | Last 3 years | Last 5 years | Last 10 years |
|---|---|---|---|---|---|
| FIPP (G) | 4.17 | 6.40 | 15.63 | 16.63 | 27.77 |
| S and P CNX 500 | −0.01 | −1.14 | 8.59 | 11.55 | 20.30 (Bench Mark) |
| HDFC Equity | −5.04 | 11.02 | 25.24 | 19.82 | 32.47 |
| HDFC CB | −5.37 | 7.93 | 20.33 | 17.26 | 27.93 |
| HDFC TS | −4.81 | 7.88 | 23.03 | 15.36 | 30.74 |
| S and P CNX 500 | −7.78 | 2.31 | 12.19 | 12.03 | 20.06 (Bench Mark) |
| HDFC Top 200 | −5.25 | 9.03 | 22.5 | 19.83 | 27.93 |
| BSE 200 | −7.96 | 2.96 | 12.08 | 12.73 | 20.53 (Bench Mark) |

Bench Mark index of S&P CNX 500 is different because FIPP and HDFC declare their annual performance in different months (July and June respectively).

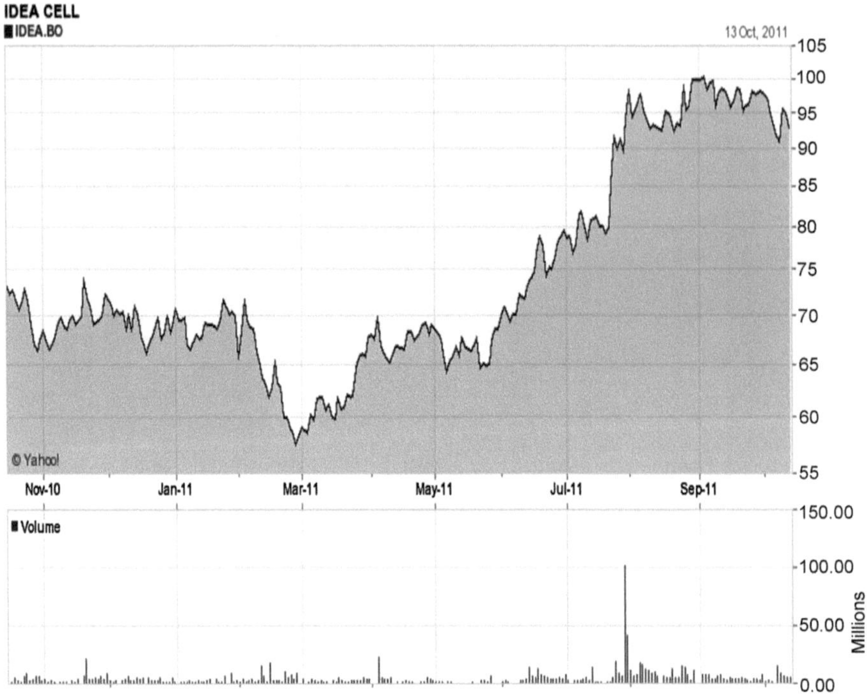


**Fig. 17.4** Daily closing values. (Courtesy: http://in.finance.yahoo.com)

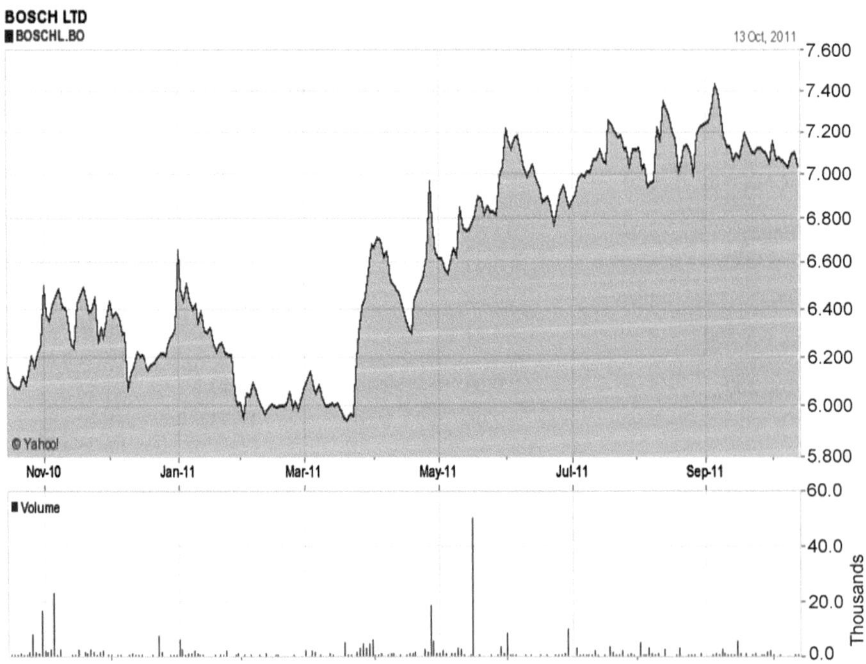


**Fig. 17.5** Daily closing values. (Courtesy: http://in.finance.yahoo.com)

## 17.6 Disclaimer

We are no way connected with the mutual funds that we have studied and the view is completely ours.

## References


1. http://web.sg.ethz.ch/wps/CCSS-09-011
2. Intouch mutually 8(12), HDFC Mutual fund, June 2011, pp 10, 11, 19
3. Franklin Templeton investments monthly fact sheet, July 2011, p 3
4. http://finance.yahoo.in

# Chapter 18
# Characterizing Price Index Behavior Through Fluctuation Dynamics

**Prasanta K. Panigrahi, Sayantan Ghosh, Arjun Banerjee, Jainendra Bahadur, and P. Manimaran**

**Abstract** We study the nature of fluctuations in variety of price indices involving companies listed on the New York Stock Exchange. The fluctuations at multiple scales are extracted through the use of wavelets belonging to Daubechies basis. The fact that these basis sets satisfy vanishing moments conditions makes them ideal to extract local polynomial trends, through the low pass or 'average coefficients'. Subtracting the trends from the original time series yields the fluctuations, at different scales, depending on the level of low-pass coefficients used for finding the 'average behavior'. The fluctuations are then studied using wavelet based multifractal detrended fluctuation analysis to analyze their self-similar and non-statistical properties. Due to the multifractality of such time series, they deviate from Gaussian behavior in different frequency regimes. Their departure from random matrix the-

P.K. Panigrahi (✉) · S. Ghosh
Dept. of Physical Sciences, Indian Institute of Science Education and Research Kolkata, BCKV Campus, PO Mohanpur, Dist. Nadia, West Bengal 741 252, India
e-mail: pprasanta@iiserkol.ac.in

S. Ghosh
Center for Quantum Technology, School of Physics, University of KwaZulu-Natal, Private Bag X54001, Durban 4000, South Africa
e-mail: 210556397@ukzn.ac.za

S. Ghosh
e-mail: say.xen@gmail.com

A. Banerjee
Dept. of Metallurgical and Materials Engineering, National Institute of Technology, Durgapur 713 209, West Bengal, India

J. Bahadur
Dept. of Electronics and Communication Engineering, National Institute of Technology, Durgapur 713 209, West Bengal, India

P. Manimaran
C R Rao Advanced Institute of Mathematics, Statistics and Computer Science, Gachi Bowli, Hyderabad 500 046, India
e-mail: rpmanimaran@gmail.com

F. Abergel et al. (eds.), *Econophysics of Systemic Risk and Network Dynamics*, New Economic Windows,
DOI 10.1007/978-88-470-2553-0_18, 

ory predictions in such regimes is also analyzed. These deviations and non-statistical properties of the fluctuations can be instrumental in throwing significant light on the dynamics of financial markets.

## 18.1 Introduction

Financial time-series which were in the past of interest to only economists, have led to considerable inter-disciplinary research due to the applicability of various physical laws and techniques in their analysis. This has led to the discovery of various new aspects like fractality [1], multifractality [2], correlated behavior [3, 4] and complex network structure [5]. The last few years has also seen a lot of activity in terms of explaining the correlations in financial markets through the Random Matrix Theory (RMT) framework [6–8].

Fractals as first predicted by Benoit Mandelbrot in the 1960s [9] have been widely applied to understand various processes in Physics [10], Chemistry [11] and Biology [12]. Mandelbrot in 1963 proposed the study of fluctuations in the market prices [1] which opened new vistas for the analysis of stock markets through statistical physics. In the recent years, Peng et al. proposed the "Detrended Fluctuation Analysis (DFA)" in 1994 [2] to study the DNA nucleotide structure as a random walk problem which was extended to study the price fluctuations in economic time-series under a mono-fractal hypothesis [13–15]. However, the inadequacy of the mono-fractal hypothesis to model the behavior of financial time series was soon pointed out and consequently, a multi-fractal model called the "Multi-Fractal De-trended Fluctuation Analysis (MFDFA)" was proposed [16]. This method used a variable window approach to calculate the local variances in the profile of the data series from the polynomial fit. Manimaran et al. in 2009 [17] building on the work by Kantelhardt et al. proposed the "Wavelet Based Multi-Fractal De-trended Fluctuation Analysis (WBMFDFA)", where, using the Multi-Resolution Analysis (MRA) capable "fractal like" kernels, the time-frequency resolution and extraction of fluctuations for multi-fractal analysis was shown to have a greater efficiency than its predecessor.

The study of the correlation matrix of the financial return series have been shown to agree well with the predictions of RMT and the nearest-neighbor-spacing of the rank-ordered unfolded eigenvalues of the correlation matrix follow that of the Gaussian Orthogonal Ensemble (GOE) [7, 18]. This behavior dubbed as an *universal* behavior of the financial return series have again, recently been studied with respect to temporal evolution of financial correlations to study the differences between the assumption of strongly correlated financial time-series and uncorrelated financial time-series [19]. In this context, it becomes important to analyze the nature of correlations in the time-frequency domain in order to ascertain the effects of non-stationarity and transience on such studies. The inefficient handling of such signals by either Fourier Transform or Short Term Fourier Transform for the purpose of a time-frequency localized study have already been established, leading to the development of Wavelet Transform [20].

In this work, we will use a wavelet based fluctuation extraction technique to study the correlations of the fluctuations at various frequency windows (called *scales* in the wavelet parlance). We will also briefly comment on the multi-fractal nature of the fluctuations and the distributions of the associated parameters: Hurst exponent and singularity strength. The organization of the article is as follows: in Sect. 18.2, we will briefly review the theoretical methods of Wavelet Based Fluctuation Extraction (WBFE), Wavelet Based Multi-Fractal De-trended Fluctuation Analysis (WBMFDFA) and the Random Matrix Theory (RMT) based method to study the time-frequency localized correlations of the fluctuations. Further in Sect. 18.3, we discuss the results obtained by the application of WBFE, WBMFDFA and correlation analysis on the price index of 196 scrips trading on the New York Stock Exchange (NYSE) between September 1984 to June 2010. Finally, we summarize and conclude with the scope for future work in Sect. 18.4.

## 18.2 Review of Theoretical Methods

### *18.2.1 Wavelet Based Fluctuation Analysis*

In the following, the analysis of a time series given by $X(t)$ is carried out by calculating the "log-normalized return series" $R(t)$:

$$r(t) = \log X(t+1) - \log X(t), \tag{18.1}$$

$$R(t) = \frac{r(t) - \langle r(t) \rangle}{\sigma_r} \tag{18.2}$$

where, $\langle \cdot \rangle$ and $\sigma_r$ are the time average and standard deviation of the log-return series $r(t)$. $\sigma_r$ is also called as the "volatility of returns". The profile $Y(t)$ is calculated by taking the cumulative sum of the log-normalized return series:

$$Y(t) = \sum_{k=1}^{t} R(k) \tag{18.3}$$

which is then subjected to the Wavelet Based Fluctuation Extraction (WBFE) [17]. The WBFE can be performed following the steps:

1. Calculate the one-dimensional discrete wavelet transform (1DWT) [20–23] of the profile:

$$Y(t) = \sum_{b=-\infty}^{\infty} c_b \phi_b(t) + \sum_{a\geq 0}^{l} \sum_{b=-\infty}^{\infty} d_{ab} \psi_{ab}(t) \tag{18.4}$$

   where $c_b$ are the "low-pass" coefficients that capture the trend or the average behavior of the signal and $d_{ab}$ are the "high-pass" coefficients capturing the local fluctuations in the signal at various window sizes $a$. The functions $\Phi$ and $\Psi$ are called the "scaling filter" and the "high pass" filters respectively. The father and mother wavelets $\phi_b(t)$ and $\psi_{ab}(t)$ are orthogonal to each other and are subjected to the admissibility conditions

$$\int \phi(t)\,dt < \infty, \tag{18.5}$$

$$\int \psi(t)\,dt = 0, \tag{18.6}$$

$$\int \phi^*(t)\psi(t)\,dt = 0, \quad \text{orthogonality}, \tag{18.7}$$

$$\int |\phi(t)|^2\,dt = 1 = \int |\psi(t)|^2 dt. \tag{18.8}$$

The scaled and translated versions of $\psi(t)$ are called the "daughter wavelets"

$$\psi_{ab}(t) = 2^{a/2}\psi\left(2^a t - b\right), \quad a \in \mathbb{R}, b \in \mathbb{Z}^+ \tag{18.9}$$

which differ from the mother wavelet $\psi(t)$ at the $a$th scale by $2^a$ in height and $2^{a/2}$ in width. $a$ and $b$ are called the scaling and translation parameters respectively and $l = \lfloor \log N / \log 2 \rfloor$ is the maximum number of scales for the profile $Y(t)$ of length $N$.

The wavelet kernel for the 1DWT should be chosen such that it captures the maximum information from the signal. For example, the Daubechies' family of wavelets satisfy vanishing moment conditions which make them blind to various polynomial trends. The wavelet Db-$N$ (with the index number $N$ being even integers between 2 and 20), has $N/2$ vanishing moments limiting the representation of a polynomial trend of $N/2$ in the signal. The Db-4 wavelet has two vanishing moments making it blind to constant and linear trends. In this work the Db-4 wavelet is employed.

2. Calculate the approximate trend $T_a(t)$ at the scale of interest $a$ and subtract it from the profile $Y(t)$ to get the fluctuations $Z_a(t)$:

$$Z_a(t) = Y(t) - T_a(t). \tag{18.10}$$

The $Z_a(t)$ obtained by this method represent the fluctuations at different frequency bands (the scale is inversely related to the frequency). Consequently, these fluctuations can be probed to analyze the behavior of the signal in various frequency bands like Fourier power law and moments of the fluctuation distribution. It has been shown earlier that the well-known $f^{-3}$ behavior of market fluctuations appear only in the low frequency or long wavelength regime [24]. Due to the varying window sizes (corresponding to different scales) and the convolution error generic to wavelet transforms, the extracted fluctuations happen to have erroneous values at the edges. These errors are corrected by performing the WBFE on the reversed profile and then taking the average of the two (forward and reversed) fluctuation series at each scale.

### 18.2.2 Wavelet Based Multi-fractal De-trended Fluctuation Analysis

These fluctuations can also be subjected to a multi-fractal analysis which is a modified form of the original MFDFA proposed by Kantelhardt et al. in 2002 [16].

The fluctuations obtained in (18.10) are further subdivided into $N_s = \lfloor N/s \rfloor$ segments of size $s$ such that $s = 2^{a-1}W$, where $W$ is the support width of the wavelet and $a$ represents the scale. Thus the fluctuations obtained at various scales can be analyzed at window sizes corresponding to the scale and the wavelet used.

Since the fluctuations are guaranteed to have zero mean, we can directly find the variance of each segment and thus calculate the fluctuation function

$$F_q(s) = \left[\frac{1}{2N_s}\sum_{k=1}^{s}\{\sigma^2(k,s)\}^{q/2}\right]^{1/q}, \quad q \neq 0. \tag{18.11}$$

$F_q(s)$ is the "$q$th order" fluctuation function, where $q \in [-m, m], m \in \mathbb{Z}^+$. The negative (positive) $q$ values capture the fractality of the broader (finer) fluctuations. It can be easily seen that at $q = 0$, the $F_0(s)$ blows up. Hence, to circumvent this issue, at $q = 0$,

$$F_0(s) = \exp\left[\frac{1}{2N_s}\sum_{k=1}^{s}\log\{\sigma^2(k,s)\}^{q/2}\right]^{1/q}, \quad q = 0. \tag{18.12}$$

The "generalized Hurst exponent" $h(q)$ can be obtained from $F_q(s)$ since, as a function of the segment size $s$, $F_q(s)$ follows a power law of the form

$$F_q(s) \sim s^{h(q)}. \tag{18.13}$$

It must be noted that at $q = 2$, this method reduces to the standard mono-fractal fluctuation analysis and $h(2)$ is the Hurst exponent. The multi-fractal scaling exponent $\tau(q)$ can be calculated as

$$\tau(q) = qh(q) - 1. \tag{18.14}$$

The singularity spectrum $f(\beta)$ is related to the multi-fractal scaling exponent $\tau(q)$ by a Legendre transform

$$\beta = \frac{d}{dq}\tau(q), \quad \text{and} \quad f\left(\frac{d}{dq}\tau(q)\right) = q\frac{d}{dq}\tau(q) - \tau(q) \equiv f(\beta) = q\beta - \tau(q). \tag{18.15}$$

### 18.2.3 Correlation Analysis of Fluctuations

The fluctuations $Z_a$ obtained at scale $a$ through (18.10) can be analyzed for correlations. The fluctuations for the entire data set of $N$ scrips can be written as a $N \times T$

matrix $\mathscr{X}$, where $T$ is the time length of the fluctuation series of each the $N$ scrips. The correlation matrix is thus given by

$$\mathscr{C}(x, y) = \frac{1}{T} \mathscr{X} \mathscr{X}^T, \tag{18.16}$$

where, $\cdots^T$ is the transposition operation. In the case where $\mathscr{X}$ consisted of $N$ mutually independent normally distributed fluctuations of length $T$, $\mathscr{C}$ could be considered to be a Wishart matrix [18, 25, 26]. Under the constraint $N \to \infty$, $T \to \infty$ and $Q \equiv T/N \geq 1$ the density of eigenvalues of the correlation matrix takes the form [27]

$$\rho(\Lambda) = \frac{Q}{2\pi\sigma_{\mathscr{X}}^2} \frac{\sqrt{(\Lambda_{\max} - \Lambda)(\Lambda - \Lambda_{\min})}}{\Lambda}, \tag{18.17}$$

with $\sigma_{\mathscr{X}}^2$ being the variance of the matrix $\mathscr{X}$ and $\Lambda_{\min}^{\max}$ given by

$$\Lambda_{\min}^{\max} = \sigma_{\mathscr{X}}^2 \left[ 1 + \frac{1}{Q} \pm 2\sqrt{\frac{1}{Q}} \right]. \tag{18.18}$$

It must be noted that under the constraint $N \to \infty$, the eigenvalues of the matrix $\mathscr{C}$ lie strictly in the range $[\Lambda_{\min}, \Lambda_{\min}]$. However, for finite sized matrices, there exists a finite probability of finding eigenvalues outside this range. Indeed, it has been shown that for financial time series, the nearest-neighbor eigenvalue spacing $\lambda \equiv \Lambda_{i+k} - \Lambda_i$, obtained through unfolding the eigenvalues of $\mathscr{C}$ follow the distribution for a Gaussian Orthogonal Ensemble (GOE) [7]

$$\rho(\lambda) = \frac{\pi\lambda}{4} \exp\left( -\frac{\pi}{2}\lambda^2 \right). \tag{18.19}$$

Here, we will investigate the scale dependence of the $\rho(\lambda)$ for the correlation of the fluctuations $Z_a(t)$ and $Z_a'(t)$ at scale $a$, where $Z_a(t)$ and $Z_a'(t)$ are different scrips in the corpus. This will give us an idea about the effectiveness of the RMT predictions in different frequency regimes.

## 18.3 Results and Discussion

### *18.3.1 Data*

We have analyzed 196 scrips trading on the New York Stock Exchange (hereafter referred to as NYSE) in the period from September the 7th, 1984 to June the 10th, 2010. The scrips analyzed in this work have been chosen so as to encompass high-cap, mid-cap as well as low-cap sections of the American stock market. This combined with the time-frame for analysis includes some of the major crashes of the

NYSE, for the example the *Black Monday* on October 19, 1987, the July 2, 1997 crash triggered by the Asian financial crisis, the burst of the *dot-com bubble* leading to a three years sluggish activities from March 10, 2000 and the two years long bear run of the market from 2007–2009. The data has been analyzed using the WBMFDFA method and also through a correlation analysis of the fluctuations in the random matrix theory framework.

## *18.3.2 Time-Frequency Localized Correlation Analysis*

Analysis and investigation of correlations of the fluctuations over different frequency windows $a$ ($a \propto 1/f$ where $f$ is the frequency) can provide us with insight into the spectral behavior of the market correlations thereby improving our understanding of the collective behavior of the market in different time spans. For example, if the correlations between different scrips representing different sectors of the market in short time windows (high frequency, low scale) is low, then this could be exploited to guard the simultaneous crashes of different sectors in the event of a crisis. However, if the scrips are correlated in the low frequency limit, then the long term correlations of the market could lead to the "healing" of the market in a systematic and efficient way after a crash. This kind of information could be very useful for the policy makers in order to both monitor the economy as well provide for safe-guards against possible crashes.

In order to understand the nature of spectral correlations in the market, we analyze the spectrum of the correlation matrices at different scales. In Fig. 18.1, we have shown the density of the unfolded eigenvalues of the correlation matrices of the fluctuations obtained at different scales. We observe from Fig. 18.1a that at lower and middle scales (up to $a = 7$), the unfolded eigenvalue distribution $\rho(\Lambda)$ is largely uniform and changes to fit (18.17) at higher scales as shown in Fig. 18.1b. Since in this analysis, we have looked at the behavior of the fluctuations over the whole time period $T = 5799$, the value of $Q = 29.58$ is very large. We could expect that the correlations of such fluctuations will show significant deviations from the RMT prediction since it fits well in short time windows [28]. We must remember that the fluctuations obtained by the WBFE method at characteristic of the frequency range (scale) under study. The spectral correlations under investigation here show that at high frequencies (low scales), the GOE fit of (18.17) is not followed by $\rho(\Lambda)$. However, at higher scales, that is at lower frequencies, the $\rho(\Lambda)$ fits reasonably well with (18.17).

In Fig. 18.2a, we have plotted the density of the eigenvalue spacing $\rho(\lambda)$ against different scales. It is clearly visible that at the lower and middle scales, the $\rho(\lambda)$ follows the RMT prediction well, which we have exemplified in Fig. 18.2b at scale 5. This is in contrast with the results in Fig. 18.1a suggesting that the eigenvalue spacing at lower scales is behaves differently from the eigenvalues themselves. This is an interesting observation which we believe should be explored in greater details.

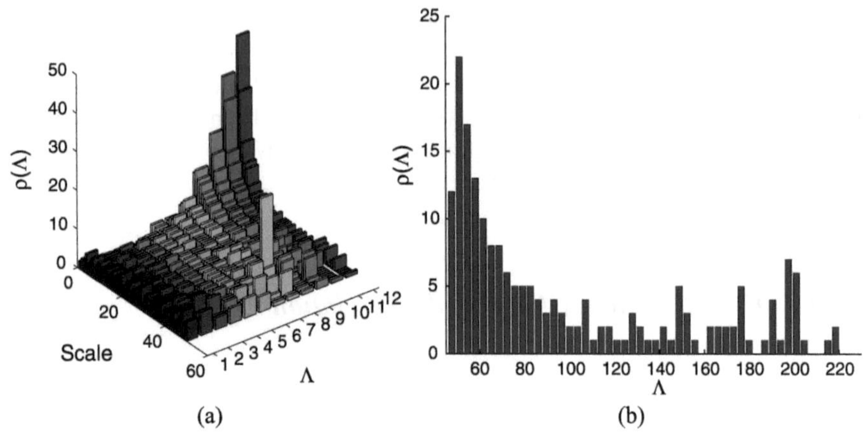


**Fig. 18.1** Plots depicting the density of unfolded eigenvalues $\rho(\Lambda)$ of the correlation matrix of the fluctuations at (**a**) different scales and at (**b**) scale 10. It can be seen the $\rho(\Lambda)$ fits with the analytical (18.17)

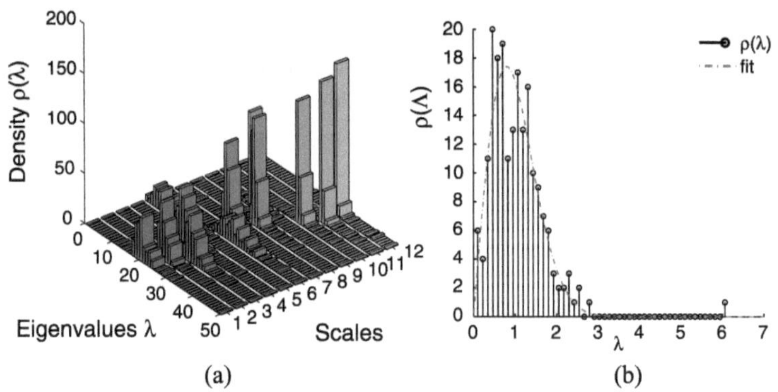


**Fig. 18.2** Plots depicting the density of nearest neighbor unfolded eigenvalue spacing $\rho(\lambda)$ of the correlation matrix of the fluctuations at (**a**) different scales and at (**b**) scale 5, $\rho_5(\lambda)$ fit with $\rho(\lambda) = a\lambda \exp(-b\lambda^2)$ where $a = 35.64$ and $b = 0.7707$ (with 95 % confidence bounds) as an illustration. We can see that though the GOE fits well in the high and mid frequency ranges, at low frequency or long periods, they do not fit with the RMT assumptions

## 18.4 Conclusion

To summarize, we have analyzed the nature of fluctuations from different sectors of the New York Stock Exchange (NYSE) through Wavelet based multi-fractal analysis and RMT based techniques. The number of companies being small, it is expected that there will be significant deviation from the RMT prediction. Interestingly, although the density of unfolded eigenvalues exhibited this behavior, the nearest-neighbor eigenvalue spacing distribution matched reasonably well with the RMT predictions at lower and middle scales. This points out that the spacing distribution can shed considerable light on the nature of the high frequency fluctuation correlation, for a smaller corpus of data.

In conclusion, wavelet analysis when combined with RMT approach can reveal considerable information about the correlations at different scales and is quite useful for modeling the behavior of high and low frequency components of physical processes.

**Acknowledgements** P. Manimaran would like to thank Dept. of Science and Technology, Government of India, for their financial support (DST-CMS GoI Project No. SR/S4/MS:516/07 Dated 21.04.2008).

## References

1. Mandelbrot B (1963) J Bus 36(4):394
2. Peng CK, Buldyrev SV, Havlin S, Simons M, Stanley HE, Goldberger AL (1994) Phys Rev E 49:1685
3. Bouchaud JP, Potters M (2000) Théorie des risques financiers, Aléa-Saclay, 1997; theory of financial risks. Cambridge University Press, Cambridge
4. Mantegna RN, Stanley HE (2000) An introduction to econophysics: correlations and complexity in finance. Cambridge University Press, Cambridge
5. Mantegna R (1999) Eur Phys J B 11:193
6. Wigner EP (1951) Ann Math 53(1):36
7. Plerou V, Gopikrishnan P, Rosenow B, Amaral LAN, Stanley HE (1999) Phys Rev Lett 83:1471
8. Plerou V, Gopikrishnan P, Rosenow B, Amaral LAN, Guhr T, Stanley HE (2002) Phys Rev E 65:066126
9. Mandelbrot B (1982) The fractal geometry of nature. Wheeler Freeman, San Francisco, USA
10. Pietronero L, Tosatti E (eds) (1986) Fractals in physics
11. Stanley HE, Meakin P (1988) Nature 335:405
12. Dewey T (1997) Fractals in molecular biophysics. Oxford University Press, USA
13. Liu Y, Gopikrishnan P, Cizeau P, Meyer M, Peng C-K, Stanley HE (1999) Phys Rev E 60:1390
14. Vandewalle N, Ausloos M, Boveroux P (1999) Physica A 269(1):170
15. Kantelhardt JW, Koscielny-Bunde E, Rego HH, Havlin S, Bunde A (2001) Physica A 295(3–4):441
16. Kantelhardt J, Zschiegner S, Koscielny-Bunde E, Havlin S, Bunde A, Stanley H (2002) Physica A 316(1–4):87
17. Manimaran P, Panigrahi P, Parikh J (2009) Physica A 388(12):2306
18. Laloux L, Cizeau P, Bouchaud JP, Potters M (1999) Phys Rev Lett 83:1467
19. Fenn DJ, Porter MA, Williams S, McDonald M, Johnson NF, Jones NS (2011) Phys Rev E 84:026109
20. Daubechies I (1992) Ten lectures on wavelets. CBMS-NSF regional conference series in applied mathematics, vol 64. SIAM, Philadelphia
21. Mallat S et al (1989) IEEE Trans Pattern Anal Mach Intell 11(7):674
22. Farge M (1992) Annu Rev Fluid Mech 24(1):395
23. Torrence C, Compo G (1998) Bull Am Meteorol Soc 79(1):61
24. Ghosh S, Manimaran P, Panigrahi PK (2011) Physica A 390:4304
25. Edelman A (1988) SIAM J Matrix Anal Appl 9(4):543
26. Baker TH, Forrester PJ, Pearce PA (1998) J Phys A 31(29):6087
27. Sengupta AM, Mitra PP (1999) Phys Rev E 60:3389
28. Kumar S, Deo N (2012). 2012.0409

# Appendix
# Discussions and Comments

The articles in the proceedings seem to hint at two aspects of systemic risk. The first one is the fact that such a risk takes place at a much larger scale than that of an individual institution. The second is that it eventually spreads to the real economy outside the financial system through various "leakage" mechanisms, of which the last crisis in 2008 has given some examples: liquidity shrinkage, fire sale of assets, drop in market value of derivatives, etc.

It is now high-time to put forward a network-based, dynamical system approach stemming from statistical physics, complex systems and chaos theory, and use it in the context of financial systems. It is clear—and many authors and observers of the markets have noted this fact—that the rapid changes in the structure of financial markets over the past two decades have had a tremendous influence on the financial system as a whole. We strongly believe that it would be hopeless to try and understand a notion such as systemic risk without connecting it to the real-life description of the complex mechanisms that control liquidity, transaction costs, derivatives valuation, electronic and algorithmic trading.

The complexity of the financial system must be explored with the eye of the physicist, the formalism of the mathematician and the toolbox of the computer scientist. Clearly, the suitable scientific framework for the understanding of systemic risk is that of complex systems and networks. Fundamental concepts such as the thermodynamic limit, mean field theory or phase transitions are well adapted when trying to understand the fundamental macroscopic features of any network, in particular the financial system. It would however be necessary to go much further than the existing attempts at understanding systemic risk by undertaking the study of non-stationary, time-dependent networks. In fact, the time evolution of networks, or to put it differently, a dynamical-system-based approach to networks, is definitely relevant in order to gain some understanding of the possible "routes to chaos", some of which have been recently witnessed. The financial system is a complex, time-dependent dynamical system with many degrees of freedom and potentially non-trivial, manifold long-time dynamics. This feature has the rather dramatic consequence that the control of the system must be thought of as a time-dependent process, rather than a stationary one. One is no more interested in the question of

F. Abergel et al. (eds.), *Econophysics of Systemic Risk and Network Dynamics*, New Economic Windows,
DOI 10.1007/978-88-470-2553-0, © Springer-Verlag Italia 2013

the system being in equilibrium, but rather, in that of knowing which "attractor" the network is prone to go to, and that of finding the available tools for forcing the system away from, or towards a particular attractor. Concepts and ideas from the control of unstable dynamical systems could be used, helping us to understand the response of the financial network to external "shocks".

Transposing some well-trodden paths to a brand new domain, it seems very natural to ask a series of questions:

(i) What are the phase transition mechanisms of financial networks?
(ii) How, if at all, does the concept of self-organized criticality apply to the financial system?
(iii) What are the graph-theoretic properties of the financial network that drive systemic risk?
(iv) How can one build a dynamical system view of the financial network?
(v) How can one control systemic risk?

Such questions may seem obvious, but it is clear that the answers to them will not be so. In addressing this task, we will need to broaden in a considerable fashion, the classical set of tools commonly used to describe the dynamics of networks, so as to, on the one hand, fully integrate ideas and concepts from dynamical system theory and, on the other hand, to superimpose to this purely graph-theoretic approach characteristic features of financial institutions: derivatives, market sensitivity, credit exposure, etc.

Hopefully, this conference proceedings will stimulate the interest of many researchers, young and old, to venture deeper into this domain. One hopes one of the future Econophys-Kolkata meetings would revisit this theme.

Photograph of the participants in the Econophys-Kolkata meeting held at the Saha Institute of Nuclear Physics, Kolkata during October 21–25, 2011

MIX
Papier aus verantwortungsvollen Quellen
Paper from responsible sources
FSC® C105338

If you have any concerns about our products,
you can contact us on
**ProductSafety@springernature.com**

In case Publisher is established outside the EU,
the EU authorized representative is:
**Springer Nature Customer Service Center GmbH**
**Europaplatz 3, 69115 Heidelberg, Germany**

Printed by Libri Plureos GmbH
in Hamburg, Germany